"He was there alone with himself, collected, tranquil, adoring, comparing the serenity of his heart with the serenity of the skies, moved in the darkness by the visible splendors of the constellations, and the invisible splendor of God, opening his soul to the thoughts which fall from the Unknown. In such moments, offering up his heart at the hour when the flowers of night inhale their perfume, lighted like a lamp in the center of the starry night, expanding his soul in ecstasy in the midst of the universal radiance of creation, he could not himself perhaps have told what was passing in his own mind; he felt something depart from him, and something descend upon him, mysterious interchanges of the depths of the soul with the depths of the universe."

Victor Hugo – from Les Miserables

The
CONSTELLATIONS
Volume II

Sky Tours For Computerized Telescopes

Enjoying the sights, history and lore of the night sky

by

P. Clay Sherrod

Printed by LuLu – Publishing and eBook Company
Copyright 2017, P. Clay Sherrod
Printed in the United States of America
All rights reserved, Published 2017

Arkansas Sky Observatories Publications

ISBN: 978-1-365-71833-5

Constellations

is dedicated to
my Grandparents
who taught me that it is okay
to use my imagination

so that I might see creatures in the sky.

"What an astonishing thing a book is. It's a flat object made from a tree with flexible parts on which are imprinted lots of funny dark squiggles. But one glance at it and you're inside the mind of another person, maybe somebody dead for thousands of years. Across the millennia, an author is speaking clearly and silently inside your head, directly to you. Writing is perhaps the greatest of human inventions, binding together people who never knew each other, citizens of distant epochs. Books break the shackles of time. A book is proof that humans are capable of working magic….."

- Carl Sagan, Cosmos

INTRODUCTION

Hopefully through Volume One of The Constellations you have enriched both your knowledge of some of the most recognizable constellations of the sky as well as your proficiency in developing an astronomical sky library for your computerized telescope, computer or smart device.

In Volume Two we explore many more constellations that can be easily seen from the Northern Hemisphere. In the two-volume set, almost all prominent constellations are covered, but not all.

Also in Volume Two we have a special bonus in the way of a discussion of how our 88 constellations of the sky originated, evolved and eventually became "locked" into the official designations, those boundaries which they have today.

Also a bonus is a discussion and extensive listing of the very curious "Carbon Stars" – some of reddish stars ever seen in the sky. When you come across, one, you will certainly know exactly what star is the carbon star.

Enjoy this guide….the historic and beautiful artwork of celestial cartographer Johannes Hevelius and the splendid photographs and sketches that make your tours enjoyable and memorable

Doc Clay - 2017

CONTENTS - VOLUME TWO

Introduction
The Origins of our Constellations – page 9
Carbon Stars –   page 87
Hercules –   page 13
Lacerta –   page 35
Leo –   page 53
Lepus –   page 71
Libra –   page 109
Lupus –   page 109
Lyra –   page 133
Ophiuchus –   page 153
Orion –   page 177
Pegasus -   page 263
Perseus –   page 289
Pisces –   page 321
Pleiades –   page 345
Sagitta –   page 353
Sagittarius –   page 379
Scorpius –   page 411
Serpens –   page 153
Taurus –   page 439
Triangulum –   page 477
Ursa Major –   page 501
Virgo –   page 521
Vulpecula –   page 353

Author Note: The Constellations guides are presented in alphabetic order for convenience of reference. However, note that many lesser constellations are included in MAJOR constellation discussions since they are bordering or very near these larger areas; hence each volume may have listings for constellations but data is not duplicated

The Origins of Our Constellations

I have always found it interesting - really - that there is not a single constellation nor bright star in the sky (nor has there EVER been) that is named on behalf of our beloved Toads or Frogs. After all, the toad and the common varieties of frogs are known worldwide and have been a staple of sound in our evening hours as we gaze upon the darkness of space. Who can ever forget the haunting sounds of the tiny peepers and green tree frogs in our remote dark sky sites as we gaze skyward in the late night cooling winds of deep summer?

But, alas, the ONLY reptilian recognition by our celestial founding fathers is that of a common LIZARD, the small northern constellation of LACERTA, perched in his threatening lizard-like manner tightly between the large constellations of Cygnus to its west and Andromeda to its east. (note that I do NOT include the Serpent, SERPENS nor the Dragon (Draco) as a true "reptile," but rather mythological in nature).

Lacerta, like the similarly small constellation of Vulpecula (*The Constellations* / Sagitta) is a very recent addition to the constellation family. Since the earliest naming of our star patterns by the Arabian and Greek stargazers, exactly one half - 44 of them - are new additions. The latest are those added by the German mathematician-astronomer ***Hevelius*** in 1687 and those include Lacerta and Vulpecula. It seems that there were a few too many "leftover" stars between the conspicuous northern cross of Cygnus and those star groups to it east and he envisioned the need for a lizard....not a FROG mind you, but a LIZARD....in that homeless groups of stars.

All of the original constellations were placed into 48 groups by the 2nd century astronomer and philosopher **Claudius Ptolemy**, who randomly selected the 1022 brightest stars and put them into these convenient groups for quick identification. You must remember, of course, that Ptolemy lived and worked in *Alexandria* of the Mediterranean, and his extensive list was comprise of ONLY star groups visible from his location and hence did NOT include any of the most recent southern constellations.

Consequently most of the newest constellations are those seen from far south latitudes. It was not until around 1603 that the first southern constellation names appeared in a star atlas by **Johann Bayer**, which added another dozen constellations to the known list. His additions were NOT from first-hand observations of these star patterns.....he depended solely on celestial navigational records by the earliest ship captains who ventured far to the south of Ptolemy's observing perch.

Three more southern constellations were added about 21 years later by yet another German, **Jakob Bartsch**. For those who have seen the beauty of the Southern Cross (Crux), we can give the little-known Bartsch credit for taking THAT group of four very bright stars out of the huge constellation of Centaurus (a Ptolemy original) and placing it as a separate constellation.

It was after this that seven MORE constellations

were "made up" from leftover stars by *Hevelius* in 1687, all in the northern skies, while far south constellations were being added yet again by *LaCaille* in 1750, who tacked on his 15 additional star patterns to the total....

.....things were quickly getting out of hand. It was perhaps the earliest example of the 20th century scam to "own a piece of Mars", or get "your own star" (for a price). The constellation numbers were adding up! Constellations were being subdivided like mobile home lots in Arkansas, and there seemed to be no control nor end in sight. Alas, there was not any "Celestial Urban Planner" around when we needed one.

But in 1928, the **International Astronomical Union** (IAU) took charge of this rampant disregard for star turf abuse and established a permanent and binding system of boundaries and identifications for the exact 88 constellations that we have today. Many constellations were thrown out entirely. Some - like ARGO NAVIS in the southern hemisphere which represented a huge ship covering way too much sky - were divided into smaller ones. Others, like Pegasus - whose "great square" (see GUIDES/Constellations/Pegasus) was transformed into a TRIANGLE overnight by the taking of *Alpheratz* to place into Andromeda's boundaries - were altered in such a way that the boundaries were either expanded or restricted more than they were originally set. Even the great benchmarks of the Milky Way - Sagittarius and Scorpius - have seen their boundaries change drastically since 1600, finally ending up permanently (we hope) as they are represented today.

A beautiful William Cunningham woodcut from 1590 showing the Ptolemaic system as known at the time. The constellations are represented in the equatorial "band" surrounding the central Earth

*   *   *

Chapter 16

HERCULES
Hanging Out With "Hercules the Hero"

This installment of the constellation HERCULES is the first tour for Volume Two: The Constellations" of our constellation study guides for all GO TO telescope users - you will learn more about the fantastic globular cluster *Messier 13*, other star clusters and many nice double stars in this large spring/summer constellation. As will all of our "GO TO guides", it features a "start" with an easy GO TO to the bright star *RAS ALGETHI* as your beginning point for a "GO TO tour

From bright starry globular clusters to beautifully-colored and complex multiple star systems, discussions along the way tell you what to expect from each telescope size and type. All objects will be discussed with exact descriptions of what the viewers of all telescopes, 3 inch to 8 inches, should expect to see...and what to NOT expect to see!

As in all Guides, useful magnifications for EACH GO TO object are discussed for certain telescopes, what type of night and conditions are needed to see certain details, double stars that can be resolved in each telescope model, and much, much more. It is your complete GUIDE for your deep sky observing pleasure and a very handy tool for use at your next star party!

Needless to say, it WILL put your computerized

telescope to work for you in a most efficient and enjoyable way!

I hope you will enjoy these comprehensive GUIDES to "Touring the Constellations", complete with diagrams, charts and illustrations. Please let us hear from you with summations of YOUR observations through these constellation tours!

Introduction

Hercules (pronounced "HER-kew -eez") is one of the **largest** of all constellations and dominates the seemingly bright-star void skies of spring and early summer. The Greek hero of the sky appears ready to fend off any celestial offenders it might encounter, the snakes, dragons, scorpions, bears, wolves and all else that the ancient Arabian and Greek skywatchers filled into our frightful night sky. Indeed, without the fearsome duo of Orion and Hercules (perhaps a good comic book venue) the darkened gloom of night with its creatures might seem a very uninviting place for the innocent likes of Cassiopeia, Virgo, Coma Berenices, Venus.....

Hercules gained his place in the heavens after a rewarding and eventful life of fighting crime and evil injustice on the mortal Earth; he subdued the *evil bull of Crete*, freed *Prometheus* from his enslavement in the Caucasus mountains, guided the famous *Argonauts* on their epic voyage, slew the *Lion of Nemea*, clubbed to death the horrible attacking *Birds of Sytmphalia*.

The Greek hero Hercules
As drawn by Johannes Hevelius 1687
*   *   *

But even Hercules can only do so much. It seems that, either from vanity or the chill of the night air when not fighting the evil forces of mythology, he donned a warm and glamorous robe which - unknown the celestial crime fighter - had been soaked in the blood of the sure-to-kill-you-no-matter-who-you-are-Centaur *Nessus*. The poison gradually pulled all of the strength, stamina and fortitude from our hero and he opted for death rather than a sure demise in shame and geekdome.

Hence, Hercules was cremated atop **Mt. Oeta** on his way to the Heavens where he gained his place among the immorals among the star group that we see today.

To even the most ancient of ancients, Hercules has

be denoted as "*the Kneeling One*," or "*kneeling man of strength*" and his place in the stars frequently depicts him in that position. HOWEVER, in Greek mythology, only after his death did Hercules realize his lifelong quest to marry the fine Princes *Here*, daughter of *Hebe* of Eternal Youth. It has always been my opinion that we are seeing Hercules in a very weak moment (perpetually through all time....) kneeling in proposition of marriage to his childhood sweetheart.

But I digress....perhaps we should remember this great man of 12 wonders more fittingly from the striking and manly figure shown above.

The click-on chart shown The star chart for Hercules shown following is for your convenient reference. Each of the reference numbers apply to the concise listing of objects found following.

Among the outlining figures within the constellation of Hercules is the somewhat famous "**Keystone**", a marker of the springtime skies. This "*asterism*" or star pattern is comprised of four stars in the traditional keystone shape marked at the corners by the fairly bright stars PI (mag. 3.4), ETA (mag. 3.6), EPSILON (mag. 3.9, and ZETA (mag. 3.0).

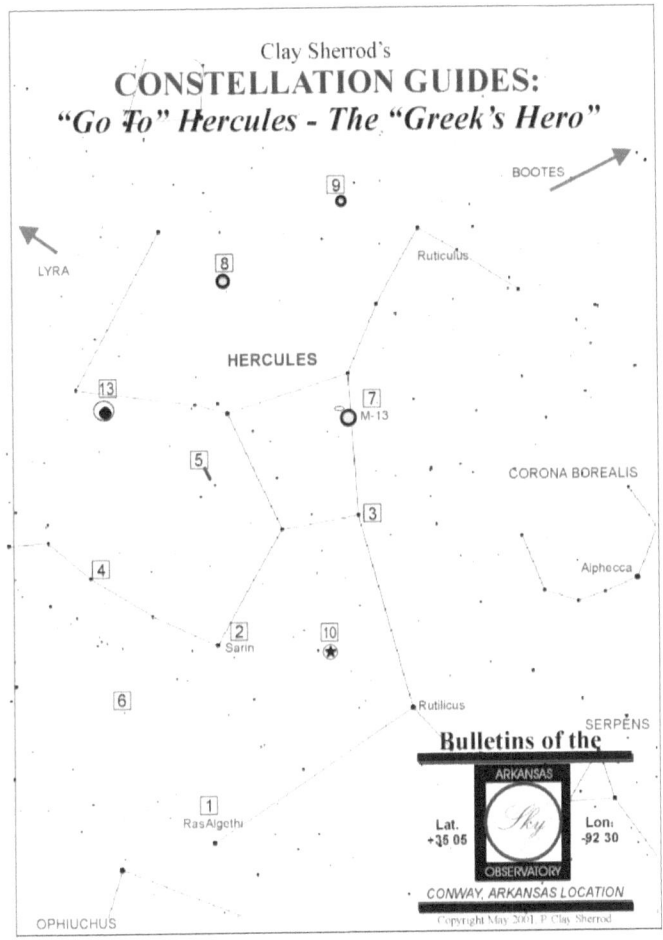

A finder chart for locating many of the GO TO objects in the constellation of Hercules; if using a computer planetarium program, you are encouraged to plot the objects on your screen for higher resolution than this chart provides.

* * *

Immediately East of Hercules is the familiar constellation of *LYRA*, the harp with its very bright white star Vega, a beacon of early summertime skies. This small constellation is featured in the ASO GUIDES/Constellation on this website as a GO TO Constellation Guide.

YOUR HERCULES CONCISE DIRECTORY

In Hercules, there are 10 nice objects in this GO TO tour; all are in reach of every telescope within 3- to 8-inches, yet each telescope will demonstrate uniquely different and challenging aspects of the objects. You will note that there is a conspicuous absence of brighter galaxies and open galactic clusters in this region of the sky. Immediately South of Hercules, we have seen that there are an overwhelming number of globular clusters in Ophiuchus and Serpens, and innumerable distant galaxies in Coma Berenices, Ophiuchus and surrounding areas (see *Constellations*, Vol. One).

As we look toward the constellation Hercules, we are looking ABOVE the galactic pole, and hence looking AWAY from the star-rich areas of our own galaxy, the Milky Way. Hence the star concentration and the number of rich star clusters diminish as we move upward, away from the "hub" of the galaxy. However, note that in the figure below, the **GLOBULAR CLUSTERS** are distributed fairly evenly around the galactic center, so there are at least two very rich such objects in our TOUR of Hercules.

As will all of our GO TO guides, I continue to recommend good a good star atlas and/or chart/and PC sky program which lists the finest objects constellation-by-constellation; if you cannot access any of these objects (or those that are not listed in this TOUR), you can access directly from coordinates - Right Ascension (RA) and Declination (DEC) of any known object via the sky program.

For Meade's Autostar, you merely need to hold down your MODE key on the AutoStar (or similar)for three (3) seconds and the RA and DEC coordinates appear for the telescope. Merely press "GO TO" and the cursor appears prompting you to enter the Right Ascension of the object if it is NOT listed among the objects in the AutoStar library; once the RA is entered, press "Enter" and the cursor once again prompts for the Declination coordinates (these coordinates for epoch 2000) are found in all good observing guides). Once those are entered, merely press "GO TO" once again and your telescope will slew to the position of the object. As discussed in Volume One of The Constellations, all GO TO telescope models may work a bit differently in terms of keystrokes or tab commands, but the process is fundamentally the same among all.

Detailed descriptions of each object of this TOUR will be described and specifics provided as to visibility of those objects in YOUR telescope model. The constellation tour star chart (click on and print to size, above) will get you started, as it demonstrates the relative positions of all 10 objects in this "Hercules tour."

Following is the summarized 10-object list for your GO TO tour of Hercules; you may wish to find the majority of the objects from the AutoStar or sky program library (for example, you can merely pull up Messiers 13 or 92 by going to "Object/Deep Sky/Messier Object/"13"....enter....GO TO" or...if you want to experiment and be a "better AutoStar user", try entering the following coordinates (provided in the list directly following) as described under MODE above.

OBJECT 1:   bright star - Ras Algethi (alpha Herculi) - R.A. 17h 12' / DEC + 14 27 - Magnitude: 3.9 (a bit variable)
OBJECT 2: "optical" double star - Delta Herculi – R.A. 17 13' / DEC + 24 54 - Magnitudes: 3.1 & 8.5
OBJECT 3: good double star - Rho Herculi - R.A. 17h 22' / DEC + 37 11 - Magnitudes: 4.5 & 5.5

OBJECT 4: nice - but tough "triple" star - Mu Herculi - R.A. 17h 45' / DEC + 27 45 - Magnitudes: 3.4, 9.8, & 10.8

OBJECT 5: U Herculi (68 Herc) nice easy variable star - - R.A. 17h 16' / + 33 09 - Mag. range: 4.6 to 5.5 (2.1 days)

OBJECT 6: nice double star - 95 Herculi - R.A. 17 59' / DEC + 21 36 - Magnitudes: 5.1 & 5.2 (color contrast??)

OBJECT 7: THE globular cluster - Messier 13 (ngc6205) - R.A. 16h 40' / DEC + 36 33 - Magnitude: 4.0 (wow!!)

OBJECT 8: globular cluster - Messier 92 (ngc6341) - R.A. 17h 16' / DEC + 43 12 - Magnitude: 6.1 ("little brother")

OBJECT 9: very faint globular cluster - ngc6229 - R.A. 16h 46' / DEC + 47 37- Magnitude: 8.9 (nice object)

OBJECT 10: faint planetary nebular - ngc6210 - R.A. 16h 43' / DEC + 23 53 - Magnitude: 9.7 (star is 12.1!)

GOOD LUCK on these objects....there are a few nice challenges ahead as described in the detail summaries for each object provided below!

A VISUAL GUIDE TO OUR OBJECTS IN HERCULES

Object 1 - Beautiful Double Star and Bright Star RAS ALGETHI (alpha Herculi)
You cannot find a more beautiful double star than *RAS ALGEHTI*. This is our starting star for our tour of Hercules and a very nice one, indeed. Comprised of it main star of magnitude 3.9 and very, very red in color, it is accompanied by an easily resolvable

companion of Magnitude 5.4 that has been described by many as "emerald green!" You don't get that color contrast too often...it is absolutely gorgeous and perhaps the best object on your tour, even rivaling the magnificent Messier 13 later on down your list!

You will never forget your medium power (about 15x per inch) view of this double star; it will be firmly implanted in your memory just like your first view of the rings of Saturn! It is quite easy in the 4-inch and larger scopes and resolves with medium power very well....higher powers are recommended to really separate the two to appreciate the wonderful color contrasts. For small telescopes, use fairly high power to separate, as they are pretty close for those scopes at only about 5" arc separation, but clearly resolvable in the smaller telescopes.

Alpha Herculi is also slightly variable, ranging from about magnitude 3.2 to around 4.0 at dimmest in a period of about 90 days. At only 420 light years away, it is a mass just about like our own sun, but about 800 times more luminous!

Ras Algethi gets it Arabic name from "*Ras al Jauthiyy*" or, "...head of the kneeling one," as described above. In ancient China it was given a bit more significance, surprisingly so for such a faint naked eye star, and was signified the "Emperor's Throne", quite a lofty perch for a fairly seemingly obscure star.

Object 2 - "Optical Double" Star SARIN (delta Herculi)

This double star is an "optical double" as has been described in previous constellation Guides. An optical double is one that APPEARS from our vantage point to be a binary, but in fact is only "lined up" to our sight as two stars seemingly very close. *Sarin*, at magnitude 3.4, has an optical companion of 8th magnitude nearly due south from the brighter star, and clearly both stars are moving perpendicular to one-another so they are NOT physically bound. Like Ras Algethi above, this double has a wonderful color contrast that can be clearly seen in 6- to 8-inch scopes, but not so easily in smaller scopes. The brighter star is somewhat a green or bluish color and the dimmer star is clearly either greyish white or slightly pinkish-white, depending on your color acuity. With a present separation of about 5" arc, this star CAN be resolved in all of our telescopes, although high magnification IS required on small telescopes to find the very faint star; be sure to look for it just west of due south.

Object 3 - A Good Double Star - Rho Herculi
This is a nice and easy double star for all of our telescopes. With an orbital period (this is a true double star system) of just more than 56 years, the apparent angular distance of these stars varies enough to allow a good separation (4.1" arc) at the present time. This is a **very close star** to Earth, at only 30 light years distant, and is a star in the "*red-giant*" making phase, very much more luminous that our Sun but about the same size in its "pre-expansion" period. This star can be best seen with about 100x in the 4-inch and about 80x in the larger telescopes...more magnification will provide better views. This object requires about 50x per inch

aperture for good resolution in the smaller telescopes.

Object 4 - A fine - but challenging - Triple Star – Mu Herculi

This is one of the most-observed of all triple star systems. It is, indeed, a physically associated triplet of stars, all orbiting one-another. Mu, the primary star at magnitude 3.4, has a VERY faint (mag. 9.8) companion nearly due west about 34" arc (slightly smaller than Jupiter appears in whatever eyepiece you are using). It is a very difficult star to find due to the brightness of Mu itself, so I highly recommend finding Mu first at low power, centering, and then increasing the magnification to about 25x per inch to locate the companion. The companion star itself is a challenge for the smaller telescopes. In the 6- and 8-inch scopes.... ONCE FOUND....increase the magnification even more (hope you are tracking well tonight!) to split the faint star into TWO stars! In a 6-inch this requires the very best of nights and I have never had a true separation, only an elongation of the image. The 8-inch will resolve the very faint star into its two red stars at an incredible 848x (the 4.7mm + barlow)!!

It will NOT resolve the star at lower magnification, so don't get your hopes up. These two are actually a pair of **RED DWARF** stars, but will appear for whatever reason a rather bright blue color! Good luck on this one!

Object 5 - A Very Good Variable Star – uHerculi (or "68 Herculi")

For small telescope users out there, this is a "DO NOT MISS" variable star for you....and for all scopes, even binoculars! This star is variable from magnitude 4.7 to 5.5 in a period of ONLY 2.1 days! This variation should be compared to other brighter stars seen around it (use the locator chart from the American Association of Variable Star Observers that I have provided the link to below) to judge its brightness during this rapid cycle and record what you see! Download and print the AAVSO chart (this is a low power, wide field "finder chart" that allows you to find relatively bright stars over a large part of the sky for comparison.....sky directions are indicated on the map):

https://www.aavso.org/apps/vsp/ . Note for these charts, simply type in the NAME of the variable at top to generate your choice of chart.

After you have observed this star in comparison to stars of equal magnitude (you will get better with practice) you can compile a graph of the brightness changes similar to the one shown. These "light curves" as they are called, are frequently included in these constellation Guides to show the remarkable changes in a star's brightness over time.

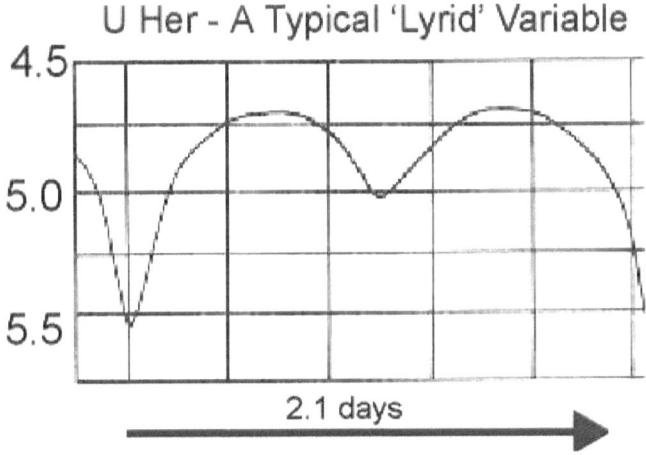

Object 6 - Another Double Star....but hold on! It's Worth Looking At!! - "95 herculi"

This magnitude 4.4 star is a famous double star and well worth looking up. Actually the total brightness of 4.4 is a COMBINED brightness of two stars, magnitudes 5.1 and 5.3 that are fairly close together, but easily resolvable in all of our telescopes. Use higher power (about 30-40x per inch) in a small scope and medium power (about 15x to 20x per inch) on the 4-inch and larger scopes to resolve. The stars, almost equal in brightness, are about 6" apart, so they should be readily seen at these magnifications.

There is a very interesting aspect of these two curious stars....they change color (or so it appears). At one point in 1856 for example the stars were both recorded as "white." In 1857, the observer, **Piazza Smyth** was amazed to see that both stars then were recorded as green and deep red! Curiously, this same phenomenon was observed by experienced and highly reputable observers WORLDWIDE during that same time period! Even more curious, is that the stars appear to have "returned" to the whitish coloration since and have not shown signs of any other colors since!! These two stars are giant stars some 400 light years distant.

MESSIER 13
Meade LX 90 @ 227x

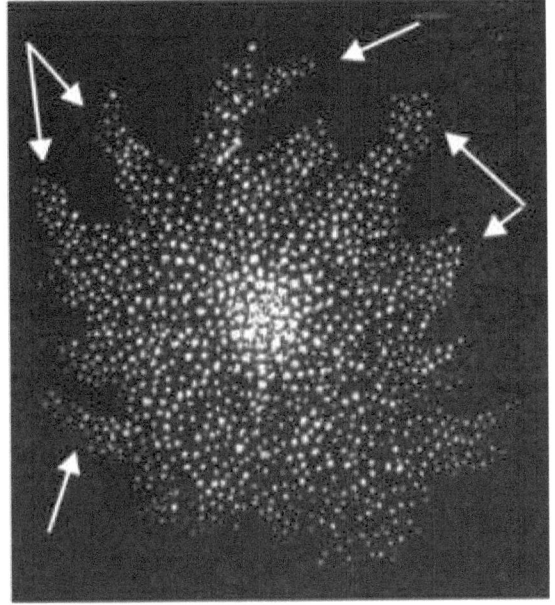

Object 7 - Without a doubt, this is the most famous of all northern hemisphere globular clusters and one of the most notable deep sky objects for all telescopes. It is rivaled in beauty only from the Omega Centauri globular and 47 Tucanae globular clusters which are much brighter and larger in telescopes. Omega Centauri is clearly visible for the lower United States observers and those south of any latitude 32 degrees north, but atmospheric extinction usually precludes the excellent views afforded by **Messier 13**.

The sketch, made in 2001 through an 8-inch telescope, shows what most visual observers are intrigued with the most about globular clusters: the obvious "strings" of stars that appear to flow out from the center as shown with the white arrows. Under very dark skies and moderate magnification, these become obvious with every large cluster. It is no coincidence....the dynamics of the gravitational forces between the stars do actually segregate many of these hundreds-of-thousands of stars into "arms" of concentration.

"Messier 13" was actually first cataloged by **Edmund Halley** (of comet fame), but credit is given some 50 years later (1764) to **Messier** as his 13th entry into his "you-may-confuse-this-for-a-comet" objects. Very high in the northern skies and conveniently located in the "Keystone" noted above, this cluster is actually visible to the naked eye and makes a nice sight even in your finderscope, the total expanse equaling about 1/3 the diameter of the moon!

The cluster is clearly visible in all of our telescopes, and can be "partially" resolved around its edges even with small APO refractors and high magnification (about 120x) on a very dark night. Many peripheral stars are seen with the 4- 6-inch telescopes begins to break down the central core into multitudes of very faint 11th magnitude stars. The view with an 8-inch and larger is incredible under medium high (about 225x) power, with star colors becoming evident with the red giant stars within the central part of the cluster. However, with any amateur telescope, there is still an unresolved "glow" of stars to the 21st magnitude which are thought to number in the 40,000's.

Be sure in all scopes under VERY dark conditions,

to look for the many "star strings", arm-like extensions of stars that are clearly evident under good conditions and medium magnification; these are visible in virtually all of our telescopes, but really begin to take on a distinct nature with more aperture. In my 6-inch there is a very pronounced "string" in a rather NW direction that almost appears as an "arm" of stars extending quite a ways from the central core.

This incredible object measures over 160 light years across, and is some 25,000 light years distant, or about hovering over the center of our Milky Way galaxy. Some outlying stars are thought to be gravitationally-bound to M-13 out to the incredible distance of 200 light years. An interesting aspect concerning ALL globular clusters is necessary here: even though, at this great distance, they APPEAR to be packed and crowded with multitudes of tightly-packed stars, in actuality they are so large that there is LESS than one star per cubic light year in each! That is barely one-half the distance of OUR closest star system, Alpha Centauri!

Object 8 - Messier 92 - Nice Globular....but overshadowed by its Big Brother to the South!
It's tough to be in competitions that you know you can't win, and little Messier 92 is clearly in that no-win situation, located just north and slightly east of its "big brother globular" Messier 13. As nice as ANY celestial object is in reality, it almost seems useless to look at Messier 92 after you have viewed the expansive beauty of the huge Messier 13.

Unfortunately for this often overlook globular, it is truly a spectacular objects - in many ways more so that Messier 13. At first glance....particularly after seeing M-13....you may not think so as this globular requires a bit more dark sky and somewhat higher magnification with all of our telescopes. The FIRST thing you should notice is that it has an incredible dense concentration of stars all across the center of the cluster, much more so (compacted) than M-13 and many other such globulars. Indeed, the object is so hard to penetrate even with the 6- and 8-inch that star resolution is confined to the peripheral parts; however even a small scope and about 120x will resolve SOME outer stars in this magnitude 6.1 cluster. At a distance of over 35,000 light years, its far distance compared to M-13 can only suggest how this cluster would appear if at the same distance! For the 4-inch and larger telescopes, observing with a very well dark-adapted eye on a dark night with about 30x per inch will actually reveal a much more incredible sight than its rival M-13!

Object 9 - NGC 6229 - A faint and distant globular cluster
If you thought that M92 was small compared to M-

13, just wait until you see this one! This very distant globular cluster is located very near the more famous pair in our skies but is a very difficult object. In the small telescope at medium power, this will appear as a very fuzzy and dim 8th magnitude star. Only the "glow" instead of a point of light eludes to its non-stellar nature in those scopes. The 4-inch does only a bit better, but with higher power, it will begin to appear as a glowing ball of light. There is no resolution of this cluster in ANY of the moderate-sized scopes. At magnitude 8.5, this tiny-appearing globular extends only 3.5' across, compared to M-13's whopping 23' and M-92's 8' arc.

Object 10 - Planetary Nebula NGC 6210
This also is a tough object but is included here because of the seeming void of any nebulae to observe in Hercules; indeed, were it not for M-13 and M-92, Hercules would be a very dull constellation other than its hundreds of striking double and multiple star systems! NGC 6210 is a very difficult, magnitude 9.7 planetary nebula, the remnants of a stellar explosion. The "remains" - a tiny 12th magnitude star at the center of this very blue disk-like object can be seen under good dark skies and about 230x with an 8-inch and can be occasionally glimpsed in the 6-inch on very steady nights at the same magnification. The disk can be seen clearly in small scopes as 20" arc elongated "dot" (about one-half the size Jupiter appears in the telescope at the same magnification). Its color makes this a "must see" for the larger telescopes.

WANDERING ABOUT....YOUR NEW "USER OBJECT" FOR HERCULES

This brief GO TO Tour of Hercules has revealed its most interesting objects. But please do not stop here. Go ahead and locate many of the hundreds of objects that are present in Hercules.....there are several more interesting double and multiple stars that are within the reach of your telescope. In addition, use this opportunity to actually LOAD ANOTHER USER OBJECT onto your sky program!

This featured user "GO TO OBJECT" is much different than all the rest of those we have loaded before. This is an object we CAN'T SEE. No, it is not a black hole....but just as good. What you are about to load is a very famous **nova**, or "exploding star" from 1934, "*Nova Herculis*" or **DQ Herculis** as it is now known. This was one of the most brilliant novae of our last century, attaining a brightness of nearly +1.0 on Christmas morning, 1934. Prior to that time, the star was captured on film as nothing more than a common 15th magnitude object in Hercules....by December 13 it had suddenly brightened to 3rd magnitude and kept increasing from there! At its peak DQ Herculis reached a brightness that would compare to 65,000 times our own sun at that distance!!

It took nearly 90 days for the star, located just on the Hercules-Lyra border, to suddenly begin fading and today it is believed to be a dwarf star. It can still be seen with some MAJOR brightness fluctuations from its average brightness of about 9.2 however, and well worth checking as often as possible....that is why I have selected DQ Herculis as your USER OBJECT in Hercules. It is visible in all of our telescopes and you never can tell when the NEXT outburst will occur. The chart link below will allow you to download and print out the appropriate AAVSO chart for DQ Herculis....other wider field and more narrow field charts are available from the same link's home page:

https://www.aavso.org/apps/vsp/ . Note for these charts, simply type in the NAME of the variable at top to generate your choice of chart.

To load this nova on your User Object on AutoStar or other sky program,, go to: "Select/Object [enter] and scroll down to "user object" [enter]. Now enter these coordinates: R.A. 18 06' / DEC + 45 51' ; under "description" enter "DQ HERC NOVA" You will now have a wonderful "cataclysmic variable star" to monitor. If you suddenly see an outburst of this star, by all means *REPORT IT* to: www.aavso.org immediately!

*   *   *

Beautiful Greek Marble Statuary
Hercules destroying the Evil Centaur Nessus

* * *

Chapter 17

LACERTA
"....leapin' LIZARDS!"
(with my apologies to Little Orphan Annie)

Featured is our twentieth Constellation Guide, "GO TO LACERTA" of the series "GO TO GUIDES for all GO TO Telescope Users". In a milestone of sorts as we edge about one-quarter the way through the 88 constellations recognized "officially" by the International Astronomical Union (I have yet to see them strike, picket or demand a wage hike), we land in the vicinity of celestial **reptiles**....actually the ONLY celestial reptile.

I have always found it interesting - really - that there is not a single constellation nor bright star in the sky (nor has there EVER been) that is named on behalf of our beloved Toads or Frogs.

But, alas, the ONLY reptilian recognition by our celestial founding fathers is that of a common LIZARD, the small northern constellation of LACERTA, perched in his threatening lizard-like manner tightly between the large constellations of Cygnus to its west and Andromeda to its east. (note that I do NOT include the Serpent, SERPENS nor the Dragon (Draco) as a true "reptile," but rather mythological in nature).

GUIDES/Constellations/Pegasus) was transformed into a TRIANGLE overnight by the taking of *Alpheratz* to place into Andromeda's boundaries –
The tiny north-south winding pattern of relatively faint naked eye stars in Lacerta DO perhaps suggest

the curious "wiggly" motion of the hyperactive lizard when scurrying across hot pavement. Likewise a SNAKE would have been appropriate, but there were already snake-like creatures (Serpens and Draco) in the sky. So Hevelius, likely after watching a writhing lizard from his hot tub in Germany, aptly appointed this small star group as "LACERTA" and the name survived the scrutiny and heavy hand of the IAU's constellation subcommittee.

Lacerta also has a couple of notable distinctions for its absences of:

1) **Messier objects** - it is one of the few constellations that has NO Messier objects within its bounds AND no deep sky object brighter than magnitude 9.0; and,

2) **Named stars** - Lacerta has NO named star within its boundary;

3) **Galaxies** - no galaxies brighter than magnitude 14.5 exist within its borders;

4) **Globular clusters** - there are no globular clusters in Lacerta.

What Lacerta DOES have are some interesting variable stars (there are 33 nice variables in the constellation), and some wonderful double/multiple stars (at least 35 for the amateur telescope). In addition there are a few select galactic clusters that are visible in small and moderate telescopes and ONE planetary nebula that "may" be glimpsed with 6-inch or larger telescope.

Lacerta is located high north of the Celestial Equator (the "0" demarcation line in declination) and hence all objects for our "GO TO" TOUR will have positive ("+") declinations. This small constellation rises far north of due east at dark (around 9 p.m. local time) in early June, but is a nearly-circumpolar constellation with its high declination and thus is visible throughout most of the year for most latitudes north of 30 degrees. It culminates (passes the meridian at midnight) each year about August 25.

Each GO TO object in Lacerta is discussed for your telescope regarding the type of conditions necessary for you to view it optimally for discern the very faintest details.........magnifications and aperture necessary for most objects, and much, much more. This is YOUR complete guide to get you on your way to exploring the best objects (albeit only a few!) in Lacerta with your computerized telescope and its GO TO function. The following listing of "BEST" objects contains the finest or most interesting.

Use the attached star chart and the following Guide as an excellent reference for your next star party itinerary, or a beginning for further study into the thousands of objects visible in this part of the sky. Truly these extensive Constellation Study Guides will most definitely put your sky program to work for you in the most efficient and enjoyable way possible! As a matter of fact, MANY telescope users are now programming their own "Tours"
based on these guides, using each constellation as a separate GO TO Tour for the AutoStar or PC sky

program library that can be added in or deleted through the main edit screen on your PC or MAC computer.

We hope you enjoy these comprehensive GUIDES to touring the constellations via your AutoStar or PC sky program and its computer-driven telescope. Each new installment is complete with diagrams, charts and illustrations that you will find nowhere else. Please let us hear YOUR feedback and your observations of each and every constellation after YOU have toured its vast reaches of our skies!

YOUR LACERTA CONCISE DIRECTORY OF INTERESTING OBJECTS –

The small constellation of Lacerta presents stark contrast to its Milky-Way-rich neighbor to the west, CYGNUS and the fall spectacular to its east, Andromeda (Both constellations discussed in Volume One). There are only six (6) deep sky objects that are within reach of even the largest amateur telescopes and two of those will likely require a 5" or greater telescope for viewing. Nonetheless, there are some fantastic double and multiple stars at your beckoning call in this tiny constellation. To find out the details and coordinates of such stars, as well as a complete listing of suitable variable stars, consult with the "*Burnham's Celestial Handbook,*" Vol. Two.

For double stars, I have always recommended the "whatever" rule: Use whatever power is necessary to cleanly and convincingly resolve the double star; on nights of poor seeing conditions less power can be effectively used, but on very steady nights, the sky is literally the limit. However, you will note that on double stars which exhibit COLOR, higher magnifications will tend to wash out color contrasts, so use only the magnification necessary to obtain good separation and optimum image scale for viewing.

I have chosen the finest (or most interesting) 11 objects in this LACERTA "GO TO" tour; as with all guides, all objects listed below will be visible in most telescopes (some naked eye) from 3-inch to 8-inch; of course larger apertures may "show" an object a bit closer and "better," but frequently a wide field and low power view is more desirable than aperture for FINDING the objects initially. Indeed, I strongly encourage you first FIND the target object, or its approximate location through your GO TO function with your lowest power and then - once IDENTIFIED positively - move up slowly in steps with magnification if necessary. Remember, not all objects "like" magnification. Sometimes better "field of view" (such as the wonderful wide fields provided by small refractors) is desired over light gathering (like the larger 8-10 inch) and magnification.

The rule for determining "optimum magnification" is that: 1) too low power results in sky background glow detracting or diminishing the contrast against the deep sky object; 2) too high magnification darkens BOTH the sky background AND the object; 3) medium magnification can be achieved at which you have MAXIMUM contrast between the object and its darkened background sky. I have found through three decades of direct observing that about 15x per inch aperture, for deep sky observing is PERFECT for most objects. That being said, always remember that DOUBLE or multiple stars require whatever power you can crank out....the seeing conditions are the limiting factor here.

For my complete and comprehensive discussion regarding seeing conditions and sky transparency, see my discussion of this topic in *GUIDES - General*, on the ASO website: www.arksky.org .

With all deep sky objects, avoid attempting to observe when the moon is in the sky, even a very thin crescent, as its brightness in the sky will overshadow the very dim contrast afforded by even the brightest deep sky object; if you see the object at all against moonlight, you will NOT see the subtle outlying areas or the full detail of what is presented.

Lacerta is dominant in late summer skies; riding mid-way on the Celestial Equator, it is suited for long-period observing for telescope users both north and south of the equator. All deep sky objects and difficult double stars are ALWAYS best observed when they are located nearly overhead (or as high in the sky as possible), thus requiring the observer to look through the thinnest portion of the Earth's "lens" of atmosphere and haze.

As with all of the "GO TO" tour constellation lists, I recommend a good star atlas and/or chart which will list all the finest objects, constellation-by-constellation. One very handy reference guide is the *PETERSON FIELD GUIDE TO THE STARS AND PLANETS*, which features complete lists with declinations, right ascensions, magnitudes, and all pertinent information for you to expand your observing horizons beyond this brief guide.

Please see Volume One of *The Constellations* for a discussion on using computerized telescope functions and sky program.

The constellation tour Star Chart above will get you started on your journey for this constellation.

With all deep sky objects avoid attempting to observe when the moon is in the sky, even a very thin crescent, as its brightness in the sky will overshadow the very dim contrast afforded by even the brightest deep sky object.

Following is the concise object list for your "GO TO" TOUR of LACERTA; you may wish to find the majority of the objects from the sky program Library (for example, you can easily go to "ngc7209" if you pull up "Object/Deep Sky/NGC/..type in '7209'...." and then press "Enter", followed by "GO TO" to access this galactic cluster. On the other hand, if you want to experiment and become a "better computer user" try entering the exact R.A. and DEC coordinates of that object as described above after holding down the MODE key (Autostar). You will find the accuracy of entered GO TO's to be somewhat better than those stored in a scope keypad , but the capability of acquiring unlisted objects is fantastic!

Of course, for direct access of the constellation itself via AutoStar or sky program, you might choose to merely key in SELECT / OBJECT /

CONSTELLATION / ....and scroll to "Lacerta." Enter and then press " GO TO" and you are off to the brightest star of lizard land!

OBJECT 1: brighter star - (alpha Lacertae) - R.A. 22h 30' / DEC + 50 02 - Magnitude: 3.9 - nice optical double

OBJECT 2: incredible multiple star - Roe 47 - R.A. 22 30' / DEC + 39 31 - Mags: 6, 10, 10, 10, 9.5, 10.5 (yep!)

OBJECT 3: quad star - 8 Lacertae - R.A. 22h 34' / DEC + 39 23 - Mags: 6, 6.5, 9.5, 10.5 - possible in all scopes!

OBJECT 4: 4-inch test double - 13 Lacertae - R.A. 22h 42' / DEC + 41 33 - Mags: 5.5 & 11, tough for small scopes

OBJECT 5: variable star - RS Lacertae - R.A. 22h 11' / DEC + 43 31 - Mag: 9.5 to 12.4 - semi regular, 234 days!

OBJECT 6: galactic cluster - ngc7209 - R.A. 22h 02' / DEC + 46 16 - Mag: 9, very large and open

OBJECT 7: galactic cluster - IC1434 - R.A. 22h 09' / DEC + 52 35 - Magnitude: 10, very small, tough

OBJECT 8: galactic cluster - ngc7243 - R.A. 22h 13' / DEC + 49 38 - Magnitude: 7.5 total, large and open -BONUS!!

OBJECT 9: galactic cluster - ngc7245 - R.A. 22h 14' / DEC + 54 05 - Magnitude: 11.3, tough, 5-inch and larger +

OBJECT 10: galactic cluster - ngc7296 - R.A. 22h 26' / DEC + 52 02 - Magnitude: 9.4, tiny, only a few stars

OBJECT 11: planetary nebula - IC5217 - R.A. 22h 22' / DEC + 40 43 - Mag. 12.2, very tough and TINY planetary

....SO LET'S LEAP ON OUR LIZARD-OF-LATE AND LOLLYGAG AWAY!! (refer to the Lacerta Star Chart for all the objects described in detail on the guide)

A VISUAL GUIDE TO OUR DEEP SKY OBJECTS IN LACERTA –

Object 1 - Brighter Star - alpha Lacertae - Also a Nice Optical Double Star
Notice that the above description indicates "brighter star," and NOT "bright star." At magnitude 3.9, this star is one of the faintest primary stars within any of the 88 constellations. It is a "solar type" star, type-A, very similar in both mass and evolution to our sun. Alpha Lacertae - also one of the FEW unnamed primary stars of any constellation - is a pretty nice optical (not physically bound by gravity) double star, also known as "Barnard 703." At a relatively far (36" arc, about the extent of Jupiter's diameter seen in the same eyepiece) is VERY faint, magnitude 12, star almost due northwest of the 3.9 magnitude Alpha. This very faint star will be best seen in an 8-inch, but can be glimpsed in a 6-inch on very dark nights; even in a 4-inch, if the brighter star is obscured with a dark crosshair of a reticle eyepiece, might reveal this faint and more distant star.

Object 2 - Roe 47 - A Complex Multiple Star
Examine the diagram I have prepared below for the interesting quadruple star "ROE 47" in Lacerta. This complex star system - just as the next object, "8 Lacerta" which is even MORE complex - is a wonderful challenge for a four inch scope, and even is a test for patience with those using larger

telescopes. Certainly the four main stars of this true physical quadruple stars are bright enough to be discerned clearly and without doubt in the 3.5" scope - magnitudes 6 (primary star), 10 (star B), 10.5 (star C) and 10 (star D). In addition the separations from the brighter 6th magnitude star are ample for a clean split.

In the past I have always discussed the location of a secondary star to its primary in terms of DIRECTION (i.e., North, South....or NW, SE, SSW, etc.). In the "real world" of astronomy, we refer to the location of one star relative to another in terms of POSITION ANGLE, as shown in the following two diagrams. Although these diagrams are oriented for YOUR telescope's view - NORTH at top and EAST to the right - they also are correctly oriented for double star plotting.

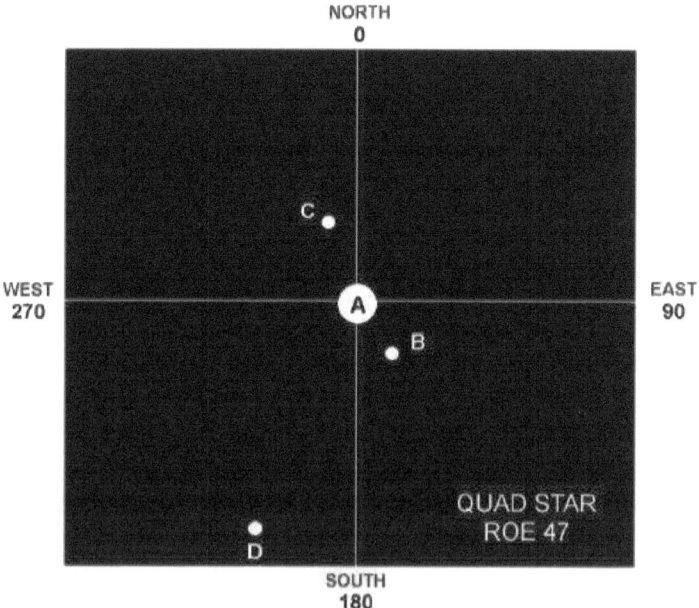

POSITION ANGLE is the measurement in degrees from 0 degrees to 359 degrees as measured FROM DUE NORTH toward EAST....east is P.A. 90 degrees....south is P.A. 180 degrees....west is P.A. 270 degrees and due north from center (the primary star again) is 360 degrees. It is very simple....you merely use this layout exactly as shown like a protractor and always measure clockwise from NORTH through EAST. So let us look at these two complex star systems beginning with this Lacerta GO TO Guide in terms of BOTH direction and position angle.

Star "B" in Roe 47 is located as you can plainly see roughly SE of the primary star "A"; its true POSITION ANGLE is 158 degrees as measured from north clockwise through east. Look for this 10th magnitude star about 43" arc (almost exactly the same distance in "arc that Jupiter's disk appears in the same eyepiece) in P.A. 158. Similar, 10.5 magnitude Star "C", at only slightly less distance is in P.A. 344 degrees, or NNW of the primary star. Tenth magnitude Star "D" is SSW of "A", at a distance of 106" (about twice the distance of the other two from "A"), and is found in P.A. 216 degrees.

This is a fun object for any four inch and larger scopes and added aperture just makes it easier and more enjoyable. Medium powers are recommended, perhaps about 15x to 20x per inch aperture for best viewing of all four stars within the same field of view. It IS possible that all four of these stars CAN be seen at high magnification (about 100x) very small telescopes.

Object 3 - "8 Lacertae" - An Even MORE Complex Multiple Star - THIS ONE IS A CHALLENGE! You will likely need a short break after attempting this one. Here is a system of SIX - yes SIX - stars that are an actually gravitationally-bound group in a common and complex orbit.

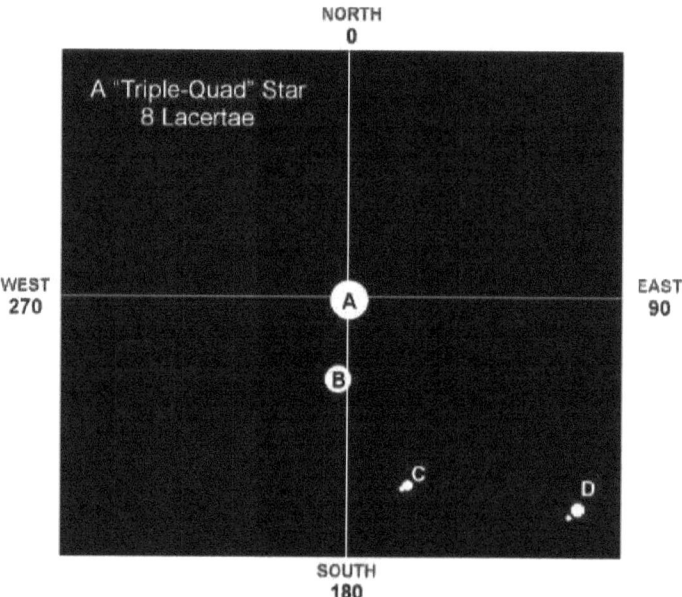

We will use again our guide from POSITION ANGLE to locate each of these stars. Note that FOUR (4) of these stars should be visible in the ETX 60 (high power and dark skies) with difficulty; the same four with a 3-inch rather easily, FIVE (5) with a good 6-inch and medium high (about 160x) magnification, and all SIX (6) with the 8-inch or larger under very dark skies and powers approaching 400x being required. The stars are again designated "A" (the primary), "B" at P.A. 186 degrees (nearly due south from A a wide 22" arc and a bright 6.5 magnitude); "C" much fainter at magnitude 10.5 in P.A. 169 degrees, again very far off at 48" arc; and "D" at magnitude 9.5 and twice as far (81" arc) as any of the rest in P.A. 145 degrees (nearly SE) from "A". Notice from the chart above how extremely close that "C" and "D" both ALSO have companion stars! The tiny 14th magnitude companion to "C" can ONLY be seen in the LX 90 and then it is a challenge....separated ONLY by 1.3" arc! (the position angle of that 14th mag. star is measured at 254 degrees from "C", and not from "A" in this case!). Star "D" has a slightly easier companion star - 13th magnitude - that can be glimpsed in the larger scopes in P.A. 226 degrees from "D" by 9.1" arc, so it should be MUCH easier to spot.

Whew!.....take a break. But DON'T MISS THIS STAR!....and "yes" this will be your "User Object" for this installment of our "GO TO" tour for Lacerta! (how could I resist!?)

Object 4 - OKAY...Let's have an EASY Double Star - 13 Lacertae - Good test for 4-inch telescope If you have recovered from that hair-raising experience with Object 3, then you are raring to get started on another double star....but this one is a piece of cake. The smaller secondary is right at the limiting magnitude of a 3-inch at about 150x; a 6-inch and larger scopes will reveal the companion easily. The star 13 Lacertae features an easy-to-find 5.5 magnitude star as the primary star and an 11th magnitude secondary star at P.A. 129 degrees and nicely spaced 15" arc away. Do you remember WHERE Position Angle 129 degrees is?

Remember the rules:

1) The primary star is the "center of a clock;"
2) Due NORTH from that star is "0" degrees and the clock moves clockwise to measure;
3) When you reach due EAST from the primary star (a "3:00 position on the clock") that is 90 degrees;
4) "6:00" on the clock face is due south from the primary and is 180 degrees P.A.;
5) As the clock "rotates" clockwise, we reach WEST at 270 degrees P.A.
6) Back to straight-up north, we are in P.A. 359 to 0 degrees!

So....based on that....have you figured out that the 11th magnitude secondary to "13 Lacertae" is just about ESE from its primary? In nearly all future constellation guides, I will be referring to double stars in POSITION ANGLE placement (with little reminders on what that means) as we all are attaining more and more proficiency with the sky! Good luck!

Object 5 - A Good Semi-Regular Variable Star, RS Lacertae

RS Lacertae is a very good variable star for observers with small to moderate telescopes, as it ranges in magnitude from 9.5 to 12.4 through its "not-so-regular" cycle of about 237 days. Within that long cycle are many "ups and downs" and sudden outbursts which may take place in only a matter of hours or days! Remember that if you are interested in serious variable star research and contributions, to contact the American Association of Variable Star Observers in Cambridge, Massachusetts via: http://aavso.org/ for a complete description of the many publications and charts as well as observing manuals that are available online.

Once you have used your AutoStar GO TO to acquire the low power field surrounding RS Lacerta, refer to the charts
https://www.aavso.org/apps/vsp/ . Note for these charts, simply type in the NAME of the variable at top to generate your choice of chart –
in which is the "a" scale finder and star comparison chart for given star magnitudes down to about magnitude 12.4; this chart should be ample for all observations with all scopes, since the star's magnitude rarely goes lower than 12.4 (there is a great "12.4" magnitude comparison star located right next to the spot for RS Lac!) A 6-inch and larger scopes will have no trouble following this star through its complete cycle; at dimmest (which does not last long), the star will be right at the limiting magnitude threshold for a 4-inch.

Object 6 - Finally! One of the few Deep Sky Objects in Lacerta - Galactic Cluster ngc7209
You can put another notch on the old blue tube assembly for this NGC object; just knowing you saw it should send chills down your spine, since you likely have not ever seen this (nor any of the other ones in Lacerta) before....and likely will not again.

NGC 7209 is a very large (20' arc...the moon is only 30'!) open cluster of about 50 stars scattered about. If you use too much magnification, you will NOT see the cluster effect at all. As a matter of fact, other than is the truly huge wide field 2" eyepieces for larger telescopes, this does not appear as a cluster at all due to its size. However, all 50 some odd stars, magnitude 9 through 12 should be visible in both of those scopes as well as most in smaller, wide field telescopes. Very low powers on very dark skies with small scopes will reveal a very faint glow, perhaps the nicest view of all. This cluster is one of many on the very eastern fringes of the Milky Way through Cygnus.

Object 7 - Another Peripheral Milky Way Star Cluster - IC1434
Located JUST inside the northwest corner of Lacerta (right inside the Lacerta-Cepheus-Cygnus borders) is this tiny, tiny - and very faint - galactic cluster of about forty stars crammed into the small space of only 8' arc; it has a total magnitude of only 10.0, so the 40 odd stars are in the 9th and 11th magnitude range, thus accessible to a 3-inch at medium high magnification; because of its small size and faint stars I highly recommend about 30x to 40x per inch aperture in all telescopes. This object will be missed in smaller telescopes.

Object 8 - Nice Galactic Cluster with a Bonus! NGC 7243 + Struve 2890 Triple Star!
This is a fairly bright and large open cluster, the same size (20' arc) as ngc7209, but somewhat brighter at a total magnitude of 8.0; its 45 stars range between 8.5 and 11, including the wonderful TRIPLE star Struve2890. The cluster itself is clearly visible in all scopes and lowest powers are best.

Look for Struve2890 DEAD CENTER of this cluster; the primary star is the brightest star of the cluster at magnitude 8.5. Look for its "B" component (mag. 8.5 also!) a full 9.4" arc away, easily split all scopes from the small APO and up! Their equal magnitudes and yellow colors provide a wonderful medium power sight. They are arranged in a near north-south orientation from one-another so it is difficult to determine "which" is the actual primary star. If you look at P.A. 278 degrees (nearly due west) from the mid-point of the two brighter stars, you will find ANOTHER easy star very far away (94" arc - two "jupiters" distant) that is magnitude 9.4 and also visible in all our scopes, although difficultly so in the small telescopes.

Object 9 - Yep...you guessed it: Another Galactic Star Cluster! NGC 7245
Just north of IC 1434 (see above) and in the northern extreme of Lacerta is a VERY small (only 3' arc!) cluster, the farthest (by double) of all the clusters in Lacerta. This tiny object is ONLY magnitude 11.5 and is not visible in a 6-inch, even though that scope's limiting magnitude should reach that far. Its 50 stars are seen in only the world's largest telescopes; in a good 8-inch this will appear as a very small fuzzy star, only the size of the planet Mercury at the same magnification!

Object 10 - Our Last Faint Fuzzy Galactic Cluster - NGC 7296
Another very small (4' arc in diameter) galactic cluster, ngc7296 CAN be glimpsed in both the ETX 90 and the ETX 125 as a VERY faint (magnitude 9.4) fuzzy object that clearly is not a star. It has only about 15 stars total, all too faint to be seen in those scopes but I have clearly seen two (2) faint stars in an 8-inch; it is resolvable entirely in the 24" at very high magnification.

Object 11 - Finally - A Planetary Nebula....but don't get too excited folks. This one is TOUGH!
NGC 5217

Okay....how about a planetary nebula that is nearly 12.5 magnitude and ONLY 12" arc. This is a large-scope object only. In telescopes 5-10 inches, it CAN be seen, but only as a roundish pale glow even at 447x under steady and dark skies. Only the imagination reveals that it "might" be nebulous similar to a planetary nebula. Larger scopes can easily reveal the central star at magnitude 14.5. Many observers with large telescope report that this central star is actually easier to make out than the faint shell of nebulosity.

WANDERING ABOUT....YOUR NEW "USER OBJECT" IN LACERTA

Here is your interesting triple star Struve2890 imbedded in the galactic cluster ngc7243! This is a double bang for your "User Object Buck!" for your Autostar library and one that you would likely overlook and never pursue if you could not GO TO it directly. Remember \ that ALL three stars will be easy (under dark sky conditions) in the telescopes larger than 6 inches.

On AutoStar, go to: "Select/Object [enter]...." scroll down to "User Object" [enter]. Now enter the coordinates given above for "Struve2890", using the number keys on AutoStar. After entering the coordinates and pressing "Enter" yet again, scroll down one and you can list the magnitude of the object as "8"[Enter]. This will be your first "double"-whammy on the USER OBJECT list, two for the GO TO of one!

Chapter 18

LEO
Entering the Domain of the "Celestial Lion"

This is the second installment, "GO TO LEO", of our constellation study guides for all GO TO telescope users. The last installment featured Ursa Major, located high in northern declinations for easy year-round viewing. This Guide will feature a constellation equally as recognizable yet one located very close to the *CELESTIAL EQUATOR*, or roughly centered on either side of "0 degrees" declination. As will all of our "GO TO GUIDES", it features a "start" with an easy GO TO to the bright star **REGULUS**, and then proceeds through the many fine examples of double and multiple stars, galaxies and interesting objects within reach of YOUR telescope! From distant galaxy groups to beautifully-colored and complex multiple star systems with discussions along the way tell you what to expect from each telescope size and type. All objects will be discussed with exact descriptions of what observers with 3- to 8-inch telescopes should expect to see...and what to NOT expect to see!

The finest stellar and deep sky objects in the constellation will be featured....and - yes - there will be something for everyone and every telescope...Even naked eye and binoculars when appropriate!

Discussed are useful magnifications for EACH GO TO object, what type of night and conditions are needed to see certain details, double stars that can be resolved in each telescope model, and much, much more. It is your complete GUIDE for your deep sky observing pleasure and a very handy tool for use at your next star party!

Needless to say, it WILL put your computerized telescope to work for you in a most efficient and enjoyable way!

I hope you will enjoy these comprehensive GUIDES to "Touring the Constellations" which will feature a NEW constellation about every two weeks, complete with diagrams, charts and illustrations. Please let us hear from you with summations of YOUR observations through these constellation tours!

Introduction

Few constellations really appear to us as what their mythological designations signify. I have never seen a "virgin" around Virgo, an "archer" in Sagittarius, or a goat herder in Auriga. But some among those 88 constellations which remain today are unmistakable in their identify. Perhaps this is why - even before communications and travel between the continents, and before cultural mythology and lore were exchanged between peoples of the world - that some constellations have always had a common association world-wide; such is the case with **Scorpius**, the Scorpion....**Orion**, a hunter or a great persona of a man....**Ursa Major** as a large bear.

And among those must be **LEO**, the celestial "lion" that roars its way into our skies each spring.

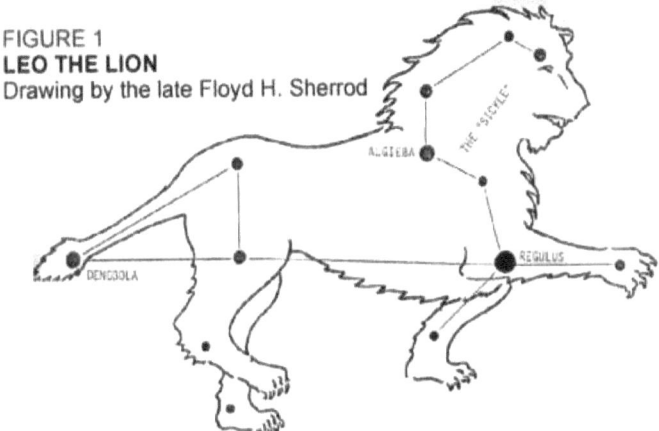

FIGURE 1
LEO THE LION
Drawing by the late Floyd H. Sherrod

No matter how you draw the lines, this figure still comes out looking like a lion. Curiously, when upside-down, as it would be viewed by our observers south of the equator, it is very difficult to even recognize, much less associated to a lion, this constellation.

Although the ancient Egyptian sky watchers worshipped this constellation as "*The House of the Sun*" to signify when the sun entered the star group, associated with the flooding of the **Nile River**, (thereby providing precious water for irrigation to the parched arid surroundings) they too perhaps signified this constellation as "lion-like" in both hieroglyphic renderings as well as a possible association to the great Sphinx of the Egyptian plains.

After the birth of Christianity, the constellation was known to early Christians as Daniel's *Lion's Den*, from the Book of Daniel. Earlier, the book of

Genesis signifies this constellation as the symbol for *Judah*.

In ancient China, star watchers were a bit different in their views of this group....to them, it was the "*Great Yellow Serpent*" or "dragon."

Note the many bright stars in the drawing (Figure 1) above that outline the head, body, tail and even feet of the great celestial lion. The stars of the lion's "mane" are also known as the "*SICKLE*" from its characteristic shape, and it is among those "sickle stars" that the famous and beautiful "headlight double" - **Gamma Leonis** is found.

As we explore Leo, we are entering the "*Realm of the Galaxies*" from our vantage point in space. Just east of Leo is the fantastic galaxy-packed constellation of *Coma Berenices* and others which exhibit hundreds of thousands of galaxies to the world's largest telescope. There are perhaps over 10,000 galaxies photographable to the world's largest telescopes in Leo alone.

The bright star **REGULUS** is so near the celestial equator (only 12 degrees north) that it provides an excellent bright white target for both star tests and for accurate telescope polar alignment using the "drift method."

YOUR LEO "GO TO" TOUR - THE CONCISE DIRECTORY

We will concentrate on 11 objects in LEO for this quick tour; all are in reach of every telescope from 3-inch to 8-inch, yet each telescope will demonstrate uniquely different and challenging aspects of the objects. I n addition to the 11 finest objects, there are MANY more exciting things to see in our celestial Lion than those given here - hundreds of double and multiple stars, hundreds of galaxies and deep sky objects, stars of curious colors and motion - that can be found with your telescope. Perhaps more than any other constellation of Spring, Leo offers a wealth of fine **double stars** to test the limits of your telescope!

I recommend good a good star atlas and/or chart/and PC sky program which lists the finest objects constellation-by-constellation; for Autostar and similar you merely need to hold down your MODE key (or other scope command) for three (3) seconds and the RA and DEC coordinates appear for the telescope. Merely press "GO TO" and the cursor appears prompting you to enter the Right Ascension of the object if it is NOT listed among the objects in the computer library; once the RA is entered, press "Enter" and the cursor once again prompts for the Declination coordinates (these coordinates for epoch 2000) are found in all good observing guides). Once those are entered, merely press "GO TO" once again and your computerized telescope will slew to the position of the object!

The following constellation guide to objects at the end of this tour will describe all the details of each object you wish to view and give specifics as to visibility in YOUR telescope model.

Clay Sherrod's
CONSTELLATION GUIDES:
"Go To" Leo - The "Celestial Lion"

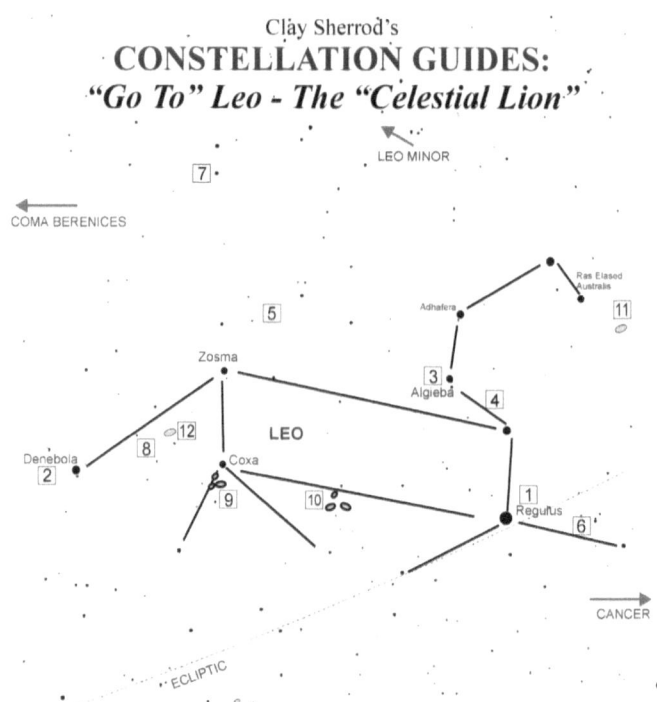

The special star chart above will get you started, as it demonstrates the relative positions of all objects in this "tour" to the conspicuous stars outlining the distinct figure of our celestial lion."

Following is the complete 11-object list for your "GO TO TOUR" of Leo; you may wish to find the majority of the objects from the AutoStar library (for example, you can merely pull up Messiers 65 and 66 by going to "Object/Deep Sky/Messier Object/M-81....enter....GO TO" or...if you want to experiment and be a "better AutoStar user", try entering the following coordinates as described under MODE above.

58

OBJECT 1: star - REGULUS (alpha Leo) - R.A. 10H 06' / DEC + 12 13 - Magnitude: 1.4

OBJECT 2: optical double star - DENEBOLA (beta Leo) - R.A. 11H 47' / DEC + 14 51 - Magnitude: 2.1

OBJECT 3: beautiful double star - ALGEIBA (gamma Leo) - R.A. 10h 17' / DEC + 20 06 - Magnitude: 1.9

OBJECT 4: multiple star + galaxy - TANIA AUSTRALIS (MU UMa) + ngc3184 - R.A. 10h 19' / DEC + 41 45 Magnitude: 3.1 (galaxy is 10.3m)

OBJECT 5: close nice double star - Leo 7704 - R.A. 10h 14 / + 17 59 - Magnitudes: 7.2 & 7.4

OBJECT 6: famous variable star - R LEONIS - R.A. 09H 45' / DEC + 11 40 - Magnitudes: 4.4 TO 11.3 (45 DAYS)

OBJECT 7: great double for ETX 60/70 - ALULA AUSTRALIS - R.A. 11H 16' / DEC +31 49 - Magnitudes: 4.4 & 4.8

OBJECT 8: double star - Leo 90 (good test for 4-inch) - R.A. 11h 32' / DEC + 17 04 - Magnitudes: 6.0 & 7.1

OBJECT 9: galaxy cluster - Messiers 65 (ngc 3623), 66 (ngc3627) and ngc3628 - R.A. 11h 16' / DEC + 13 23

OBJECT 10: galaxy cluster - Messiers 95 (ngc 3351), 96 (ngc3368) & 105 (ngc3379) - R.A. 10h 41' / DEC + 11 58

OBJECT 11: galaxy - ngc2903 - R.A. 09h 29' / DEC + 21 44 - Magnitude: 9.1

*   *   *

A VISUAL GUIDE TO OUR OBJECTS IN LEO

<u>Object 1</u> - Bright Star Regulus (Alpha Leonis)
This bright white star is known throughout the world as "*The Heart of the Lion,*" and fittingly so. It is a "B" type very bright star, shining at about 160 times the light of our own sun. Even though it is white, it is a very "late" star, having depleted much of its hydrogen-rich fuel and now exhibiting fusion through helium.

In ancient middle eastern cultures, Regulus was one of the four (4) "royal" stars: *Regulus, Aldebaran, Fomalhaut* and *Antares*. The Babylonians before them depicted Regulus as "*the King*", with no mention as to a reference to a lion, only to royal association. Because of the seasonal rains that bring life-giving nutrients to the arid regions, many such cultures associated this star - as did the Egyptians the entire constellation (see above) - to great floods and a time of increased agriculture.

An interesting aspect of Regulus is its location so close to the ecliptic (see the constellation chart above); what this means to amateur astronomers is that many times this bright star is **OCCULTED**, or "covered up" by the moon as the earth's satellite slowly moves eastward through the ecliptic. Such occultations are spectacular when involving a bright star. In 1959 Regulus was actually occulted by the planet VENUS, an event so rare that many centuries will pass before anything close occurs again.

Look for a very difficult (because of the brightness of Regulus) companion star at a Northwest angle from Regulus at a distance of 180" arc, or about 5

Jupiter widths away; medium-to-high power is recommended to see its companion. This 8th magnitude star is, itself, double, but it requires at least a 16-inch telescope to resolve.

Object 2 - Bright Optical Double Star DENEBOLA (beta Leonis)
This bright yellow star, only 43 light years distant, signifies the "Lion's Tail" of Leo; also Arabic is the name "Serpha" which translates "keeper of the weather," perhaps also a reference to the flooding of the Nile river when the sun moves into the boundaries of Leo.

There is a significant difference between an "optical double" and a "physical double" star; the optical double, like this target star, Denebola, only APPEARS to be two related stars when, in fact, they are seen at two greatly different distances. In fact, Denebola has MANY optical companions; the one visible in the 4-inch and larger telescopes is nearly due south, about 4' arc, from the bright yellow main star.

Object 3 - Bright Star Double Star **ALGEIBA** (gamma Leonis) "*The Auto Headlight Stars*"
This star, folks, is a keeper. If you look at NOTHING ELSE in Leo, you MUST see this incredible sight. Not only is this one of the finest - and easiest resolvable - double stars in the sky, but it shows a remarkable brilliant deep yellow coloration with both components, and they appear as to automobile HEADLIGHTS coming right at you out of the sky.

This is an excellent test star for a smaller telescope

but easily resolvable at medium magnifications in larger scopes. In my telescopes, I frequently have my best views of this double at about 400x with a high quality eyepiece. The figure below shows the incredible and unforgettable sight of this star seen at very high power.

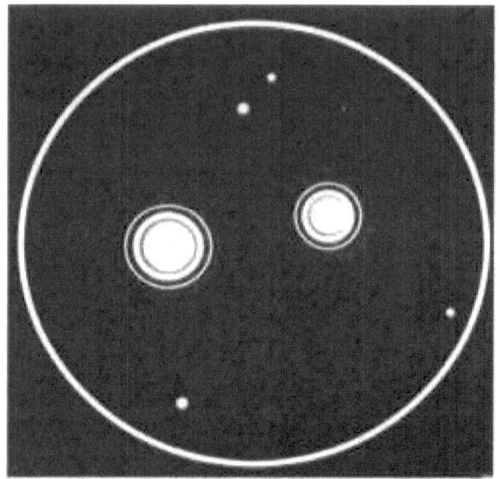

FIGURE 2
Drawing of the beautiful double star Algeiba through the ETX 125 @ 404x

When highest in the sky, steady air and high magnification shows Algeiba with two distinct yellow Airy disks surrounded by two very nice rings of light if your system is perfectly collimated. There is about a magnitude difference (2.1 and 3.4) between the two of them.

The name "Al Jabbah" appropriately signifies the "*Lion's Mane*" from the original Arabic designation.

After you have "looked it all up", turn your telescope for a moment away to the northwest from Algeiba about 2 full degrees and put in your lowest power eyepiece (small, wide field scopes will have a great benefit here). The incredible LEONID meteor show (peaks November every year) actually appears to emanate from this exact location as the Earth turns into the swarm of debris left from **Comet Swift-Tuttle** in 1866!

Object 4 - Double Star 7704 Leonis
This will be a good test for you to enter your own R.A. and DEC coordinates on the AutoStar or similar device since it is not part of the AutoStar library of double stars or named stars. Enter the coordinates (listed above) as described earlier via the MODE key and once entered, merely push "enter" and the telescope will slew to this very, very nice star! This is a TEST for your ETX 125....even though both stars are 7.5 magnitude, they are only 1.2" arc apart, theoretically resolvable. I have actually been able to see some "elongation" in a 3-inch of these two stars, but no clean separation; however, both the 6- and 8-inch scope will definitely split it. High magnification is a MUST. Note: make sure you have the right star....there are several of that similar magnitude nearby!

Object 5 - Great double star "54 Leonis" - A super double star test for smaller APO refractors
Again, use the "mode" key on this one and enter your coordinates. This star is not so difficult to find, however, as both are brighter - magnitudes 4.5 and 6.3! They are separated by over 6" arc so the ETX 60/70 with high power can plainly show this very pretty pair. Use low powers on the 6- to 8-inch scopes....medium magnification (120x) on a 3- inch is ideal.

Object 6 - Variable Star **R LEONIS**

This is our very first variable star on the Constellation Tour Guides, and it is one of the most famous and easy-to-monitor for all of our scopes. The entire cycle of brightening (to magnitude 4.4, naked eye!) and dimming (to 11.3, right at the threshold of a 3-inch on a dark night) can be followed in a very short period of only 44.9 days! You can keep a log and monitor this red "*Mira-type*" star as it goes through its pulsating increase in size to maximum and slow (yes, actually shrinking) to minimum. Once your data is collected for several cycles, you can compile a "light curve" exactly like the one shown below!

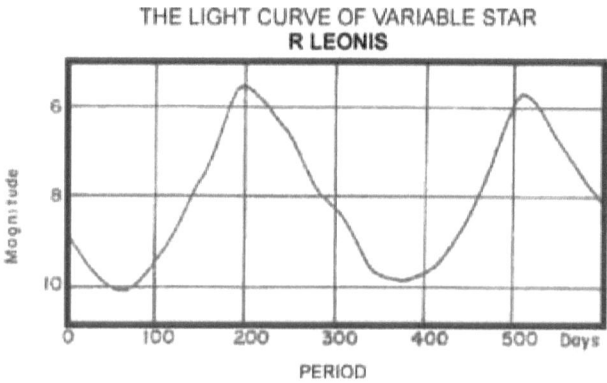

R Leonis is known as a "long period variable" and perhaps one of the best for getting started in variable star observing (see my Guide to Variable Stars.... here on this web site under "Observational Guides...."). Below is a click-on for you to obtain the wide-field chart for R Leonis from the American Association of Variable Star Observers; this chart is suitable for observing magnitudes when the star is brightest....access www.aavso.org for a full set of downloadable charts to watch this wonderful star through its dimmest part of the cycle.

R Leonis Finder/Comparison Chart link: (note that "reversed" charts are available for catadioptic and all variations of telescopes) https://www.aavso.org/apps/vsp/ . Note for these charts, simply type in the NAME of the variable at top to generate your choice of chart.

Object 7 - Double Star for Small Scope Test – Beautiful object for all!
This star, ALULA AUSTRALIS, is actually in the constellation of Ursa Major (one of its back paws) and was featured in the previous constellation TOUR; if you missed it, check it out while you are here. This is a fine double star that is challenging for the smaller telescope at medium power and a beautiful sight in all telescopes.

Object 8 - Nice Double Star 90 Leonis - key in your coordinates to find this one!
This double star is an excellent test object for the 4 inch, with the two stars (magnitudes 6.1 and 7.1) separated by about 3" arc, which should be easily resolvable on a steady night with high (about 160x+) magnification. The 6-inch will easily split these two components at 73x, as will the 8-inch at 77x. The two stars will appear very close together, nearly equal in brightness, with the dimmer of the two a bit West of due SOUTH from the brighter star. If you cannot immediately resolve, check to make sure you have the correct star located via the "Mode" key to enter your own R.A. and DEC. Also, increase your magnification if the star does not readily split into its two components.

Object 9 - Galaxy Group - **Messiers 65, 66** and **ngc3628**

This is one of the nicest and easiest-to-locate of your GO TO objects; both **M-65** and **M-66** will be in the same lowest power field of view (use a quality low power eyepiece); with the many other stars in the field, it is a remarkable sight, with M-65 appearing distinctly oval in shape and brighter than M-66 to its "left" in the field of view.

Messier 65 photographed from
Arkansas Sky Observatories
* * *

Both galaxies are oval in shape as seen in your scopes, with M-65 seemingly more cigar-shaped but with M-66 showing a bit more detail on its southern perimeter to the larger 6-inch and 8-inch scopes on VERY dark nights. This detail is actually extensions of a prominent galaxy "arm" and associated star clusters within that arm. M-65 is a uniform, featureless glow in all scope sizes.

Messier 66, in Leo
From Arkansas Sky Observatories
*   *   *

Just to the north of these two, and still within the same fields of the smaller telescopes, is the wonderful ngc3628 galaxy, seen nearly edge-on with a very pronounced dark "lane" across its edge. This dark "streak" can be seen clearly with the 4-inch on a very, very dark night, but NOT with anything smaller; the galaxy, however, CAN be seen with these scopes. The 6-inch and larger scopes reveal a great amount of detail on ngc3628 under perfect dark sky conditions with magnifications from 150x to 210x. You might be able to distinguish the dark lane as it "separates" one half of the galactic hub (it will appear brighter north of the dark lane and slightly more narrow and dimmer below, or south, of the dark lane). These are but a few of the remarkable collection of galaxies throughout this region of sky.....use some good reference books and the next clear moonless night to peer deeper into space that you likely have ever gone!

Object 10 - Galaxy Group - Messiers **95, 96** and **105**

Like the group just described, this is yet another close assemblage of fairly bright galaxies; all should be visible in the same field of view with the 4-inch, although vague; the 6- and 8-inch telescope will likely require (unless a really wide field eyepiece is used) you to move a bit to the northeast from M-96/95 to access M-105. The three galaxies are about 8' arc apart, making a pretty equal triangle. M-96 is actually a full magnitude brighter than M-95 (magnitude 10.1), although it is somewhat smaller. Messier 105 (magnitude 9.5) is also brighter than M-95, but can easily be mistaken for a "fuzzy star" as it is very small (only 2' arc across) and very round. It is very difficult to distinguish in the small telescopes and requires medium power in the larger telescopes.

Object 11 - Galaxy ngc2903

This bright (magnitude 9.3) oval-shaped galaxy will look quite symmetrical in a 4-inch and larger scopes, appearing elongated almost due north-south; it is very bright toward the center and then falls off rapidly toward its large 11' arc length. Because it is so large, its brightness is scattered over a very large area and thus appears dimmer than it actually is. The larger aperture telescopes will clearly show some "rough" mottling across this face-on galaxy, with many dark and light zones appearing where star clouds and dark matter actually exist in this nice galaxy.

WANDERING ABOUT....YOUR NEW "USER OBJECT" FOR LEO

This brief GO TO tour of Leo has revealed its most interesting secrets. But please do not stop here. Go ahead and locate many of the hundreds of other ngc galaxies that are present in this constellation.....there are several more interesting double and multiple stars that are within the reach of your telescope. In addition, use this opportunity to actually LOAD ANOTHER USER OBJECT onto your sky program!

Your friends, family and fellow stargazers would be delighted if you could show them the "Automobile Headlights in Leo" in your computerized telescope; merely go into "Select/ Object/ User Object....[enter]" and then add in the coordinates (R.A. and DEC given above) and a name/description of the beautiful yellow "headlights" of gamma Leonis and press "enter" to store your information into the Sky library!

*   *   *

DID YOU KNOW....

The famous artist **Vincent Van Gogh**
Was fascinated by the night time and its stars?

In addition to his representations of the starry night through his art, here are a few of his memorable quotes:

"Be clearly aware of the stars and infinity on high. Then life seems almost enchanted after all."

***

"I don't know anything with certainty, but seeing the stars makes me dream."

***

"I often think that the night is more alive and more richly colored than the day."

***

"At present I absolutely want to paint a starry sky. It often seems to me that night is still more richly coloured than the day; having hues of the most intense violets, blues and greens. If only you pay attention to it you will see that certain stars are lemon-yellow, others pink or a green, blue and forget-me-not brilliance. And without my expatiating on this theme it is obvious that putting little white dots on the blue-black is not enough to paint a starry sky."

***

Chapter 19

LEPUS
The Hare Beneath the Feet of a Great Hunter

BONUS: *Your "GO TO" Guide to Observing CARBON STARS"*

With this constellation of LEPUS, are now deep into the winter skies of the northern hemisphere in our "GO TO" tours for constellation studies with computerized telescopes of all types. This installment of our *Constellation* guides features an oft-ignored little hare of a constellation resting beneath the feet of Orion, the mighty hunter. Orion was well covered in our last three installments of the "GO TO" tours, being a comprehensive three part series. LEPUS, immediately south of the Great Hunter, is often skirted over quickly after observers "ewww!" and "ahhhh" over the splendors of the magnificent bright and young stars and nebulae of Orion. However, there is a wealth of fine objects in Lepus and.....well....while you are in the neighborhood anyway....

Let's explore the relatively small but object-rich confines of LEPUS, the rabbit or "hare."

Oddly, *LEPUS* has little mythological reference or history. Indeed, there is no reference to the "hare" in the Greek, Arabic, nor Chinese records of antiquity. The typical assumption is that the constellation was placed - and named - where it is in reference to the Mighty Hunter, Orion. For the pitiful and fragile rabbit is poised beneath the firm footing of Orion (the bright stars Rigel and Saiph) to the north, and running away from (westward) the pursuits of Orion's hunting dogs, CANIS MAJOR and CANIS MINOR, noted to the east by the very bright stars "Sirius" and "Procyon", in Canis Major and Minor, respectively. It is interesting to me that LEPUS was included in **Ptolemy's** original 48 constellations established in the 2nd century A.D., for typically only the 12 constellations of the Zodiac and those with strong heritage and Greek mythological ties were included in his early list.

At any rate, the rabbit does not look one bit happy in **Bode's** 1603 star atlas rendering reproduced below:

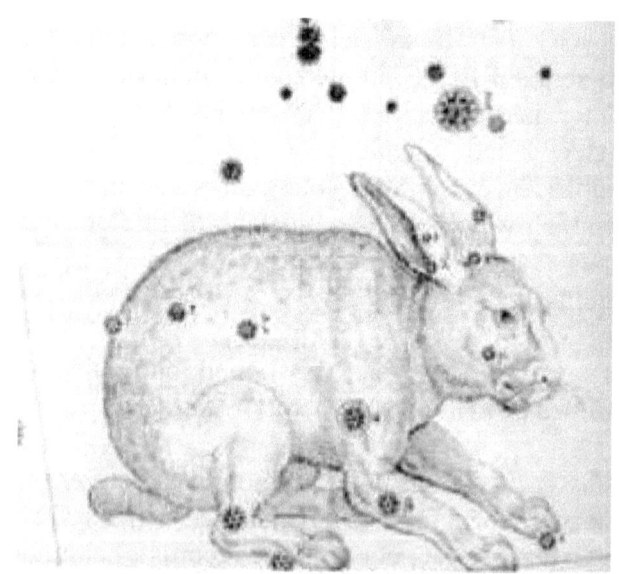

LEPUS THE RABBIT (Hare)
from
Bayer's 1603 Star Atlas

In virtually all cultures the small groups of stars has remained associated with a cowering rabbit beneath Orion's feet. Early German astronomers referred to it as the "*Hase*," the Portuguese "*Lebre*," the Italians called the stars "*Lepre*" and the French saw "*le Lievre*." The Arabian starwatchers, most of the stars being named by them, imagined the four stars, alpha, beta, gamma and delta as the "*Chair of the Giant*" , or a place in which the great *Orion* could rest. Al Sufi, a 10th century A.D. Persian astronomer/astrologer referred to the four bright stars of Lepus as the "*Thirst-slaking Camels*", because of their position near the meandering Celestial River, the constellation of Eridanus (see: ASO *Constellation* Vol. One: *Eridanus*).

Early Egyptians knew Lepus as the "*Boat of Osiris*", the god who was one of the most widely worshipped deities in Egypt and the "*Judge of the Dead*" and "*King of the Underworld*". The early Chinese sky watchers were a lot more practical with many of their star associations. Looking at these four, plainly spaced and rather symmetrical stars in a mostly void portion of dark sky, they knew the same found stars simply as Tsih, or "the *Shed*".

Within Lepus are NO star clusters of note with amateur instruments, one Messier Object (Messier 79), one very faint planetary nebula with a notably bright central star, about five spiral galaxies and very few bright stars. Indeed, there are TWO named stars in this rather faint constellation and one rather interesting "circlet of stars" as will be described following. In addition there are some fine double and multiple stars that we shall explore and at least one VERY fine variable star that is my favorite colored star in all the sky....."*Hind's Crimson Star.*"

Indeed, as a "bonus" for this "GO TO" TOUR, a selected list of the 75 most famous and "reddest" of all stars - "**CARBON STARS**" is provided, giving the coordinates, the periods (yes, they are ALL variable stars!), the magnitude range, and the extreme "red end" spectral type of each star. Note that all the stars provided in our bonus listing are within range of almost all amateur telescopes!

Lepus (pronounced "LEAP-us") is contained within a box-like border that is barely one hour of right ascension wide (from about 5h R.A. to just past 6h R.A., some 17 degrees of sky) and barely the same extent in north-south directions. All objects within Lepus are SOUTH of the celestial equator (see the chart below) and thus have negative ("-") declinations. Lepus is due west of the much brighter (and more in the star-rich fields of the winter Milky Way) constellation Canis Major....another reason that it is often overlooked by stargazers.

For a complete cross reference of named stars, Bayer and Flamsteed designations, SAO numbers and double stars along with their coordinates, I refer you to the wonderful tables at:
http://www.peripatus.gen.nz/astronomy/DatInd.html
This particular URL is specifically targeted for every star and object of the sky is so cataloged in this wonderful web site.

Clay Sherrod's
CONSTELLATION GUIDES:
"Go To" LEPUS : the "Hare, or Rabbit"
Ehhhh...what's up Doc?

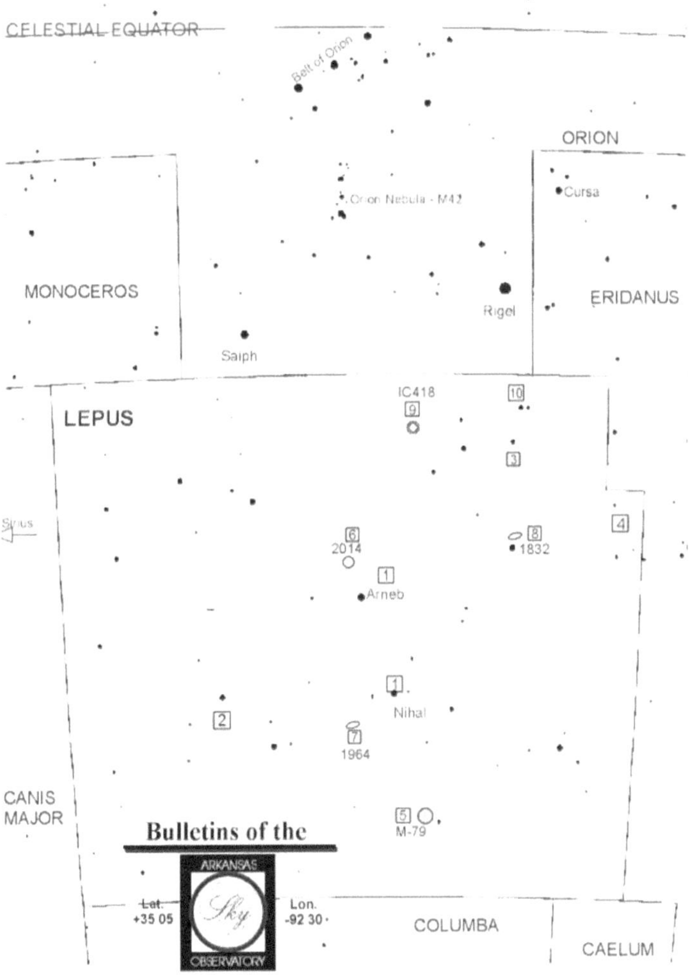

A finder chart for locating many of the GO TO objects in the constellation of Lepus; if using a computer planetarium program, you are encouraged to plot the objects on your screen for higher resolution than this chart provides.

THE STARS OF LEPUS –

Take the time to fully explore this fantastic and user-friendly constellation cross-references. From mid-northern latitudes, the stars of Lepus rise only slightly later than do those of Orion; wherein Orion rises pretty much east for latitudes 35 degrees and below, Lepus is far to the southeast and its faint stars are not visible until much later in the evening than those of Orion. For late November, Lepus rises about 9 p.m. and is located south of overhead (culmination on the celestial meridian) the same as those of Orion, or about 1 p.m. Midnight culmination of the middle of Lepus occurs each year around December 15.

Because of the stark absence of spectacular (or even "reachable") deep sky objects in Lepus, observers are encourage to visit this celestial rabbit with "double stars in mind." There are many splendid double and multiple stars, dozens of which are observable in much detail in telescopes from 3" to 8". Unfortunately I am able to only select a few of the finest double and multiple stars to discuss in our Eridanus "GO TO" Guide; I very much encourage all telescope users to obtain the three-set copy of *Burnham's Celestial Handbook* for reference on each and every constellation; there is no finer reference work for deep sky viewing to be found. The discussion on the constellation LEPUS will be found in Volume Two of the Handbook.

Also note that there are very useful programs (please do a computer search) now available for you to install on your computer to CONVERT the epoch 1950-1960 coordinates listed in the Burnham reference tables DIRECTLY to epoch 2000 are beyond which you can install within minutes....all you must do is merely type in the coordinates exactly as they appear in Burnham's and enter, with the resulting epoch conversion done instantly for use with your Autostar or computer controller!

Once scanning the pages of Volume 2 of Burnhams, you will quickly see the "best multiple" and unusual stars for observations. There are no less than 100 fine multiple stars within reach of most amateur telescopes, many of which are fascinating objects. A few have been selected here as described following, some of which are wonderful and challenging targets in selected telescope sizes.

OBSERVING THIS CONSTELLATION WITH BINOCULARS –

For those who wish to explore the regions of Lepus in binoculars I highly recommend a standard good quality 7 x 50 or 10 x 50 glass used in very dark, moonless skies away from artificial lighting. Remember to let your eyes become "dark adapted" for at least 15 minutes prior to searching out fainter objects. For a wonderful selection of binocular tour objects, visit
http://www.dibonsmith.com/lep_con.htm;
a tremendous guide by **Richard Smith** from his web page entitled "The Constellations."

GETTING STARTED

As with every "GO TO" TOUR guide, each GO TO object in LEPUS is discussed for your telescope regarding the type of conditions necessary for you to view it optimally for discern the very faintest details.........magnifications and aperture necessary for most objects, and much, much more. This is YOUR complete guide to get you on your way to exploring the best (and few!) objects in this HUGE constellation. The chart provided above from the Arkansas Sky Observatory and the subsequent detailed listing of "BEST" objects contains the finest or most interesting from my own observing experience and preference.

Use the attached star chart shown above and the following Guide as an excellent reference for your next star party itinerary, or a beginning for further study into the thousands of objects visible in this part of the sky.

OBSERVING TIPS –

Every deep sky object and every double/multiple star will have a "PERFECT MAGNIFICATION" .....this is the magnification that you should use that will show the object as bright and with as much as detail with possible and still increase its size appreciably so that you can view it comfortably and unmistakably. The rule for determining "optimum magnification" is that: 1) too low power results in sky background glow detracting or diminishing the contrast against the deep sky object; 2) too high magnification darkens BOTH the sky background AND the object; 3) medium magnification can be achieved at which you have MAXIMUM contrast between the object and its darkened background sky. I have found through three decades of direct observing that about 15x per inch aperture for deep sky observing is PERFECT for most objects. That being said, always remember that DOUBLE or multiple stars require whatever power you can crank out....the seeing conditions are the limiting factor here.

For my complete and comprehensive discussion regarding seeing conditions and sky transparency, see the *ASO Guides General,* www.arksky.org .

With all deep sky objects, avoid attempting to observe when the moon is in the sky, even a very thin crescent, as its brightness in the sky will overshadow the very dim contrast afforded by even the brightest deep sky object; if you see the object at all against moonlight, you will NOT see the subtle outlying areas or the full detail of what is presented.

For detail descriptive lists of the great double and variable stars within LEPUS, and as with all of the "GO TO" tour constellation lists, I recommend a good star atlas and/or chart which will list all the finest objects, constellation-by-constellation. One very handy reference guide is the *PETERSON FIELD GUIDE TO THE STARS AND PLANETS*, which features complete lists with declinations, right ascensions, magnitudes, and all pertinent information for you to expand your observing horizons beyond this brief guide. For the many double and multiple stars, I again urge you to refer to the indispensable "*Burnham's Celestial Handbook*", Volume 2 for a complete abbreviated listing.

Truly these extensive *Constellation* Study Guides will most definitely put your PC sky program to work for you in the most efficient and enjoyable way possible! As a matter of fact, MANY users are now programming their own "Tours" based on these guides, using each constellation as a separate GO TO Tour for telescope library that can be added in or deleted through the main edit screen on your PC or MAC computer.

We hope you enjoy these comprehensive GUIDES to touring the constellations via your AutoStar and its computer-driven telescope. Each new installment is complete with diagrams, charts and illustrations that you will find nowhere else. Please let us hear YOUR feedback and your observations of each and every constellation after YOU have toured its vast reaches of our skies!

YOUR LEPUS CONCISE DIRECTORY OF INTERESTING OBJECTS –

To discover more information, and learn of many, many more double and multiple stars that are within reach of your telescope, turn to your Burnham's Celestial Handbook. For a brief discussion on double star observing and their "Position Angles" refer to my brief overview in the "GO TO" tour guide for Lacerta in these *Constellation* Vol. Two..

The most interesting 9 targets (as well as your "GO TO" GUIDE of 75 CARBON STARS throughout the entire sky) in the constellation have been chosen for this LEPUS "GO TO" TOUR; as with all GUIDES, all objects listed below will be visible in most telescopes (some naked eye) in telescopes from 3" to 8"; of course larger apertures may "show" an object a bit closer and "better," but frequently a wide field and low power view is more desirable than aperture for FINDING the objects initially. Indeed, I strongly encourage you first FIND the target object, or its approximate location through your GO TO function with your lowest power and then - once IDENTIFIED positively - move up slowly in steps with magnification if necessary. Remember, not all objects "like" magnification. Sometimes better "field of view" (such as the wonderful wide fields provided by smaller telescopes) is desired over light gathering and magnification of larger telescopes.

Note that your PC sky programs may NOT have every object listed on every constellation GO TO tour....this is intentional. You can access some of the most interesting objects of the sky directly from their coordinates. It is quite simple as you merely enter these coordinates as described (please consult with your program's HELP section) process for the Autostar (other scopes' controls are similar).

The constellation tour Star Chart above will get you started on your journey for this constellation.

Following is the concise object list for your "GO TO" tour of LEPUS; you may wish to find many of the objects from your sky program/Autostar Library (for example, you can easily go to the globular cluster Messier 79, if you pull up "Object/Deep Sky/Messier/..then type in '79'...." and then press "Enter", followed by "GO TO" to access this nice and easily resolvable star cluster. Also note that Messier 79 has the designation "NGC1904", so you can also access it that way, by keying in: "Object / Deep Sky / NGC / [type in "1904"] and press enter and then GO TO.

You will access your FIRST GOTO target - (usually the brightest star in each constellation) - via the command "SETUP / OBJECT / STAR / NAMED....and scroll to "**Arneb**," then press "Enter" and subsequently "GO TO" to move to one of these bright stars.

You may also access the constellation by: SETUP/OBJECT/CONSTELLATION/"Lepus".....Enter....GO TO, which will slew your telescope very near the cartographic center of this sprawling star group.

For a full and comprehensive listing, including coordinates, magnitudes, sizes and more concise information about the many NGC objects contained within the boundaries of Lepus, go to the web site https://en.wikipedia.org/wiki/List_of_NGC_objects_%281%E2%80%931000%29
which provides wonderful tabular data on all deep sky and stellar objects in every one of the 88 constellations.

TOUR OBJECTS FOR LEPUS

OBJECT 1: brighter stars - ARNEB (alpha Lep) - R.A. 05h 33' / DEC -17 49 - Mag: 2.6 - "YES!....It IS on your Autostar!" /
NIHAL(beta Lep) - R.A. 05h 28' / DEC -20 46 - Mag: 2.9 - ALSO on your Autostar and a very tough double, faint companion!
OBJECT 2: triple star - Gamma Leporis - R.A. 05h 45' / DEC -22 27 - Mags. = 3.5, 6.2 & 11 - very nice color contrast, 3rd star pretty tough

OBJECT 3: nice double - Kappa Leporis - R.A. 05h 13' / DEC -12 57 - Mags. = 4.4 & 8.2 - very nice star for 3" and 4" scope test!
OBJECT 4: + "bonus Carbon Star Directory!" Hind's Crimson Star - R Leporis (variable, famous "carbon star") - R.A. 04h 59 ' / DEC -14 49 - Mag. 5.9 to 11, 432 days! Wow! What color!
OBJECT 5: difficult globular cluster - Messier 79 (ngc1904) - R.A. 05h 24' / DEC -24 31 - Mag. 8.4 - tough for northern observers!
OBJECT 6: cluster? multiple star? - ngc2017 - R.A. 05h 39' / DEC -17 51 - six stars mag. 7 to 10.2; nice loose multiple system
OBJECT 7: spiral galaxy - ngc1964 - R.A. 05h 33' / DEC -21 57 - Mag. 11.7, very low in southern sky; difficult except in 8" and larger
OBJECT 8: spiral galaxy - ngc1832 - R.A. 05h 12' / DEC -15 43 - Mag. 12.0 - very faint, compact....close to mag. 10.5 star
OBJECT 9: planetary nebula - IC418 - R.A. 05h 28' / DEC -12 42 - Mag. 11.9 - central star is mag. 10.8; good object for larger scope
OBJECT 10: bright double star to end on! - Iota Lep (#3 Lep) - R.A. 05h 12' / DEC -11 52 - Mag. 4.2 & 10.2 - BONUS variable star RX Leporis!

YOUR VISUAL GUIDE TO DEEP SKY OBJECTS IN LEPUS

Object 1 - Our "Starting" Brighter Stars - ARNEB (alpha Lep - pronounced "ar-NEEB") and NIHAL (beta Lep - pronounced "KNEE-hal")
ARNEB:
Also known in Arabic as "*Arsh*", this 2.7 magnitude star is nearly 920 light years distant and far more luminous than our own sun. This rather

uninteresting star is very close to another object on our "GO TO" TOUR, ngc2017 (below) which is officially cataloged as a galactic cluster, but is in reality nothing more than a 6-component multiple star system. Arneb is a solar-type yellow star and there is a very faint (magnitude 11.1) star some 36" arc southeast (Position Angle 156 degrees); this is a tough target for most telescopes even with that distance (comparable to almost the size of Jupiter in the same medium power....about 15-20x per inch aperture), since the brightness of the primary star outshines the very faint one. This star is most likely NOT a true gravitational companion to Arneb, but merely a more distant object aligned to our line of sight. Look for the nice color contrast: Arneb is distinctly yellow, while the faint star almost appears a blue-green color at medium-high power. At third fainter (12th magnitude) star is located about THREE TIMES the distance between these two; look for the third "optical double" nearly in line with Arneb and the other star, but slightly more to the south, three times the same distance as the 11.1 star.

NIHLA:
This is a MUCH closer star to us than is Arneb. At only 115 light years it is obviously not as luminous a star as Arneb, since its magnitude is nearly the same: 2.8. This is a true TRIPLE star, but is among the most difficult of all for amateur telescopes. It is comparable to attempting to see the companion to the bright star Sirius (see my discussion about attempting to see such doubles at *ASO Guide to Observing Companion to Sirius* under ASO Observational Guides on at www.arkskyk.org), where the primary star is so overshadowingly bright

that the companion star is masked from the glare. This is particularly true with this star. However, I can barely see the stars if I use a crosshair eyepiece fitted into at least an 8" scope (I think that a keen-eye person should be able to spot this in a 5" or 6" on a steady clear night); I use one crosshair to "hide" the bright star just enough to expose the elusive fainter star!

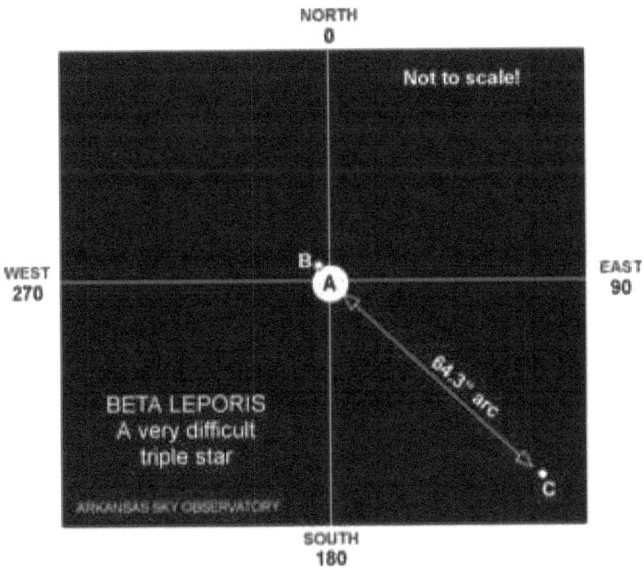

However, the first companion, magnitude 11.1 is VERY close...about 2.5" arc, (see the chart above) and just northwest (Position Angle of 336 degrees) of the bright star; so in the Maksutov or Schmidt-Cassegrain telescope with the diagonal in place, the finder chart above is oriented correctly for your field of view. This will be a challenge for ALL telescopes and observers to see this star. If you give up, try the THIRD member of the group, another 11th magnitude star exactly southeast (Position Angle 145 degrees) of Nihla quite a distance (64"

arc) away. This faint star is most definitely "seeable" with the 4" and larger aperture telescopes.

Object 2 - The Fine Double Star Gamma Leporis
This is a quite nice double star with enough magnitude difference in the two stars to "make it interesting" for all size telescopes. With a whopping 1.2' arc (that's "minute"....not "second") this is a fabulous star for small telescopes. The primary star is magnitude 3.6 and its companion - just a bit west of due south from the brighter star - is a much fainter 6.2. However, with this nice spacing between the two stars, the glare from the brighter star does not prevent even the smallest of telescopes form enjoying this fine double. This is one of the closest stars to our own solar system, only 26 light years away, and it moves through space somewhat in concert with the brightest star of the sky, nearby Sirius. Because this star rests so close to (just west of) the winter Milky Way, the star field surrounding Gamma Leporis is a spectacular sight for low power instruments; even at the lowest magnification look for the distinct difference in color of these two stars: Gamma itself is a true yellow "F" spectral type star, while the companion is anything but. Color estimates for the fainter star, even with experienced observers, range from blue-green, the ivy green, to purple! Take a look and see what color YOU see....to me in smaller telescopes the color appears somewhat grayish green. In slightly larger telescopes look for yet a third star in this collection, but this one is NOT actually a gravitationally-bound member of the Gamma Leporis group. This 11th magnitude star is almost exactly in the same line as the two stars, but located almost exactly half-way between Gamma and its 6th magnitude

companion...quite a sight!

Object 3 - Another Very Nice Double Star - Kappa Leporis.....tough on a 3" scope but "do-able!" Like Gamma, Kappa Leporis offers a wonderful - though more difficult - contrast in magnitudes. Kappa is a relatively bright magnitude 4.2 while its much fainter companion (and MUCH closer than that of Kappa!) is only magnitude 8.1. Now that is plenty bright enough to spot in a 3" or 4" telescope, but it is not that easy; since the two are separated ONLY 2.5" apart, the star is a tough one for small telescope, due primarily to the overpowering glare of the 4th magnitude star; use medium (15x per inch aperture and up) power and my "trick" of putting a crosshair eyepiece to position one reticle as a mask to block the light of the brighter star. Then look for its faint companion exactly DUE NORTH from the brighter star! This is a nice target for larger telescopes; Kappa looks bright white to me, while its faint companion is distinctly yellowish.

Object 4 - A Colorful Variable Star - "HIND'S CRIMSON STAR" - R Leporis - YOUR CARBON STAR "GO TO" DIRECTORY All of the "carbon stars" - such as the remarkably blood-red "Hind's Crimson Star" in Lepus - are variable stars; incredible and interesting red stars with spectral types that indicate the "latest" of all stars known, perhaps spending what precious little fuel remains unused and that in the form of atomic Carbon. It is likely that all of the primordial hydrogen and helium that first powered these scarlet stars is now spent, combined from nuclear fusion first from hydrogen....to helium...to lithium....and ultimately to Carbon.

The Carbon Stars and all variables are named according to variable star nomenclature. The first variable that was discovered in a constellation was named "R," followed by the name of the constellation in "genitive" form. The more "common" form of the name, "R Lepus," is NOT used.....rather it is called "R Leporis" in a more gentlemanly form. The second variable in every constellation is named "S," then T,U,V,...Z. Then the naming continues with RR, RS, RT,...,RZ. It continues with SS (not SR to keep things from getting confusing!) through SZ, TT through TZ, and continuing in this fashion through ZZ. This standard method of variable star designations covers the first 334 variables that were cataloged in all 88 constellations....but obviously there are more than that: thousands more. Most such variables after the "first 334" are named V335, then V336, V337. In addition, many of the brighter variable stars are known by a common name, such as Polaris and Alcor, and thus are merely referred to by the more common name and no "variable star designation."

This variable star is different than nearly all of the "carbon stars": these very "late" R- and N-type spectral "cold" stars are normally part of the interesting and yet-to-be-understood "Semi-Regular" stars; such stars demonstrate a "suspected or slight" regular period between brightening, dimming and then re-brightening again, but are characterized during that period by sometimes major fluctuations....unexpected dimming or brightening....long periods at maximum or minimum....very rapid brightness increases followed perhaps by extremely slow fading....and so pure "irregular" and change brightness without pattern nor notice! Hence the need for YOUR observations of such stars. They should be observed at least once weekly when possible. For a complete discussion on this classification of variable and methods for observing variable stars, see my guide at the *ASO Guide to Observing Variable Stars* on this website GUIDES link.

R Leporis on the other hand demonstrates a fairly regular period of 432 days, fluctuating from a fairly bright magnitude 5.9 to a dim 11.2 to even 12.0 in half that period. Many times it will fluctuate near halfway from minimum to maximum (see the graph below) and is very interesting to monitor during that period.

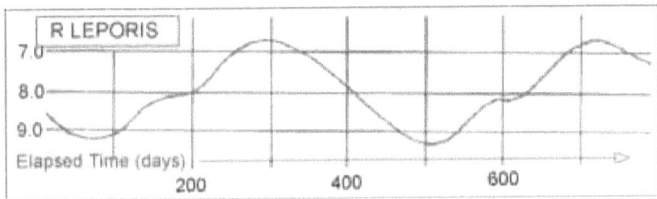

This is an incredible star just to "look at," as the British **J.R. Hind** agreed in 1845 when he first happened upon this apparently unseen object until then. Although the star raised considerable attention among astronomers at the time, many of them English noblemen and not "professional" as we think today, it was not until seven years later that anyone even noticed that this blood-colored star was even variable! Indeed, its variations are so great that within one year, the star will nearly appear to the naked eye for nearly 40 days and slowly fade to such a faint object that Mr. Hind's modest telescope would likely not be able to even find it!

The carbon stars are the most intensely RED of any know, and for good reason. They have spent all their lighter element nuclear fuel: hydrogen, helium and lithium and have fused what is left into an atomic carbon form, burning very slowly and very cool compared to younger and more energetic stars. The color of the star is much like the color of a flame here on Earth: WHITE is the hottest (brightest) as we would expect from the flames of the efficient and incredibly hot acetylene torch; next would be BLUE, like the intensely hot flame from natural gas; YELLOW is a cooler flame that the first two, much like the "gentle heat" given up by a rapidly burning log fire or the wick of a healthy candle; the flame (and star) gets cooler as it burns on less efficient fuel, gobbling its resources in a most efficient way and rendering an ORANGE flame; then of course - like the RED coals simmering in the bed of your campfire from the previously hotter and yellow-colored flames - the flame is the coolest and will provide a much "slower burning" environment with a less energetic output of heat and light.

It is just the same for the colors of stars. R Leporis is about 1500 light years away, curiously at the same distance as the many bright, young and conversely HOT stars of the constellation of Orion!

Be sure to observe this star when at MAXIMUM to appreciate its brilliant scarlet color...it is like nothing else in the sky; although still appearing red even when at faintest, the color is truly remarkable when at its peak. The 4" telescopes can handle this star all the way through its 432 changes in brightness; however, the 3" and smaller scopes can appreciate it at all times except during its "valley" at dimmest. I have noticed that long focal ratio telescopes (such as f/10 to f/15) will show the color of R Leporis far more intensely than very short Newtonian (f/4 to f/8) instruments. Likewise, the color is enriched in a traditional refracting telescope over compound telescopes of any type.

R Leporis is an excellent target for continued wintertime viewing for all observers....it is likewise very easy to locate. For free observing and finder charts for ALL variable stars, observers need only go to the American Association of Variable Star Observers (AAVSO) web site at: www.aavso.org for complete observing information and packets to get started; in addition nearly every major variable star "worth observing" is mapped in their convenient variable star charts. You can find the proper chart (low power and wide field, but to magnitudes less than 11) at https://www.aavso.org/apps/vsp/ . Note for these charts, simply type in the NAME of the variable at top to generate your choice of chart. The reversed image is for users of Maksutovs and Schmidt Cassegrains with the field like a mirror image, or with NORTH up and EAST to the right in your field of view.

A wonderful observing project for astronomers with small, medium and large instruments is to monitor the complete "GO TO" directory of Carbon Stars provided below. The coordinates provided are epoch 2000, so you need only key in the RA and DEC of each star to locate. However, if you want your "telescope to do the walking" you can download the wonderful tour: http://groups.yahoo.com/group/lx90/files/Autostar%20Guided%20Tours/CarbonStars.mtf which provides these in R.A. sequence and in the proper format ready to load onto your Autostar in its Tour Library!

One caution about observing the variability of Carbon Stars and actually MOST variable star since they typically are leaning into the "red" spectral range. There is a well known and documented effect, known as the "*Purkinje Effect*" in which two mis-estimations can occur when looking at very reddish stars:

1) the observer with more red-sensitive eyes will always note a reddish star brighter than it really is; and,
2) the longer ANY observer stares at a reddish object, particularly a star, the brighter its light will appear to the receptors of the eye.

CARBON STARS DIRECTORY:
STAR NAME
COORDINATES / MAG. / PERIOD/ SPECT-TYPE

Star	RA	Dec	Mag	Period	Spect
VX And	00:19:51	44:42:00	8.0-9.5	367	N7
T Aql	18:45:42	08:44:14	8.8-10.0	Irr	M5
V Aql	19:04:22	-05:41:27	6.6-8.1	350	N6
Y CVn	12:45:09	45:26:35	5.0-6.4	158	N3
W CMa	07:08:02	-11:55:33	7.0-8		

	Irr	C6		
R Cap	20:11:18	-14:16:00	9.4-14	
345	Ne			
RT Cap	20:17:02	-21:19:40	6.5-8.1	
395	N3			
X Cas	01:56:40	59:15:40	9.5-13	
423	Ne			
WW Cas	01:33:33	57:45:24	9.1-11.7	
Irr	N?			
Mu Cep	21:43:32	58:46:49	3.7-5.0	
Irr	M2e			
S Cep	21:35:14	78:37:31	7.4-12.9	
487	N8e			
V CrB	15:49:29	39:33:55	6.9-12.5	
358	N2			
V Cyg	20:41:17	48:08:33	7.8-13.8	
420	Npe			
RR Cyg	20:46:04	44:52:07	6.6-9.4	
417	Npe			
RV Cyg	21:43:16	38:00:48	7.1-9.3	
300	N5			
RY Cyg	20:10:23	35:56:57	8.5-10	
Irr	C4			
TT Cyg	19:40:55	32:37:04	7.8-9.1	
118	N3e			
WX Cyg	20:18:34	37:26:27	8.8-13.2	
411	N3e			
V460 Cyg	21:42:01	35:30:45	6.1-7.0	
Irr	N1			
V778 Cyg	20:36:10	60:05:17	9.4-11	
?				
SY Eri	05:09:46	-05:30.16	9.0-10	
96	N0			
R For	02:29:14	-26:05.39	7.5-13.0	
387	Ne			

Name	RA	Dec	Mag	Period	Type
VW Gem	06:42:08	31:27:04	8.7-9.1	Irr	C5
RT Gem	06:46:34	18:36:54	9.9-15	350	N?
TU Gem	06:10:55	26:01:19	7.5-8.4	230	N3
CR Gem	06:34:23	16:04:36	8.5-9.5	Irr	C8
U Hya	10:37:34	-13:22:38	4.7-6.2	Irr	N2
V Hya	10:51:38	-21:14:58	6.5-12	533	N6e
Y Hya	09:51:07	-23:01:06	6.9-9	303	N3
CZ Hya	10:27:21	-25:33:20	8.5-14	442	Ne
TV Lac	22:56:08	54:13:04	9.9	11	N3
R Lep	04:59:35	-14:48:34	5.9-11	432	N5e
T Lyn	08:22:40	33:31:20	8.0-12	419	N0e
T Lyr	18:32:20	36:59:58	7.5-9.3	Irr	R6
U Lyr	19:20:09	37:52:40	8.3-13	457	N0e
W Mon	06:52:20	-07:08:42	9.0-13	Irr	C4
RV Mon	06:58:23	06:09:53	7.0-8.9	132	C4
BG Mon	06:56:24	07:04:01	9.2-10.4	30	C5
GY Mon	06:53:11	-04:34:46	7.7-8.9	Irr	N3
V Oph	16:26:44	-12:25:43	7.3-11.5	298	N3e

TY Oph	18:31:23	04:22:12	9.5-11.0
Irr	N?		
R Ori	04:59:01	08:07:30	9.0-13.5
378	Ne		
W Ori	05:05:23	01:11:02	6.5-10
210	N5		
RT Ori	05:33:12	07:09:03	8.0-8.9
320	N3		
BL Ori	06:25:28	14:43:30	6.3-7.0
Irr	N3		
GK Ori	06:17:42	08:31:06	9.5-11
236	C4		
V431 Ori	05:15:57	11:57:48	9.3-11.1
122	C5		
RX Peg	21:56:25	22:51:19	8.0-9.5
630	C4		
RZ Peg	22:05:55	33:30:40	7.7-13.5
439	C9		
Y Per	03:27:43	44:10:25	8.1-10.9
252	N?		
SY Per	04:16:35	50:37:25	9.5-12.5
476	Ne		
19 Psc	23:46:22	03:29:42	5.5-6.0
Irr	N0		
Z Psc	01:16:08	25:45:51	7.0-7.9
144	N0		
RT Pup	08:05:19	-38:46:38	8.5-9.2
100	C6		
RU Pup	08:07:29	-22:54:55	8.9-11.1
425	C5		
AC Pup	08:22:43	-15:54:42	8.9-10
Irr	C5		
X Sge	20:05:06	20:38:37	8.7-9.7
196	N3		
BF Sge	20:02:24	21:05:27	8.5-10
Irr	N3		

SS Sgr	18:30:24	-16:53:54	9.0-10
Irr	N3		
SZ Sgr	17:44:57	-18:39:12	9.0-10
73	C7		
AQ Sgr	19:34:16	-16:22:24	6.6-7.7
200	N3		
V1942 Sgr	19:19:10	-15:54:26	6.7-7.1
Irr	C6		
SU Sco	16:40:38	-32:22:47	8.0-9.4
414	N0		
SX Sco	17:47:28	-35:42:02	8.5-9.5
Irr	N3		
R Scl	01:27:00	-32:32:26	6.1-8.8
363	N3p		
S Scl	18:50:19	-07:54:27	7.3-9
148	N3		
T Sct	18:55:25	-08:11:05	8.9-10
122	N3		
RX Sct	18:37:00	-07:36:25	9.0-11
Irr	N3		
W Sex	09:50:57	-02:02:06	9.0-10
40	C6		
Y Tau	05:45:41	20:41:09	7.1-9.5
241	N2		
TT Tau	04:51:33	28:32:02	8.0-10
166	N3		
RT UMa	09:18:23	51:24:21	8.6-9.6
Irr	C4		
VY UMa	10:45:51	67:24:19	6.0-6.6
Irr	N0		
SS Vir	12:25:16	00:46:22	6.0-9.6
355	Ne		
BD Vul	20:37:18	26:28:48	9.3-12.7
430	Ne		

* * *

Object 5 - A Far South and Difficult Globular Cluster - Messier 79

This is a very difficult object to resolve in any size amateur telescope from the northern hemisphere except perhaps Florida, and the southern tip of Texas. At the far-south declination of nearly minus 25 degrees, the atmopshere plays havoc with your telescope's ability to resolve this otherwise bright cluster into the hundreds of stars that should be visible. It was discovered by the French comet-hunter **Pierre Mechain** in October of 1780 when he recorded in his observing log: "...a nebula without star, situated below Lepus, and on the same parallel (Right Ascension) as a star of sixth magnitude..." By December, **Charles Messier**, the countryman and somewhat adversary of Mechain, located the object saying:

"....this nebula is most beautiful, the center brilliant, the nebulosity a little diffuse, its position being determined by Epsilon Leporis....a star of the fourth magnitude."

John Herschel, observing from his huge observatory near Cape Town, South Africa years later noted that Messier 79 was ".....a globular, pretty large, extremely rich; extremely compress and well resolved." in his giant 20-foot reflecting telescope. On the other hand, you can see that this is not a very spectacular not large globular from the photograph below taken with the 200" Palomar telescope, providing photographs normally commanding a gasp of air or a sudden swoon.

Without a doubt Messier 79 is the most difficult globular cluster for discerning viewers of all the Messier objects. This is due to two factors: 1) the cluster is located very far south in the skies for northern hemisphere observers and atmospheric extinction blocks much of the subtle detail and contrast of the very faint stars within it; and, 2) the stars within this cluster are primarily magnitude 14 and fainter and thus beyond the resolution of smaller amateur instruments. On very dark nights an observer with an 8" telescope might begin to see some stars sprinkled around the perimeter of this small cluster. A 12" has difficulty with individual stars on average nights but present a very beautiful view of "stardust" throughout the cluster when the skies are deep and dark. NEVER hesitate to use high magnification on nights of very good steadiness. For a complete discussion regarding how "seeing" and air transparency affects your viewing, go to my discussion in the *ASO Guides* to *Magnification and Seeing Conditions*. _ I routinely use over 50x per inch aperture to observer planets and doubles stars when the conditions are favorable; the old myth that higher powers are useless on telescopes are not true with today's Null-figured optics and excellently corrected eyepieces.....it is the AIR around us that limits the "power potential" of your telescope. I have learned that higher powers (up to around 200x to 250x) are not necessarily detrimental to observing the MOST in globular clusters! Through the power of space and an unfettered view through the emptiness of the cosmos, NASA astronauts aboard the Shuttle were able to capture the beautiful infrared image shown below:

The magnitude of the cluster is about 8.9, and its size measure nearly 8' arc; hence its distance of over 50,000 lights years is partly responsible for the poor views afforded of this cluster. However, it is also a physically smaller cluster than many of the showcase objects such as Messier 13 and those throughout Ophiuchus.

Object 6 - NGC 2017 (Herschel's 3780) –
You tell ME what this is.... a nice multiple star or a very poor galactic cluster?
Here is an object that has (probably in error) been classified as a very sparse and loose galactic star cluster. Six Stars....galactic cluster. So it's not the richest star-packed cluster in all the sky...but it IS six stars all at the same distance bound to one-another gravitationally, so is it actually a "cluster" or a very healthy "multiple star?" This is much like the now-famous "is Pluto a planet?" debate.

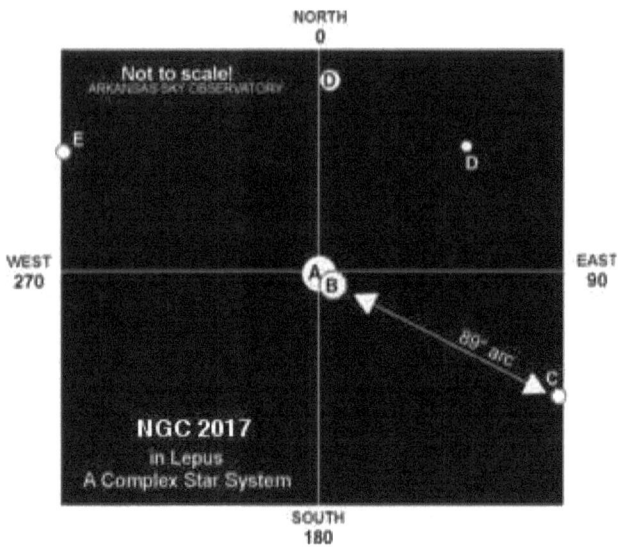

Let's study the small double star locator graphic I have prepared above. This chart has NORTH up and EAST to the right, which is the proper field orientation of your Maksutov or Schmidt Cassegrain telescope with the diagonal mirror installed. The primary brightness of this object ("cluster") is about magnitude 7, yet the object is NOT commonly referred to in many catalogs of NGC objects. Nonetheless, it IS designated as "2017" while William Herschel merely noted it as "object 3780" and did not distinguish as anything but a celestial object. Most of the brightness is the combined light of two very close stars - "A" and "B", magnitudes 7.1 and 8.2, respectively - which are only 0.9" arc apart. This equal-magnitude double can make a wonderful test for the "perfect" 5 inch or 6 inch telescope, but it will be quite a challenge with either; I have found most times it is difficult to split with even an 8", although the 10" and 12" size telescopes does a nice job showing black sky between two very nice disks. The lesser star, "B", is located SOUTHEAST (Position Angle 147 degrees) from the brighter star; make sure to use at least 25x per inch aperture to even get close to resolving these first two members of the group.

Now look for star "C" almost in a STRAIGHT LINE from these two, also further southeast from "A", but a considerable distance....89" arc, or two Jupiter diameters. Nonetheless, this star should be quite easy to identify, as it is only slightly dimmer (magnitude 9.3) than star "B" and in the same direction (SE) from the brightest star....so this "string of stars" should be quite a nice sight for larger telescopes. Even if you do not see "B", be sure and look for "C"; in fact, I have often used "C" as an angle marker denoting the exactly location "next to" the brighter star where the star "B" might be found.

Nearly due NORTH and a bit to the east (Position Angle 7 degrees) is star "D" of this six-star system, almost the same distance from "A" that is "C". This star is brighter than "C", at magnitude 8.4.

The fifth star ("E") of the group is about the same magnitude as "D" (8.3) but oriented almost exactly northwest (Position Angle 330 degrees) from the bright star "A". It is nearly 2/3 farther away in your field (129" arc) than are either "C" or "D". This star, likewise, is an easy target in even small telescopes.

The last star is "F" and it is a bit harder to see as it is only magnitude 10.1; it is still visible in the 3" and 4" telescope range however; look for "F" about

60" arc to the northeast (Position Angle 49 degrees) from the brightest star.

All-in-all, this is a nice object for all telescopes to study. It is a pretty sight and a challenge for medium size instruments if you want to see all six members; "A" and "B" will always be a challenge, but a total of at least five stars should be detected in most telescopes. Except to split the A-B pair, use medium-low (about 10x per inch aperture) to view this nice little group against a beautiful winter star field.

Object 7 - A Faint but Observable Spiral Galaxy - ngc1964

The "ringed effect" of this distant galaxy, magnitude 11.6 visually, is clearly shown in the 200" Palomar photo shown above; with telescope of 10" or less, all that is visible is a very tiny star-like smudge of light that measure 5' x 1' arc at best. It is a very difficult object, but a 5-6" scope on a very dark night and about 25x per inch CAN see the diffuse nature of this tilted spiral galaxy. In a 12" on the darkest of nights, some suggestion of the "ring" (actually merely clumping of stars within the galactic arms) can be glimpsed using averted vision. This ring is distinct and very interesting in detail in a 24" scope.

Object 8 - Another Faint Galaxy - ngc1832
Normally I would not even put a 12.2 magnitude galaxy in here as a "GO TO" tour object, but this one might be interesting to test your telescope's light gathering ability. Many observers with small-to-medium sized telescopes simply "refuse" to search out faint galaxies and for a justifiable reason: many times an "impression" or suggestion of an object might be there and the observer is wondering whether or not the object was actually SEEN or IMAGINED. If you are like me, this leaves more frustration than satisfaction. NGC 1832's image is very "tight" and compact in a telescope, the light of this galaxy compressed nearly all into its very central region that measure only 2.1' by 1' arc across; thus, even though it is a visual 12.3 magnitude, all of its light is fairly packed into a small area....a 4" or 5" telescope CAN see this galaxy, but barely, given perfect dark sky conditions. With ngc1832, you have an excellent "guide" to this star, for in the same low power field of view, just about due south of this tiny smudge of light is the bright star "mu Leporis (see the Palomar photo following, digitally re-oriented to match the reversed view of your telescope). If you will look about 1/2 degree NORTH and just a bit west of mu (magnitude 3.3) , you should be able to lock onto this galaxy. If you have dark enough skies and at

least a 4" telescope when Lepus is highest in the southern sky, then YOU can claim bragging rights to a 12.3 magnitude galaxy among your observing trophies!

NGC 1832
Mu Leporis at bottom

Object 9 - A Faint Planetary Nebula with a Brighter Central Star - IC 418
This one is not in the custom "Planetary Nebula" tour that I provide, nor is it listed among your Autostar Library NGC objects....it doesn't even have an "NGC" designation! IC 418 is a very difficult nebula to make out...BUT the central star that exploded millions of years ago to create it is still there at magnitude 10.8, visible in even a 3" scope, but perhaps more easily in the 4" and larger apertures. The "ring" of stellar gases in this planetary nebula has a brightenss of 12.0 and with its tiny size of 14" x 11" arc, the concentrated brightness is easy to spot in a 5" and certainly an 8"

telescope. However, note that the star itself is far brighter than the planetary nebula which is many times the case in some of the fainter and more compact objects of this class. On a very dark night, medium power (about 20x per inch minimum) to search for the faint cloud around the star; the keen-eyed should spot this very subtly surrounding this star in a good 4" glass.

Object 10 - Taking You Out on a High Note - The Challenging and Bright Double Star - Iota Leporis ("3 Lep") and the Variable RX Leporis

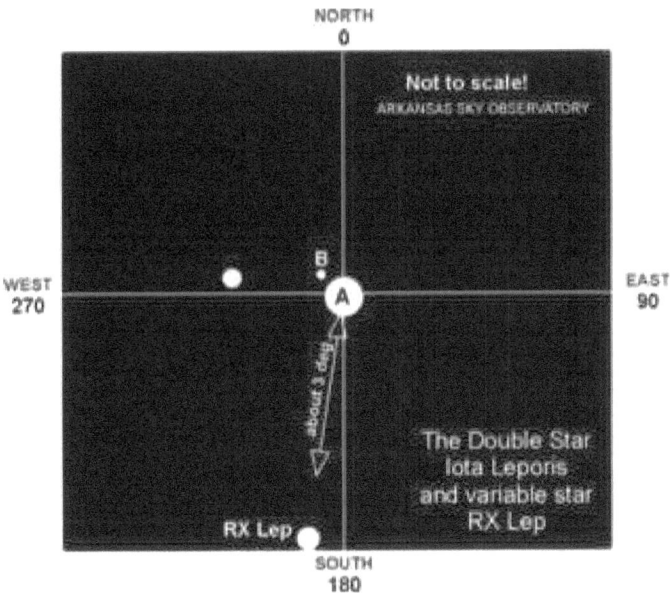

Here is a nice way to end out "GO TO" list of objects in Lepus....normally I try to end the segments with some spectacular deep sky object, but since you have suffered through the miserably difficult and faint objects in this "wascally-wabbit", this should come as a breath of fresh air. The bright

(magnitude 4.2) star Iota Leporis has a very faint (magnitude 10.4) companion star AND a nice brighter (5.7) VARIABLE STAR immediately south of it! Iota Leporis is about 277 light years distant and the faint star can be found (note the finder chart above) nearly northwest of the primary star about 12.8" arc; the pair should be able to be seen in a good 4" telescope and possibly in a 3" under very dark skies and medium-high magnification (about 25x per inch aperture). This pair has a period of rotation around one-another about every 33 years (we think!). Note that there is a nice and relatively bright (magnitude 6.1) star immediately WEST of Iota which makes a very nice grouping in any telescope.

Only 3 degrees to the south and a bit west of Iota Lep is the variable RX Leporis, an irregular variable star with a brightness range from magnitudes 5.7 to a bit less than 11, with NO particular cyclic period. Thus, the entire range of light changes in this unpredictable star can be monitored easily in a 4" and larger telescope. Note its location in my locator chart above which has a field of view oriented properly for your Maksutov or Schmidt-Cassegrain telescope. Unfortunately there is no AAVSO locator (not needed here!) nor comparison star chart available, but observers can make relative estimates via the 8th, 9th and 10th magnitude stars stars within the field of view. It is worth keeping up with, as are any irregular variable stars!

WANDERING ABOUT....YOUR NEW "USER OBJECT" IN LEPUS

There is simply no other option to consider here.

Since this is a remarkable object to view over and over again and a wonderful star to share with those at public star parties or within your observing group, NO telescope should be without "Hind's Crimson Star" on its User List. Not only is it a splendid sight like none you have ever seen, but it is a great variable for nearly all telescopes to monitor!

This wonderful object is the Rosette Stone among the Carbon Stars featured previously; it should be in every computerized sky library without question.

On AutoStar, go to: "Select/Object [enter]...." scroll down to "User Object" [enter]. Now enter the coordinates given above for R LEPORIS using the number keys on AutoStar. After entering the coordinates and pressing "Enter" yet again, scroll down one and you can list the magnitude of the object as "6" [Enter].

Remember that each computer program or keypad control varies from model-to-model between telescopes. Always consult your user guide or the HELP portion of the program for keystrokes and the use of tabs for GO TO operation. Although all programs work essentially the same way, they do vary and you should strive to master your individual sky program.

*   *   *

Lepus is often overlooked because of the dominance of the bright starred and popular Orion, to its north. Be sure to visit the tour of the stars of ORION later in this *Volume Two*; this constellation is a trip not only into space, but into a realm of mythology and memories of your own as we explore the wondrous boundaries of the constellation ORION, the Hunter.....the Hero....the Legend....the sparkling revelation of stars that virtually announce the grandeur of winter and its crisp and cold skies. ORION is NOT to be taken lightly, both the legend of the man nor the constellation pattern and the boundaries that contain it. For within the man is mythology that encompasses all cultures, all times and all of mankind. Within the bright pattern of stars are virtual stellar nurseries that harbor the first life of newborn stars, vast clouds of nebulous gases that hold the keys to the very advent of our sun, our planet and our lives. Countless beautiful and colorful multiple stars grace the constellation as the intriguingly mysterious winter clouds of the Milky Way sink from the hunter's raised club in hand, past the fiery red signal of Betelgeuse and down into the realm of the Diamond of the Sky, *Sirius*.

Because of the countless objects....the far-reaching and fascinating mythology and legends....the beauty of nearly each and every naked eye star and asterism... the "GO TO" tour version of the Orion Constellation guide is presented in THREE parts:

ORION: Part I - The Constellation, Making a Mark in the Sky and in Legends for Mankind
ORION: Part II - Observing and Understand Messier 42 - The Great Orion Nebula
ORION: Part III - Revealing the Treasures Within - Your "GO TO" tour.

* * *

Chapter 20

LIBRA
*(and some strokes for **Lupus**, the Wolf)*
....merely a question of balance.

This is the fifteenth Constellation Guide, "GO TO LIBRA" of the series "GO TO GUIDES for all GO TO Telescope Users". In case you have never looked into this rather vague constellation - the "7th Sign of the Zodiac" - you perhaps have no idea how difficult it is to make this interesting as a GO TO constellation study! Nonetheless, there ARE some interesting attributes of this obscure area of the sky. Included in this featured "GO TO" TOUR is the more southerly constellation **LUPUS**, the Celestial Wolf.....somewhat of a newcomer to the 88 constellations recognized by the **International Astronomical Union** (IAU).

Since Lupus is so far south of the celestial equator, it is a difficult constellation for observers in the northern hemsphere to concentrate on, save those in southern states of the United States and southern Europe. Nonetheless, this canine of a constellation deserves some attention as there are some interesting objects within grasp of our telescopes.

Libra, like Lupus is ALSO somewhat of a "newcomer" as far as named constellations go. It is ONLY because of its location within the ZODIAC - the twelve star patterns near the celestial equator through which all the planets, sun and moon appear to pass as we observe from Earth - that Libra is denoted as a "stand-alone" constellation at all!

THE ZODIAC CIRCLE OF TWELVE - CONSTELLATIONS
Archives of the Arkansas Sky Observatory

The only four brighter stars of Libra - alpha, beta, gamma and iota - for a small square shape in the sky, but at one time the star group was actually part of SCORPIUS (see *Constellations /Scorpius* on this ASO website), forming the "claws" or front pinchers of this venomous arachnid. It was the Egyptian priest-astronomers who first isolated the larger Scorpius into two separate constellations, although the reasoning is NOT all that clear for doing so.

There is speculation however as to the importance of this dull star group to these early people. It is theorized from much later hieroglyphics from those of the great Pyramids that the pattern of obscure stars symbolized the EQUALITY of day and night.....exactly what happened at the spring and autumn equinox as the SUN entered this now-constellation some 2000 years ago. Presently, through precession of the Earth, the autumnal equinox - where the sun crosses the celestial equator at the beginning of autumn has "migrated" from Libra westward into Virgo (see my Virgo Guide *Constellations* /Virgo I this Volume Two).

From the year 2000 and beyond for our lifetimes, the sun goes through Libra in November, passing bright *Antares* (in Scorpius) in the first week in December as it approaches Winter Solstice.

Now - a lesson in changes of history, and how most things really make sense if you know all the facts. Remember the ZODIAC? The name is from the Latin as the "*zone of animals*," signifying the nature of the constellations that primarily the sun passed through to early stargazers. The ONLY constellation of the Zodiac that is NOT an "animal" is......LIBRA. Why? Because Libra has not always "been there..." It originally was part of the 6th "sign" of the Zodiac: Scorpius.

This explanation as to the Egyptian importance of making this a separated entity in the sky ALSO explains very well why we envision a "scale" or "balance" of all things for this constellation. It seems somewhat out of place, doesn't it? Well, when you consider the importance of the fall equinox and the position of this constellation 2000 years ago as winter months approached....perhaps it quickly explains both the significance and labeling of this inconspicuous star pattern.

Although now a stand-alone ("isolated" is a good word for Libra) constellation of its own, Libra still has strong ties to its original affiliation with Scorpius through the original star names given to its brightest stars by the earliest Arabian skywatchers. We often overlook the beauty and creativity of their insight into the wonderful star names that we still preserve (thankfully) today.

ARABIAN ASTRONOMERS - namers of the stars
from a 1531 woodcust - Archives of the Arkansas Sky Observatory

The principal stars (although very faint to the naked eye) all have some association with CLAWS of the dreaded Scorpion! "*ZUBEN* ....." referred to the appendages of this creature to the Arabian astronomers/astrologers. Thus, we have the brightest (alpha) star named *ZUBEN EL GENUBI* ("the Southern Claw"), beta being *ZUBEN ESCHAMALI* ("the Northern Claw"), *ZUBEN EL AKRAB* (gamma), *ZUBEN EL AKRIBI* (delta), and

ZUBEN HAKRIABI (upsilon), as shown in the chart below.

Aren't these simply wonderful star names?

After the wonderfully galaxy-rich tours of Coma Berenices (*Constellations /Coma Berenices*), this one might seem pale by comparison....indeed after a quick flurry of nearly 80 galaxies combined in the previous two guides, we do not discuss even ONE galaxy for your Libra / Lupus tour!

Note from the sky chart included here that the *CELESTIAL EQUATOR* passes just NORTH of Libra, and thus all angular measures (declinations) are south of the celestial equator and negative ("-"); hence you will see references in this "GO TO" GUIDE to "(-)" declinations for celestial objects in both Libra and Lupus.

Also note that the *ECLIPTIC* (the band in which all the planets, sun and moon appear to move relative to Earth, on which the "Zodiac", above is centered) passes from southeast to northwest through Libra. Thus, there are on some occassions (not anytime soon for the early decade of the 21st century!) bright planets that linger from time to time in this star-poor region of sky.

As with every "GO TO" guide, each GO TO object in Libra and Lupus is discussed for your telescope regarding the type of conditions necessary for you to view it optimally for discern the very faintest details.........magnifications and aperture necessary for most objects, and much, much more. This is YOUR complete GUIDE to get you on your way to

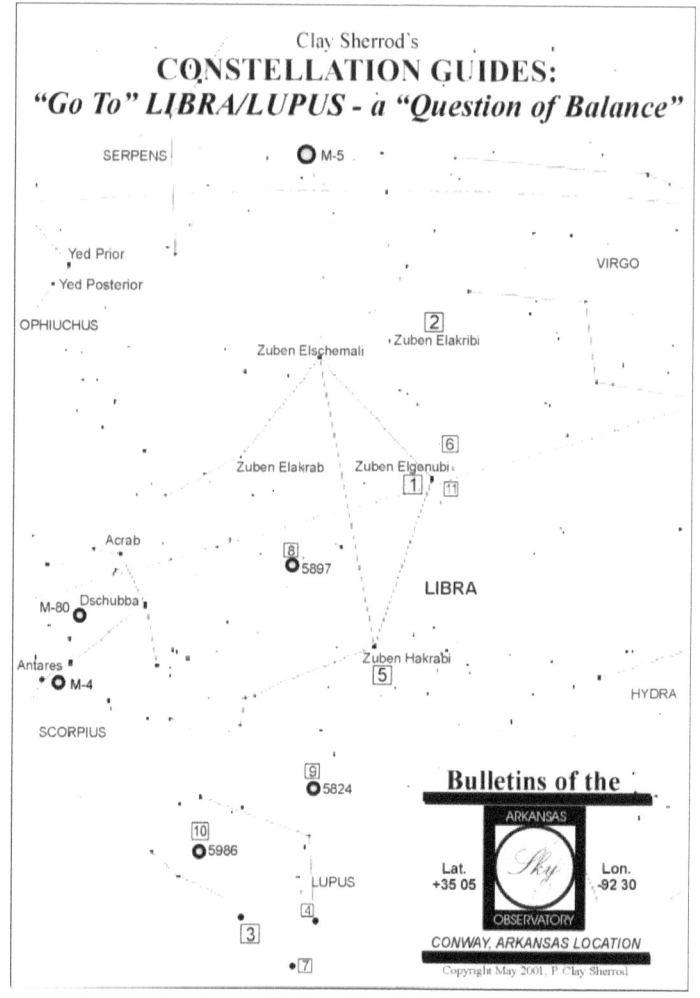

exploring the best (and few!) objects in these two constellations. The following listing of "BEST" objects contains the finest or most interesting from my own observing experience and preference.

Use the attached star chart and the following guide as an excellent reference for your next star party itinerary, or a beginning for further study into the thousands of objects visible in this part of the sky. Truly these extensive *Constellation* study guides will most definitely put your sky program to work for you in the most efficient and enjoyable way possible! As a matter of fact, MANY AutoStar or PC sky program users are now programming their own "Tours" based on these guides, using each constellation as a separate GO TO Tour for the PC sky program library that can be added in or deleted through the main edit screen on your PC or MAC computer.

We hope you enjoy these comprehensive guides to touring the constellations via your AutoStar or PC sky program and its computer-driven telescope. Each new installment is complete with diagrams, charts and illustrations that you will find nowhere else. Please let us hear YOUR feedback and your observations of each and every constellation after YOU have toured its vast reaches of our skies!

YOUR LIBRA / LUPUS CONCISE
DIRECTORY OF INTERESTING OBJECTS

Certainly there is slim pickings for real "show-stopper" celestial objects in both Libra and Lupus. Nonetheless, there are many objects that make our visit rewarding and worthwhile!

I have chosen the finest (or most interesting) 11 objects in this LIBRA / LUPUS "GO TO" TOUR; as with all GUIDES, all objects listed below will be visible in most telescopes (some naked eye) from the small telescopes to about 8 inch aperture; of course larger apertures may "show" an object a bit closer and "better," but frequently a wide field and low power view is more desirable than aperture for FINDING the objects initially. Indeed, I strongly encourage you first FIND the target object, or its approximate location through your GO TO function with your lowest power and then - once IDENTIFIED positively - move up slowly in steps with magnification if necessary. Remember, not all objects "like" magnification.

The rule for determining "optimum magnification" is that: 1) too low power results in sky background glow detracting or diminishing the contrast against the deep sky object; 2) too high magnification darkens BOTH the sky background AND the object; 3) medium magnification can be achieved at which you have MAXIMUM contrast between the object and its darkened background sky. I have found through three decades of direct observing that about 15x per inch aperture for deep sky observing is PERFECT for most objects. That being said, always remember that DOUBLE or multiple stars require whatever power you can crank out....the seeing conditions are the limiting factor here.

For my complete and comprehensive discussion regarding seeing conditions and sky transparency, see my discussion in the GUIDES tab: GUIDES/ General on the ASO web: www.arksky.org .

With all deep sky objects, avoid attempting to observe when the moon is in the sky, even a very thin crescent, as its brightness in the sky will overshadow the very dim contrast afforded by even the brightest deep sky object; if you see the object at all against moonlight, you will NOT see the subtle outlying areas or the full detail of what is presented.

The convenient sky placement of Libra during spring and early summer lends itself well to very good and long-period observing for ETX and LX 90 users both north and south of the equator. When rising about dark in the east (mid-spring) it will remain in the sky throughout the night, transiting the meridian just before midnight during that season. All deep sky objects and difficult double stars are ALWAYS best observed when they are located nearly overhead (or as high in the sky as possible), thus requiring the observer to look through the thinnest portion of the Earth's "lens" of atmosphere and haze.

As with all of the "GO TO" TOUR constellation lists, I recommend a good star atlas and/or chart which will list all the finest objects, constellation-by-constellation. One very handy reference guide is the *PETERSON FIELD GUIDE TO THE STARS AND PLANETS*, which features complete lists with declinations, right ascensions, magnitudes, and all pertinent information for you to expand your observing horizons beyond this brief GUIDE. For the computer or media device, my favorite quick reference and very comprehensive descriptions of objects are the many Smart Phone APPS that are available such as Voyager, etc.. Note that this very affordable software package ALSO allows very quick and accurate construction of "Go To" Autostar tours for these constellations as well as allows a very nice laptop control of any telescope via Bluetooth or other wireless connection!

The constellation tour Star Chart will get you started on your journey for this constellation.

Following is the concise object list for your "GO TO" tour of LIBRA and LUPUS; you may wish to find the majority of the objects from the AutoStar Library (for example, you can easily go to NGC 5897 if you pull up "Object/Deep Sky/NGC/..then type in '5897'...." and then press "Enter", followed by "GO TO" to access this remote globular cluster (Messier 13 it is NOT!). On the other hand, if you want to experiment and become a "better computer user" try entering the exact R.A. and DEC coordinates of that object as described above after holding down the MODE key. You will find the accuracy of entered GO TO's to be somewhat less than those stored in AutoStar, but the capability of acquiring unlisted objects is fantastic!

OBJECT 1: brighter star - ZUBEN EL GENUBI (alpha Librae) - R.A. 14h 48' / DEC (-)15 50 - Magnitude: 2.8

OBJECT 2: naked eye variable - ZUBEN EL AKRIBI (delta Librae) - R.A. 14H 58' / DEC (-) 08 19 - Mag: 4.7 to 5.9

OBJECT 3: nice double star - eta Lupi - R.A. 15h 19' / DEC (-) 44 31 - Mags 3.9 & 6.0 great 3-inch test, tough!

OBJECT 4: (Lupus) supernova remnant! - "Nova 1006" - R.A. 15h 00' / DEC (-) 41 42 - First EVER recorded! Visit its "grave"

OBJECT 5: (Libra) nice double - ZUBEN HKRABI - upsilon Librae - R.A. 15h 34' / DEC (-) 27 58 - Mags: 4 & 11.5, nice test!

OBJECT 6: (Libra) nice double star - mu Librae - R.A. 14h 47' / DEC (-) 13 57 - Magnitudes: 5.5 &

6.5, great for 3-inch+ +

OBJECT 7: (Lupus): southern double - pi Lupi - R.A. 15h 02' / DEC (-) 46 51 - Magnitudes: 4.7 & 4.8 - great star for 3-inch +

OBJECT 8: (Libra): globular cluster - ngc5897 - R.A. 15h 14' / DEC (-) 20 50 - Magnitude: 10.8, very loose, nice cluster!

OBJECT 9: (Lupus): globular cluster - ngc5824 - R.A. 15h 01' / DEC (-) 32 53 - Magnitude: 9.5, bright and very small - nice!

OBJECT 10: (Lupus): globular cluster - ngc5986 - R.A. 15h 43' / DEC (-) 37 37 - Magnitude: 8, very bright and resolvable!

OBJECTS 11: (Libra): nice variable stars - V Librae (long period) / UW Librae (semi regular) - R.A. 14h 38' / DEC (-) 17 27

A VISUAL GUIDE TO OUR DEEP SKY OBJECTS IN LIBRA

Object 1 - Our "Starting" Brighter Star - *"ZUBEN EL GENUBI" (*alpha Librae) - Also a nice double! Our starting point for every "GO TO" TOUR is always (or usually!) the brightest star of the constellation or region and Libra holds no exception. This is by far the DIMMEST bright star we have used as our "starting target" for any constellation "GO TO" TOUR, being only magnitude 2.8, with no other conspicuously bright stars anywhere close.....until looking far to the southeast at rising Scorpius and all of its bright and colorful Milky Way stars.

Denoting the "Southern Claw" of what used to be Scorpius (see above), Zuben El Genubi is a close star, only about 65 light years away....it is a highly evolved star, with spectrum indicators of later evolutionary stages. In even a small telescope, look about only 4' arc from Alpha and find star "8 Librae", a true double companion to Zuben El Genubi at magnitude 5.2. This is one of the nicest pairs of all for the wide fields of view of these low power refractors, and is a very pleasing sight in all telescopes larger using very low magnifications.

Object 2 - "Zuben El Akrisi" - (delta Librae) - A Very Good Naked Eye / small scopes Variable Star Remember the star "Algol" listed on your AutoStar and sky programs? You can dial in under EVENTS and find out the exact closest time of minimum brightness of this ECLIPSING BINARY star.

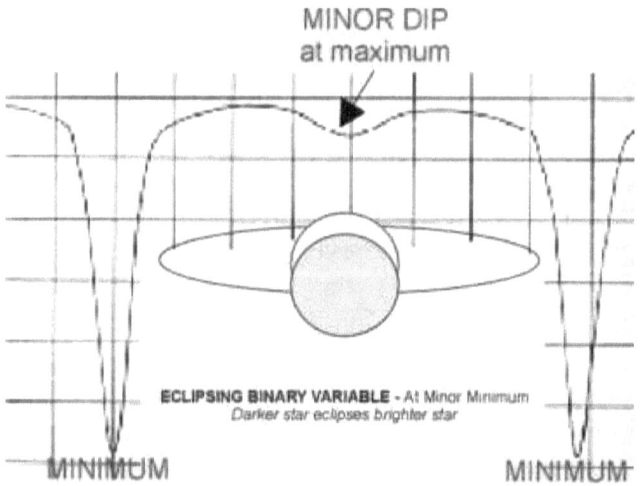

Algol-type variable stars are very simple; they have two stars (one brighter than the other) which orbit one-another and - from our line of sight - occasionally one ECLIPSES the other, or "covers it up." From our distance, we see the brightness of TWO very close stars normally; however, when the second (darker) star orbits in front of the first star, we see ONLY the brightness of the darker star in the foreground. In the diagram above, you can clearly see the nature of the eclipsing binary-type stars. If it is the brighter star that is in front, the minimum is NOT as dim as just stated.

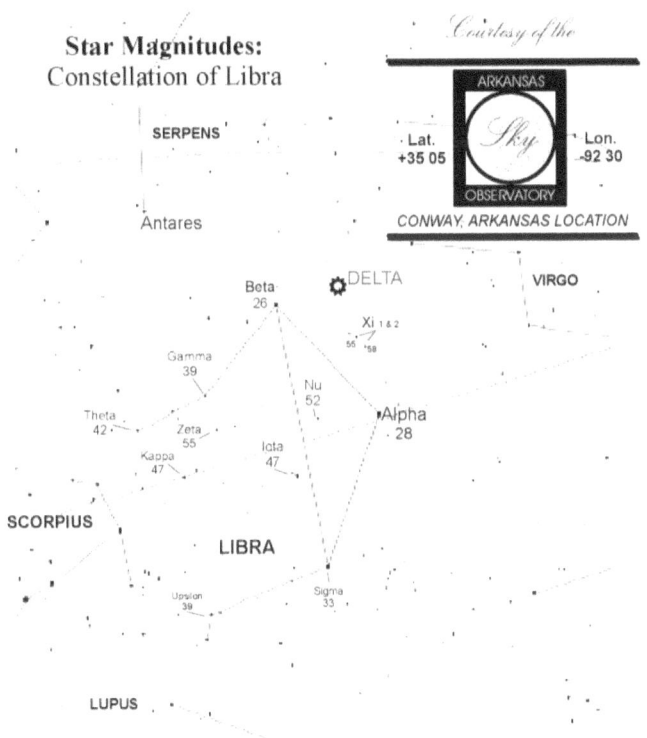

Delta Librae - ZUBEN EL AKRISI - is just such a star, with the combined brightness of 4.7 eclipsed every 2.32735 days into the brightness of ONLY star "B", or about 6.0. Although this cycle can be seen clearly with the naked eye on very dark nights all the way through minimun (mid-eclipse by the darker star), this phenomenon is best seen with the small APO refractors with their wonderfully wide fields of view; binoculars also afford an excellent way of monitoring the cycle of these two stars. An AAVSO star chart is NOT available for this bright and rapidly-changing variable, but you can use the brightness of the major Libra stars as guides as shown in my special bright star magnitude chart for Libra below. Also see previous chart.

alpha -	2.75
beta -	2.74
sigma -	3.31
upsilon -	3.78
gamma -	4.02
theta -	4.34
iota -	4.66
48 librae -	4.85
18 librae	5.95
47 librae -	6.00
mu -	6.10
5 librae -	6.50

Object 3 - A Beautiful Double Star - Eta Lupi –
YES, you can see this one from most of the United States!
Here is a great southern double star that can be located even in the northern hemisphere observing latitudes. Eta Lupi is a beautiful and bright yellow pair of stars, both of equal 5.5 magnitude brightness and only 10.5" arc separation. Thus, these can be clearly resolved with even small scopes at medium power and the stars make a gorgeous pair in the larger telescopes at low and medium-low powers. The stars are oriented in a NE-SW fashion and orbit one-another in a period of 51 years. At southern declination -33 degrees, this star should be accessible for observers south of the Mason-Dixon

line in the United States and from the southern latitudes of Europe.

Object 4 - A Place Where a Star *ISN'T!* – The Location of **Nova 1006** in Lupus

Turn your telescope via your sky program to this spot and you will be peering at a celestial graveyard, a place that dominated the nighttime sky in the year 1006, the event being the very first historically recorded supernova on record. Today, nothing brighter than magnitude 17 appears at this exact location (**14h 59.6m / -41 degrees 42m**), but in its heyday, *Nova Lupi 1006* was a terrifying and brilliant sight. The finder chart below from my Observatory denotes the exact position near Beta Lupi where the Nova was thought to have been centered.

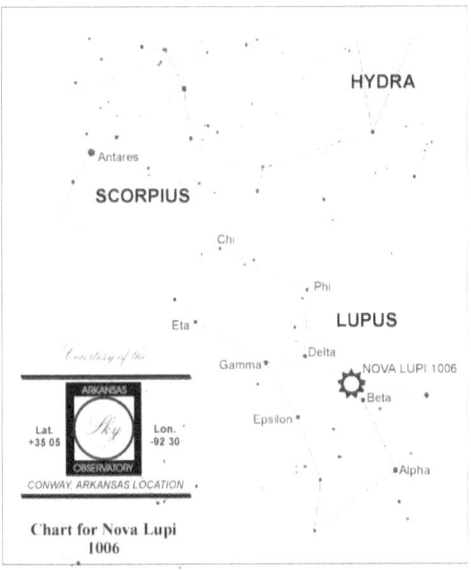

Chart for Nova Lupi 1006

The star was recorded in Medieval Europe, China, Arabia, Egypt and Japn in that year in the spring of

1006 as an object that might have been FOUR TIMES BRIGHTER than the brilliant planet Venus and almost half as bright as the moon! The Egyptians recorded the object (outburst?) as a "comet" or "Nayzak," but there is no mention of the characteristic tail. Arabian astronomers noted that the "rays" of this huge star appeared to illuminate the ground of Earth much like those from the light of the moon! Many of the same sages of the region noted that the appearance of this sudden and bright "new star" occurred, not coincidentally in their minds, with "...strong winds, bursts of lightening, and shattering thunder with no rain." Chinese astronomer/priests recorded very accurately the date of first appearance as May 1, 1006. It was seen incredibly as very bright all the way until it disappeared below the western horizon that following September! No trace of the star - now totally converted to energy save a very small portion of mass - remains as with some supernovae....no diffuse nebula, but perhaps only a tiny ring of gas that is seen surrounding a faint 17th magnitude star near that position. However, **Radio Source #1459-41** only 1.5 degree NE of Beta Lupi is likely the remaining energy of that primordial explosion.

Object 5 - Another Good Double Star - "Zuben Hakrabi" - (upsilon Librae) –
Your 3-inch "supertest"
This is a GREAT star to try on either the 4 or 6-inch scope.....it might give some 8-inch users a bit of frustration as well! This is a wonderful pair with strongly contrasting brightness, but both are
accessible in a 3-inch and above scopes. The primary star is a bright magnitude 4, while the

orbiting secondary star is a faint 11.5! Thus, it is right at the limit of the 3.5 inch scope. Look for the faint star nearly due SOUTH of the brighter one, only 3.3" arc separation between the two, also attainable with the -90 and larger scopes. However, the glare from the brighter star will make the target all the more difficult to see. Both stars are very RED stars, and the color of the brighter one should be distinctive in the 6- to 8-inch range. The secondary star will not be visible in the smaller telescopes.

Object 6 - A Great Double Star - Mu Librae - 6-8 inch telescopes only please! Tough for the 6-inch!
Here are two totally different stars as a pair when compared to upsilon above. This is a pair of stars, magnitudes 5.5 and 6.5 that are VERY close (only 1.8" arc) together and should make a very tough test for the 6-inch, even though that spacing is well within its limit. Theoretically, the 3-inch (with a 1.3" arc Dawe's limit) SHOULD be able to separate these two stars....good luck! Look for the slightly fainter star "just" southwest of the brighter one, but almost touching. Once you have found the correct star in low magnification, begin moving up in steps higher and higher until you can begin to "elongate" the image. It might take as much as 150x in the LX 90 and 200x in the 6 and 8-inch scopes.

Object 7 - A Double for "Southern Folk" - Pi Lupi
Here is another one like the one just preceding. Equal magnitude stars (5th and 5th) and separated by ONLY 1.4" arc, only 0.1" wider than the theoretical limit of a three-inch.....so how good is YOUR telescope? I suspect that this will be VERY tough even for a 6-inch. It is fairly easy in an 8-

inch, however, and I have "elongated' it in my 5" several times. At the very low declination of minus 46 degrees, this is an object only accessible to those lucky observers south of latitude 25 degrees north.

Object 8 - NGC 5897- A "Real" Deep Sky Object - Most Unusual Sparse Globular Cluster

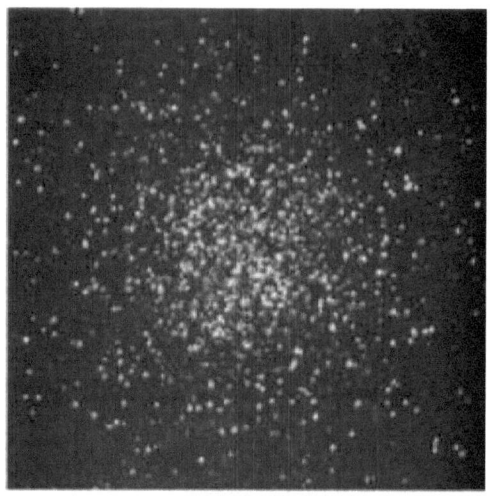

NGC 5897
A Sparse Globular Cluster in Libra
Palomar 200" Photograph

Don't expect a Messier 13 out of this globular, but it is an interesting - but difficult - object. It is likely to be missed in 3-inch and smaller telescopes, since this is a very big (8.1' arc - 1/3 the size of the moon's disk) and very dim (10th magnitude) globular. Hence, its brightness is spread out over a very large area of sky. The photo above, taken with the 200" Palomar telescope, will give you some idea of how faint the stars comprising this cluster are....and how FEW of them there are when compared to other clusters. Expect NO resolution of stars - they are much too faint - in even the 8" scope. At best, the 6-inch and 8-inch will perhaps "hint" at a very faint and large glow in this position....it is a very tough object. Using averted vision (not looking directly AT the object) and moving the telescope slowly side-to-side will assist in verifying that you do, indeed, see it! It will appear large (about 1/4 FOV) but VERY faint in about 75x to 90x in these telescopes. This cluster appears a sort-of cross between a very dense galactic cluster (like M-11 in Scutum - see my *Constellation* Guide in Volume One) and an extremely loose globular cluster like M-80 in Scorpius (Constellation Guide Scorpius). Actually M-80 (just southeast of ngc 5897) is very similarly located in distance as this fainter globular, and both are comprised of about the same number of apparent stars.

Object 9 - Very Nice Globular Cluster in Lupus - NGC 5824
For our observers Down Under and those with very low northern latitudes (from about 35 degrees North and below) this is an excellent star cluster. You will need an unobstructed southern horizon from the U.S. however, as this is low in the sky at about -33 degrees declination. However it IS worth looking at if you can go there, at magnitude 9.3, making it a very bright globular. This object is clearly visible in the small refractor but appears very small, as it is only 3.2' arc in diameter; the stars cannot be resolved in these nor the 3-inch telescope. I can see a few peripheral stars (from latitude +35 05 north) with the 6-inch and pretty good resolution across the globular with the 8-inch; even the 24" from this latitude does NOT show any more stars than seen

with the 8" scope. This is a very compact cluster and has a very rich and dense central region that almost appears star-like in larger telescoes on a very dark night.

Object 10 - Southern Globular - Spectacular Object for Low Latitudes - NGC 5986
Also in Lupus, this is a fine globular cluster and clearly visible from lower latitudes in the northern hemisphere, with a declination of about -37.5 degrees. This is a fairly large and star-dense cluster, measuring a bit over 5' arc in diameter with MANY stars of about 12.8 to 13.0 magnitude, making this a wonderful target for the 6-inch scope which shows clear star images all the way across the bright globular. It is spectacular in the 8" scope, even from 35 degrees north. It is VERY similar in appearance to the bright Mesier 4 in Scorpius (see my Scorpius tour in this *Constellations* volume Two) just to its east and appears to be at exactly the same distance as ngc 5897 and ngc 5824. M-4, on the other hand, is about three times closer than these three distant clusters. This is a great object for the 3-inch, showing good star resolution around the perimeter of the globular, and high powers on a small refractor will hint at some detail as well, although NOT as good as seen with M-4 and M-13 in Hercules.

Objects 11 - A Good Semi-Regular Variable Star - UW Librae / Long Period Variable V- Librae
Semi-regular, irregular and "cataclysmic" variables are my favorites; the "semi-regular," like this on - UW Librae - are very interesting because, even though they some"regular" cycle of dimming and brightening, there are many irregular periods of

behavior with these stars. Most such fluctuations are still beyond complete explanation so observations of these minor outbursts and erractic behavior is much needed by the scientific community. YOU can make a difference in our understanding of these stars by logging onto the American Association of Variable Star Observers (http://aavso.org/) for full information about observing and reporting these stars. In addition free star comparison and locator charts are available through this site. HOWEVER, there is not a dedicated chart for this star, but the chart for "V Librae" (R.A. 14h 38' / DEC (-)17 27) will get you close and the "b" wide field chart will show suitable comparison stars. You can download this chart, save to file, resize and print from: https://www.aavso.org/apps/vsp/ . Note for these charts, simply type in the NAME of the variable at top to generate your choice of chart.

V Librae is a long period variable, ranging in magnitude from 9.0 to 14.5; most stages can be observed by this star through the larger telescopes, and the brightest portion through the 3-inch. The cycle of this star requires 255 days to complete. For a more complete discussion of variable stars and getting started in their study, see my *Observing Variable Stars Guide* under the tab GUIDES/Observational Guides.

WANDERING ABOUT....YOUR NEW "USER OBJECT" IN LIBRA

Now here's a "GO TO" twist....what may seem like a very mundanely dim (magnitude 4.7) star in Libra is actually quite interesting. And....we've never put just a "star" on our GO TO User Object Library directory.

On your sky program (example here for for Autostar), go to: "Select/Object [enter]...." scroll down to "User Object" [enter]. Now enter the coordinates R.A. 15h 09' / DEC (-) 19 degrees 36' for "IOTA LIBRAE", using the number keys on Autostar. After entering the coordinates and pressing "Enter" yet again, scroll down one and you can list the magnitude of the object as "4.7"[Enter].

This star is actually NOT mundane at all...nor is it a single simple star. Iota Librae is a "silicon star", meaning that much of its nuclear fuel has evolved from hydrogen and helium through nuclear compression into heavier silicon for its nuclear fusion processes to continue; although an "A" type star, not totally unlike our sun spectrally, this star is about 60 times more luminous. In addition what you see as a 4.7 magnitude star is actually two VERY close stars that CANNOT be separated except in the world's largest telescopes. It was discovered SPECTROSCOPICALLY, meaning that two sets of spectral "fingerprints" were photographed for this star rather than just one.

In addition to this "out-of-reach" component, Iota Librae has ANOTHER actual physical star associated with it that you CAN see with the ETX 90 and larger telescopes! Discovered by **William Herschel** in his great "40-foot reflector" (http://www.britannica.com/biography/William-Herschel), there is a faint 9.7 magnitude star about "one Jupiter diameter (51" arc)" almost due east from the spectroscopic double; in ADDITION, this faint star is ALSO DOUBLE! It is actually two more stars, both magnitude 9 only about 2" arc away, in a NW-SE orientation. This is an excellent test object, even for the 6-8 inch and I would like to hear from those of you who can actually see this faint fifth star to the Iota Librae system! So you have just loaded your first "silicon-based Quintuple star" into your User Object database! It can't get any better than that for a show-stopped at the next star party!

* * *

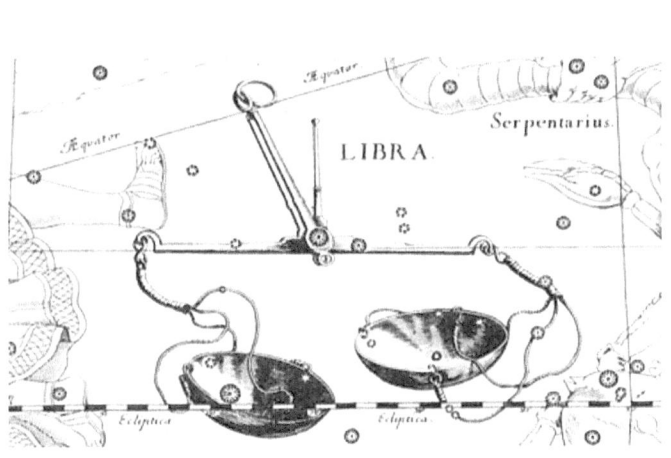

The scale of Libra
As drawn by celestial cartographer
Johannes Hevelius
Firmamentum - 1690

* * *

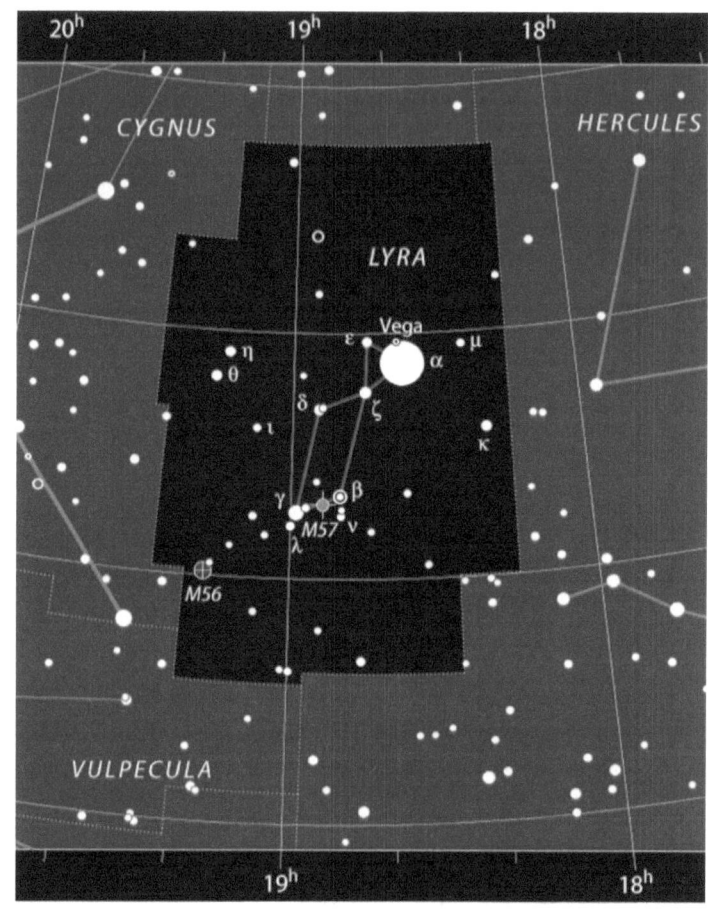

Next stop:
The Celestial Harp, Lyra
Chart Courtesy
International Astronomical Union

Chapter 21

LYRA

*Honing in on the Heavenly Harp
and its Hidden Treasures!*

This Constellation guide, "GO TO LYRA" of the series for all GO TO telescope users explores one of summer's "early arrivals," the constellation of Lyra, heralded by the bright diamond among stars, VEGA. As with all constellation guides, you will start your "GO TO" with a routine and easy slew to a bright star, in this case **Vega**. From there you will proceed to explore this small but wonderful constellation, a fantastic "planetary nebula" that remains from a massive stellar explosion, a fine globular cluster, and some beautiful and challenging multiple stars. Along the way, you will be treated to short descriptions of the mythology of this wonderful constellation and its inhabitants, astronomical history and interesting tidbits of information to make YOUR explorations of the heavens much more enjoyable and diversified..

As always, all objects will be described - first in a concise list with exact coordinates and magnitudes and then individually - with specific details of what YOUR SCOPE.....probably a computerized GO TO model....can expect to show you....and what you can expect to NOT see. So many times casual observers are disappointed when they attempt to find an object beyond their telescope's capability....or perhaps let down from what the photographs of the great observatory telescopes have shown them in the textbooks, magazines and advertisements.

Each GO TO object is discussed for your telescope regarding the type of conditions necessary for you to view it optimally for discern the very faintest details....double star challenges for each size telescope .....magnifications and aperture necessary for most objects, and much, much more. This is YOUR complete guide for deep sky observing enjoyment and satisfaction. In addition these guides make an excellent reference for your next star party itinerary!

Truly, as reports continue to come in on these extensive *Constellations* study guides, they will most definitely put your AutoStar or other computer sky program to work for you in the most efficient and enjoyable way possible!

We hope you enjoy these comprehensive GUIDES to touring the constellations via your AutoStar or sky program and its computer-driven telescope. Each new installment will appear frequently, complete with diagrams, charts and illustrations that you will find nowhere else. Please let us hear YOUR feedback and your observations of each and every constellation after YOU have toured its vast reaches of our skies!

Introduction

The constellation LYRA (pronounced "LIE-rah") ranks among the smallest of the entire regime of official **88 recognized constellations** seen from both the Northern and Southern hemispheres. Though small in size, it is indeed "Mighty" in memories and imagery....as well as ancient and even more modern history.

As mentioned in the Draco installment of the *Constellations* "GO TO" tour guides, the constellation of Lyra will hold special meaning in 15,000 or so years, not that it does not already to many people. As *Thuban* (alpha Draconis) was the Pole Star (the "north star") for the Egyptian pyramid-builders some 5,000 years ago, so will the bright star VEGA (alpha Lyrae) **be in 15,000 years**!

This is due to the Earth's PRECESSION (wobbling on its axis as it moves through space, like a top askew from its perfect spin) over tens of thousands of years. At the present time, of course, POLARIS (alpha Ursa Minoris) just happens to be lined up in such a way as we see it superimposed over the distant line transcribed into space by the Earth's axis of rotation.

The tiny but beautiful constellation of Lyra represents the *LYRE OF HERMES*, fashioned from the shell of a tortoise, according to Greek legend; interestingly earlier Persian sages also denoted the star pattern as *ZURAH LYRE*, an association that survived middle eastern conflict and religious persecution all the way through the 15th century when the "*harp of Phoenicia*" shone brightly in the skies over Arabia.

In later Greek myth, the lyre fell into the hands of noted *Orpheus* who turn his stringed song into a magic spell of sorts that mesmerized all - including animals and insects - who fell upon its melody. It was with Hermes musical instrument that Orpheus was able to enter the famed "underworld" attempting (without success) to rescue his sweetie,

Eurydice from the forces of darkness.

As you gaze at the beauty of such a small gathering of stars, it is quite easy to see Hermes' little harp suspended majestically among the stars; to me it is a wonderful reminder of the warmth, the smells and the wonderful innocent traditions of an impending childhood summer....

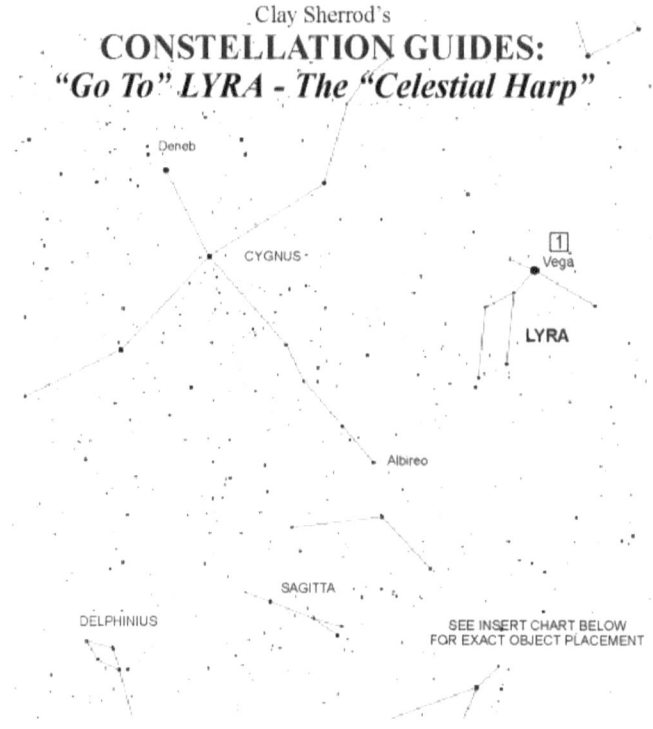

Clay Sherrod's
CONSTELLATION GUIDES:
"Go To" LYRA - The "Celestial Harp"

THE "SUMMER TRIANGLE" –

Perhaps the real beautify of this small constellation is in its association with so many other striking star patters of the late spring and early summer skies. As it heralds in the great bright clouds of the Milky

Way through Cygnus, Aquila, Sagittarius and Scorpius, the first sighting of Vega each year signals - at least to me - the advent of exciting explorations of our Milky Way galaxy!

Note from the sky chart above the arrangement of the bright stars *VEGA*, *DENEB* (alpha Cygni) and *ALTAIR* (alpha Aquilae); all are very bright stars, with Deneb and Altair appearing distinctly yellowish and Vega its brilliant blue-white color. Together, these three stars form an "asterism" in the sky known as the "Summer Triangle," a very conspicuous (particularly from lighted city and suburban sites) isosceles triangle in the night sky.

Please see the guide to Cygnus for more Summer Triangle discussion!

YOUR LYRA CONCISE DIRECTORY OF OBJECTS –

I have chosen the finest nine (9) objects in this tiny constellation to showcase for your "GO TO" tour of Lyra; in this case, all objects listed below will be visible in all telescopes from the 3-inch to 8-inch; of course larger apertures may "show" an object a bit closer and "better," but frequently a wide field and low power view is more desirable than aperture. This is the case for MANY of these objects (as well as scores of fantastic views in Lyra that are not included here), as Lyra skirts the outer periphery of the fantastically star-rich summer Milky Way skies.

The constellation's high declination in northern skies allows for ideal viewing during dark sky conditions as objects pass high in the sky, rather than low to horizons as does Scorpius or other southerly constellations for northern hemisphere observers.

As with all of the "GO TO" tour constellation lists, I recommend a good star atlas and/or chart which will list all the finest objects, constellation-by-constellation. One very handy reference guide is the *PETERSON FIELD GUIDE TO THE STARS AND PLANETS*, which features complete lists with declinations, right ascensions, magnitudes, and all pertinent information for you to expand your observing horizons beyond this brief guide.

For a discussion on keystrokes and the use of specific sky programs, please see Volume One for your particular sky program.

The constellation tour Star Chart following will get you started on your journey for this constellation. That and the "close-up" insert for Lyra (click on the small chart icon below, save to file, open, resize it to fit your page and print!) demonstrates the relative positions for all objects listed on this tour and the conspicuous stars outlining the distinct shape of Lyra the Celestial Harp.

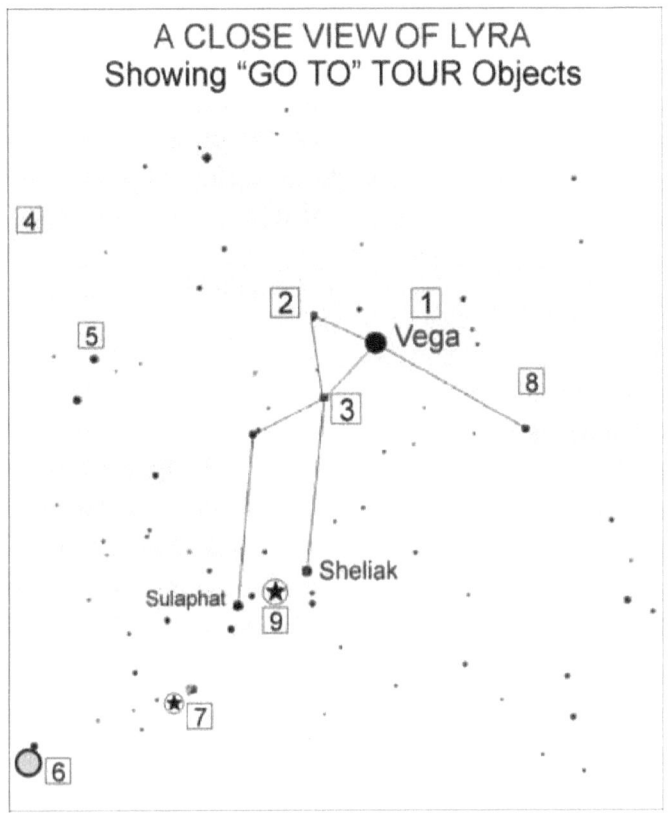

Following is the complete 9-object list for your "GO TO" TOUR of Lyra; you may wish to find the majority of the objects from the sky program Library (for example, you can easily pull up "Object/Deep Sky/Messier Object/..57...." and then press "Enter", followed by "GO TO" to access the famous Ring Nebula. On the other hand, if you want to experiment and become a "better computer user" try entering the exact R.A. and DEC coordinates listed below as described above after holding down the MODE key. You will find the accuracy of entered GO TO's to be somewhat less than those stored in your sky program, but the capability of acquiring unlisted objects is fantastic!

OBJECT 1: very bright star - Vega (alpha Lyrae) - R.A. 18h 35' / DEC + 38 44 - Magnitude: 0.04
OBJECT 2: multiple star - Epsilon Lyrae - the famous "DOUBLE-DOUBLE" - R.A. 18h 43' / DEC +39 37 - 4 stars
OBJECT 3: nice low-power double - Delta Lyrae - R.A. 18h 53' / DEC + 36 54 - Magnitude: 4.5
OBJECT 4: variable star - R Lyrae (a good beginner's variable) - R.A. 19h 24' / + 42 41 - Magnitude range: 3.9 to 5.0
OBJECT 5: nice test double - Eta Lyrae - R.A. 19h 12' / DEC + 39 04 - Magnitudes: 4.5 & 8.7, wide double
OBJECT 6:   nice globular cluster - Messier 56 (ngc6779) - R.A.19h 15' / DEC + 30 05 - Magnitude: 8.2
OBJECT 7: very difficult planetary neb. - ngc6765 - R.A. 19h 09' / DEC + 30 38 - Magnitude: 11.1 (tough!)
OBJECT 8: nice variable star - W Lyrae - R.A. 18h 13' / DEC + 36 39 - Magnitude range: 7.3 to 13.0 (196 days)
OBJECT 9: best planetary there is! - Messier 57 (ngc6720) - R.A. 18h 52' / DEC + 32 58 - Magnitude: 9.3

A VISUAL GUIDE TO OUR DEEP SKY OBJECTS IN LYRA –

Object 1 - Very Bright Star VEGA (Alpha Lyrae)
At 27 light years distant, this brilliant blue-white orb is the fifth (or some say fourth) brightest star of the night sky. It is frequently called the "Harp Star"

from its association with the mythological Hermes' lyre. However, "Vega" gets its name from the Arabic skywatchers of thousands of years ago who called it "Alnasr al Waki" or "*The Bird of Prey*", signifying a hunting eagle, falcon, or vulture as was the entire constellation known for those early skygazers. Personally, I prefer the more modern and mundane "harp" to a blood-thirsty raptor. Early Chinese astronomers knew Vega as "*the girl who weaves*" who was always accompanied by her friend Altair, "*the boy who herds.*"

Vega, at magnitude 0.04 has a more brilliant blue color than even bright Sirius, whose light is more pure white. It is only 58 times brighter than our sun would appear at the same distance, yet is about 2.5 times larger. Although Vega is somewhat variable, it is not of interest to common telescopes for that reason.

As mentioned, this star will be the brightest of all possible northern pole stars in about **12,000 A.D** where it will be farther from the true pole - some 4.5 degrees - than Polaris is today. Historically, you might want to remember for your next "teaching guide" at the local star party that Vega was the first star to be actually PHOTOGRAPHED! (boy have we come a long way!). In 1850 **William Bond** and his assistants used the "*Daguerreotype Process*" to capture its brilliance in a 1.7 minute exposure. The 15-inch telescope at **Harvard University** in Cambridge was then the largest refractor in the world, and is still in use at the same location today.

Object 2 - The Famous "Double-Double Star (Epsilon Lyrae)....a fantastic object for all scopes!

This is a FAVORITE at all star parties, and one that - no matter how many times you have seen it before in no telling how many different telescopes -you keeping coming back to view time and time again! It is THAT good.....if you have not seen this remarkable star system you have either been living in a cave or merely did not know it was there. Epsilon Lyrae - the "*double-double star*" is located very close to Vega and is the closest bright star to it with the unaided eye. Your finderscope will begin to show TWO stars of equal magnitude....the 8 x 50 finder will clearly show both stars at only 8 power. Using the standard 26mm eyepiece in virtually all telescopes will reveal that EACH of those two stars is ALSO TWO STARS! A bit more magnification is required for the small telescopes, but these scopes will show this beautiful system clearly. So, you have two bright stars that orbit one-another; those two stars are actually also two stars that orbit each other; thus, you have two stars revolving around a common center that are synchronously revolving around yet ANOTHER pair doing the same thing! Wow. Look for all four stars to be pretty much "white" in color, as they are ALL type-A stars. The two brighter pairs are actually separated in space by about only 5 trillion miles.

Object 3 - Nice Double Star With Color Contrast - Delta Lyrae

This is a good objects for all telescopes. With a huge separation of 10' arc (1/3 the size of the moon at the same magnification), this pair can be seen easily with a small refractor and larger scopes at very low power. The dimmer star ("Delta-1") is magnitude 5.5 and the brighter star ("Delta-2") is magnitude 4.5. At 800 light years, it appears that these two stars are ACTUALLY the two brightest members of a very sparse and loose cluster of stars of magnitudes 7 to 10; so, likely the stars that you see with your wide field view around these two brighter stars are, indeed, part of a loose cluster of stars. Be sure and note the interesting contrast of colors of these two stars, even though both are thought to be the same age. ALSO....scan this area around Delta Lyrae with very low power and a wide-field eyepiece (here is where the small APO refractors have the "big boys" beat!) for some of the most gorgeous views of the rich star fields imaginable. It is truly a beautiful sight, but one that must be enjoyed at the darkest sky location possible.

Object 4 - Good Beginner's Variable Star....R Lyrae
Here's one you can actually use your finderscope on....or try a pair of binoculars....it is ideal for the wide fields of the ETX 60 and 70. R Lyrae is a fairly bright (magnitude 3.9 to 5.0) star with a short period of only 46 days. It is ideal for beginning a pursuit in understanding recording techniques of variable stars! Variable star observing requires only that you routinely observe the brightness of a given star over time and record what that brightness is.....this is ONE area that the amateur astronomer can greatly contribute to the body of scientific knowledge that the professional astronomer has neither the time nor equipment to do. Stars like "long period variables" have very long light curves (see my *OBSERVING GUIDE: VARIABLE STARS* on the ASO website under the GUIDES tab for a complete description) and do not need to be observed every night. They are so predictable that observing them once a week or so is fine; on the

other hand there are "cataclysmic variables" that are violent and tend to have unannounced outbursts....observing these as often as possible! Stars like R Lyrae, with a short period of only 46 days should be "looked at" every time you are out since you do not even really need a telescope.

A great comparison star chart from the American Association of Variable Star Observers (AAVSO) can be accessed, saved to file, resized and printed for outdoor use from the following link: https://www.aavso.org/apps/vsp/ . Note for these charts, simply type in the NAME of the variable at top to generate your choice of chart.

In only three months you will have recorded two full cycles of this interesting star; use this experience to begin accessing more and more charts via these free and useful links to the AAVSO; there are hundreds of charts for all constellations and star types as well as important and helpful information about variable star observing!

OBJECT 5 - Nice very wide Double Star; great wide field views of the sky! Eta Lyrae.
This is a beautiful field surrounding the star Eta Lyrae. The double is a very wide star, equally about one-half the diameter of Jupiter as seen with the same magnification, from the brighter (magnitude 4.5, easy in your finderscopes) to the companion star (magnitude 8.7). This is a good test for your visual acuity with the telescope and the star is NOT as easy as you might think based on the separation. The tiny 8.7 magnitude star tends to get "lost" in the field if too low power is used...on the other hand, if too high a magnification is attempted, the glare of

the 4th magnitude star will obliterate the tiny image. This is a true challenge for a double with the small telescope. It begins to show more easily with the ETX 90 almost DUE EAST of the brighter star and is fairly simple to find in the 6- and 8-inch scopes. Both stars are very young "B" types stars and are true white in color.

Object 6 - Messier 56 - A very nice Globular Cluster

If it were not for the fact that SO MANY globular clusters are scattered in this region of sky (see my guide *OBSERVING GLOBULAR CLUSTERS* here on the ASO website under GUIDES) Messier 56 would be an outstanding deep sky objects. However, rivaled by larger and closer M-13 and M-5 in Hercules and Ophiuchus, respectively, this is an often-overlooked compact and interesting globular. Look for this object under your "Object/Deep Sky/Messier...." library of your sky progrram, or find it about half way from Gamma Lyrae to Alberio (beta Cygni).

The first person to actually resolve this "faint ball of light" as it was known to Messier was *Sir William Herschel*; he realized that the object was comprised of thousands of 11th and fainter magnitude stars. This object can be seen clearly as a small and compact "ball of light" in the smaller telescopes with no resolution; however, magnifications of about 156x are ideal for beginning to see actual stars around the edges -NOT the center - of this cluster with the 6 inch and larger scopes. Even on the best nights, only a couple of brighter members toward the center can be glimpsed with the 8" scope; otherwise it appears as a uniform glow surrounded by "star dust" in the larger apertures. This is an extremely distant globular; hence its small size and low brightness....it is about 46,000 light years - very close to the center hub of our Milky Way galaxy - from the Earth.

Object 7 - NGC 6765 - Faint Planetary Nebula
This is one of two planetary nebula located in Lyra, the other being the famous and un-challengeable "Ring Nebula", Messier 57. This small and faint object is a very difficult challenge for a 6-inch and is barely visible to the 8-inch telescope even on the best of nights. See if YOU can see this distant object. It is only one-half the size of the Ring Nebula and requires averted vision and VERY dark skies to see. You must let your eye adapt very long to see this one. If you do not first see it in the larger apertures, don't stare....merely scan the field of view at about 156x and eventually your averted vision will spot the object.

Object 8 - A Fine Variable Star - W Lyrae

Now that I have gotten your feet wet as a "veteran variable observer" with R Lyrae above, how about trying out one of those "long period variables?" These stars required that you aim your scope at them every two weeks or so and compare the brightness to some very similar nearby star. With your telescope's GO TO function, this makes easy work out of a valuable observation.

W Lyrae can be seen throughout its entire cycle of 196.4 days with scopes 6-inches and larger as it varies in brightness from a relatively bright magnitude 7.3 to a very dim 13.0. When at its brightest period and for three weeks either side of that, the smaller telescopes can keep up with the light changes easily. For the 3-inch, observers can watch about two thirds of the interesting cycle provided that good finder and comparison charts are at hand. The link below to the AAVSO provides a finder chart for W Lyrae; when it is dimmer, you must also download the more narrow field and fainter comparison star chart to follow this star through its dimmest periods. Like many long-period variable stars, R Lyrae is a supergiant very large and red star that pulsates through diameter to result in the light changes that we see.
https://www.aavso.org/apps/vsp/ . Note for these charts, simply type in the NAME of the variable at top to generate your choice of chart.

Object 9 - And here is the grand finale: Messier 57 - the **RING NEBULA** - Planetary Nebula
It may not be the largest that we can see....and it certainly is NOT the closest. It is by far not the most curious as to shape and morphology....but it is a "ring" and looks like a "ring" and has a cute

name. So it is the most famous of all planetary nebulae. A PLANETARY NEBULA is a (usually) spherical shell of gas (no, not a "ring") that is ejected from a star that has violently "shed" its outer layers of gas. As the star ages, the pressures inside build until a minor collapse of the outer layers plummets downward, increasing the pressure until the star "can't take it anymore....." and it hiccups.

The ring nebula is the hiccup of a tiny star (see the ASO photograph following, that attempts to match the visual impression in moderate telescopes) that still remains in the center of this beautiful object, some 1800 light years distant. In a smaller telescope Messier 57 is clearly visible. It appears as a very dull ghost-like planet (hence the name "planetary nebula"....it has nothing to do with planets other than that appearance). The central region of M-57 are very difficult with the low power instruments.

The 4-inch begins to reveal a bit of the darkened center of the nebula however and with the 6-inch much detail can be seen with magnifications between 150x and 200x. Note for reference the very faint star just to the right of the Ring Nebula that is seen in the photo. This is a very blue dwarf star, very close to the end of its "life." In neither the 5 or 8-inch was the central star seen. Although the star is referenced at "magnitude 14.2" it is very difficult because of the bright cloud of gas that you can clearly see in the drawing. I have never been able to see it with a 14" to 16" telescope, but can clearly make it out in the 24" at about 300x. With a CCD camera and 10 inch telescope the central star pops out in 25 seconds!

Remember when looking at this object, total magnitude 9.3, that it is NOT a ring of gas...it is a spherical SHELL of gas. You are merely looking through a thinner portion of gas toward the center; the perimeter appears brighter and "ring-like" because you are looking horizontally through a much more extended cloud of gas. NOTE that in most telescopes, particularly larger ones, the Ring Nebula has been described as "bluish." I have never, ever seen this coloration; indeed, if anything it has always (even back in my youth with good eyesight) appeared a rather brownish-gray color. I would be very interested in hearing of YOUR impressions of color of this fantastic object!

WANDERING ABOUT....YOUR NEW "USER OBJECT" IN LYRA

There are not many "showcase objects" in this small constellation of the northern skies, but what there

IS, is very impressive indeed. We have rarely entered a "double star" into our "User Object" list, so in Lyra we will take that opportunity with a fairly obscure double known only as "17 Lyrae," located at: R.A. 19h 06', DEC + 32 degrees 25 minutes.

I have selected this star because of three reasons:

1) it is likely that you might never search it out if it were not in this guide....and NOW ready to enter into your user list;
2) the star pair can be seen and resolved (though very difficultly in the small refractors because of magnification limits!) in all of our telescopes, and appears pretty in all of them as well;
3) it actually is a very nice and often overlooked classic double star.

THE DOUBLE "17 Lyrae" is magnitudes 5.0 and 9.4, and thus is easily accessible via the brighter star in even the finderscopes of the 3- to 5-inch scopes; the GO TO will get you very close and you can determine the actual star from there. Once found, increase the magnification in the small telescope to just over 100x (closer to 120x is ideal) and you should begin to split this star with a hint of a thread of dark sky between them. The 3-inch will do the same thing at 83x (the 15mm eyepiece); the 6-inch requires only 75x using the a good Plossl. The same is true for the 8-inch. The separation of 17 Lyrae is about 3.7" arc so you would think this would be easily resolvable in all of our telescopes.....but the fact of the matter is that the brighter star outshines the dim one so greatly that it actually is NOT an easy object!

SO LET'S LOAD THIS CHALLENGING STAR INTO OUR USER OBJECTS! Take your Autostar and let us key in the particulars for "17 Lyrae." This will add yet another multiple star system to your sky program library of interesting and unusual "User Objects."

Using AutoStar, go to: "Select/Object [enter]...." scroll down to "User Object" [enter]. Now enter the coordinates just given above for 17 Lyrae, using the number keys on AutoStar. After entering and pressing "Enter" yet again, scroll down one and you can list the magnitudes of both stars [Enter]. One scroll more after that and you may type in your brief description of the object....but look at it first! [Enter again].

Remember! Sky program vary in terms of accessing and entering information for GO TO telescopes; always consult your User Guide or HELP tab for specifics regarding your computerized sky library.

*   *   *

"If the stars should appear one night in a thousand years, how would men believe and adore; and preserve for many generations the remembrance of the city of God which had been shown! But every night come out these envoys of beauty, and light the universe with their admonishing smile."

Ralph Waldo Emerson

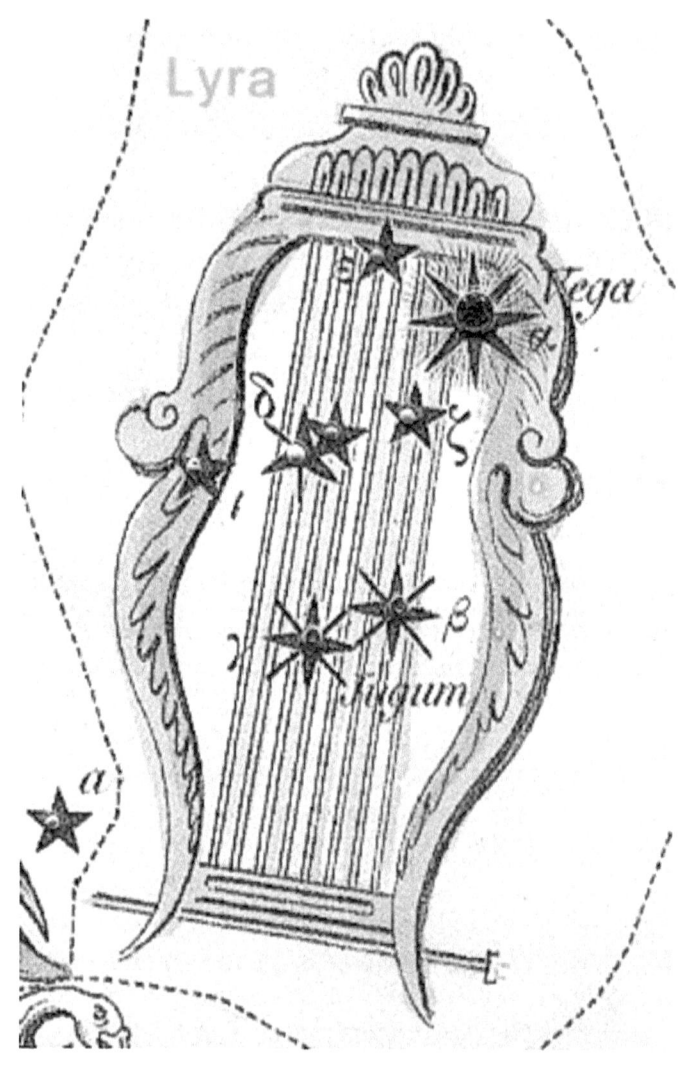

An example of early woodcut constellation art.
Lyra, the Harp, showing the star Vega

Chapter 22

OPHIUCHUS (with discussions of objects in Serpens)
Slithering With "The Serpent Handler"

In this *Constellations* guide installment, "GO TO OPHIUCHUS"- of our constellation study guides for all GO TO telescope users - we will explore a fabulous selection of fine globular clusters, and our first "dark nebula" of the deep sky.

As will all of our GO TO guides it features a "start" with an easy GO TO to the bright star **RAS ALGHAGUE**, and then proceeds through the many woonderful *globular clusters* and interesting objects within reach of YOUR telescope! This guide encompasses both Ophiuchus and the "snakes" of our night skies....*SERPENS CAPUT* and *SERPENS CAUDA*, to the west and east, respectively, of Ophiuchus.

From bright starry globular clusters to beautifully-colored and complex multiple star systems, discussions along the way tell you what to expect from each telescope size and type. All objects will be discussed with exact descriptions of what all observers should expect to see...and what to NOT expect to see!

As in all Guides, useful magnifications for EACH GO TO object are discussed for certain telescopes, what type of night and conditions are needed to see certain details, double stars that can be resolved in each telescope model, and much, much more. It is your complete GUIDE for your deep sky observing pleasure and a very handy tool for use at your next star party!

Needless to say, it WILL put your computerized telescope to work for you in a most efficient and enjoyable way!

I hope you will enjoy these comprehensive guides to touring the Constellations which feature most constellations, complete with diagrams, charts and illustrations. Please let us hear from you with summations of YOUR observations through these constellation tours!

Introduction

Ophiuchus (pronounced "O-fee-ew-cus") is one of the largest of all constellations and dominates the seemingly bright-star void skies of mid spring at late night. The mighty "*Serpent Handler*" of our skies would not be complete without his two accompanying "snakes," *Serpens Caput* (to the west of Ophiuchus) and to its east, *Serpens Cauda*.

Ophiuchus once took up much more of the starry sky than it does today, but for thousands of years has been known as a "snake charmer," serpent handler, or the "snake man" to ancient cultures. Although the early Greek philosophers clearly recognized him through the Arabic origins, it is interesting to note in the very early and picturesque French planisphere shown above that the constellation is referred to as "*Serpentarius*", albeit a rose with a different name.....is still a serpent handler.

The click-on chart shown below can be downloaded to a file on your computer, resized in that file and printed as a complete star reference chart for these four constellations. Each of the reference numbers apply to the concise listing of objects found following.

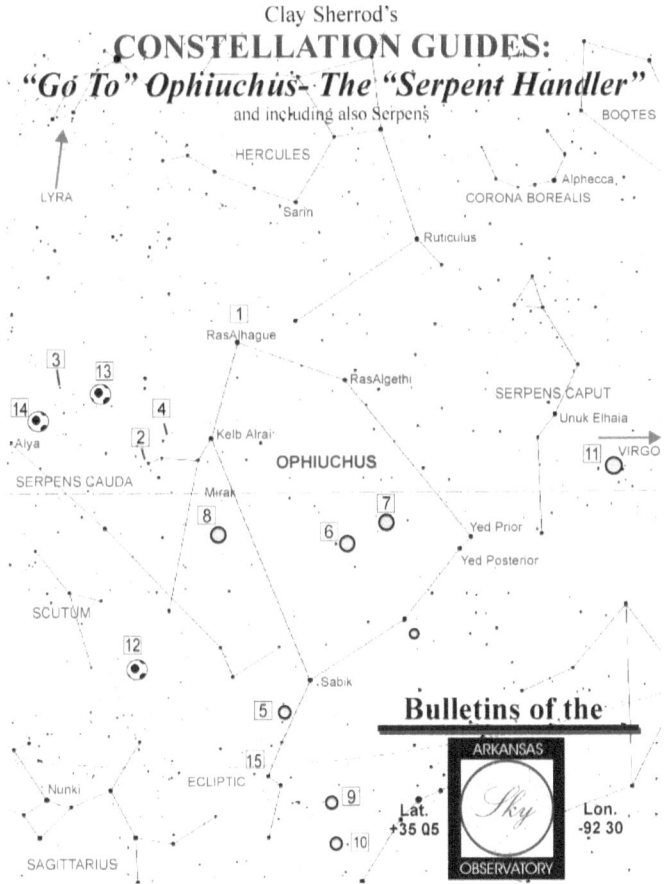

Perhaps my readers will have a bit easier time than do I realizing a depiction of a "snake handler" among this widely scattered star group. Note that our **celestial equator** ("0" degrees declination)

passes through the southern part of this huge constellation and that the ecliptic (the "zone of animals" in which the sun, moon and planets appear to progress) is arced to the south of this springtime constellation.

If you can visualize Ophiuchus as a person handling snakes, then you surely can make out the two snakes that he has in his hands, Serpens Caput grasped tightly in his left (west) hand and another seized serpent held firmly in his right hand, Serpens Cauda.

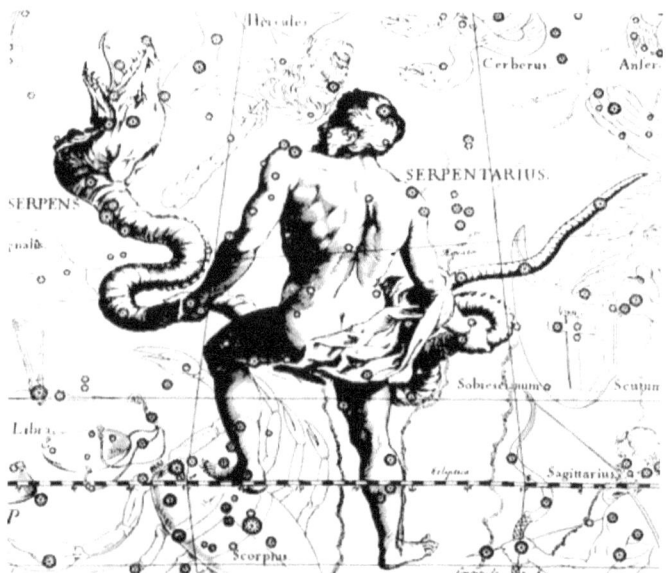

The Snake Handle Ophiuchus by Hevelius

It is among these three constellations that many of the finest **globular clusters** visible from Earth are found, as well as very rich star clouds of the Milky Way in eastern Ophiuchus and Serpens Cauda. This is an area so rich in star clouds, nebulosity, and dark nebula that everyone should take time out away from their high-powered and limited, narrow field telescopes and scan these constellations slowly with a good pair of binoculars on a very dark sky away from city lights. It is truly a spectacular sight in the eastern realm of the constellations.....we are approaching the center of our own Milky Way galaxy as we progress eastward toward Sagittarius.

Although often overlooked because of the impending arrival of the brighter Lyra, Aquila and Cygnus constellations of late night spring or summertime, this area of sky is very rich in galaxies, clusters, dark obscuring dust (one such item is included in the GO TO list that follows), and many wonderful double stars that are both optically appealing and challenging to test the precision and alignment of your telescopes as well.

YOUR OPHIUCHUS / SERPENS CONCISE DIRECTORY

There are 15 objects in this constellation GO TO tour; all are in reach of telescopes 3-inch to 8-inches in aperture, yet each telescope will demonstrate uniquely different and challenging aspects of the objects. In addition to the 15 finest objects, there are literally hundreds of wonderful double stars that are visible in most telescopes as well as many fainter globular clusters (check your sky program library for "ngc" listings!), stars of curious colors and motion. It is in this general area, the constellation of OPHIUCHUS that offers our best vantage point for observing the absolutely incredible selection of beautiful globular clusters.

As will all of our GO TO guides, I continue to recommend good a good star atlas and/or chart/ and computer sky program which lists the finest objects constellation-by-constellation; if you cannot access any of these objects (or those that are not listed in this TOUR), you can access directly from coordinates - Right Ascension (RA) and Declination (DEC) of any known object via the keypad or program. For Autostar, as an example, you merely need to hold down your MODE key on the AutoStar for three (3) seconds and the RA and DEC coordinates appear for the telescope. Merely press "GO TO" and the cursor appears prompting you to enter the Right Ascension of the object if it is NOT listed among the objects in the AutoStar library; once the RA is entered, press "Enter" and the cursor once again prompts for the Declination coordinates (these coordinates for epoch 2000) are found in all good observing guides). Once those are entered, merely press "GO TO" once again and your telescope will slew to the position of the object!

The following constellation guide to objects at the end of this TOUR will describe all the details of each object and provide specifics as to visibility of that object in YOUR telescope model.

The constellation tour star chart here (click on and print to size, above) will get you started, as it demonstrates the relative positions of all objects in this "tour" to the conspicuous stars outlining the distinct figures of our three constellations of this TOUR.

Following is the complete 15-object list for your

"GO TO TOUR" of Ophiuchus and Serpens; you may wish to find the majority of the objects from the AutoStar library (for example, you can merely pull up Messiers 9 or 12 by going to "Object/Deep Sky/Messier Object/M-9....enter....GO TO" or...if you want to experiment and be a "better AutoStar user", try entering the following coordinates (provided in the list directly following) as described under MODE above.

OBJECT 1: bright star - Ras Alhague (alpha Ophiuchii) - R.A. 17h 33' / DEC + 12 36 - Magnitude: 2.1

OBJECT 2: tough double star - 70 Ophiuchii - R.A. 18h 03' / DEC + 02 03 - Magnitudes: 4.2 & 5.9

OBJECT 3: good variable star - X Ophiuchii - R.A. 18h 36' / DEC + 08 47 - Magnitude range: 6 to 9, 334 days

OBJECT 4: Barnard's Star! - "the runaway star" - R.A. 17h 55' / DEC + 04 24 - Magnitude: 9.5 (chart attached)

OBJECT 5: a globular cluster - Messier 9 (ngc6333) - R.A. 17h 16' / -18 28 - Magnitude: 9.0

OBJECT 6: globular cluster - Messier 10 (ngc6254) - R.A. 16h 55' / DEC -0402 - **Magnitude: 7.9**

OBJECT 7: globular cluster - Messier 12 (ngc6218) - R.A. 16h 45' / DEC -01 52 - Magnitude: 8.0

OBJECTS 8: globular cluster - Messier 14 (ngc6402) - R.A. 17h 35' / DEC -03 13 - Magnitude: 9.5

OBJECT 9: globular cluster - Messier 19 (ngc6273) - R.A. 17h 00' / DEC -26 11 - Magnitude: 8.1 (nice object)

OBJECT 10: (actually in Scorpius) bright globular - Messier 62 (ngc6266) - R.A. 16h 58' / DEC -30 03 - Magnitude: 6.6 (very nice globular)
OBJECTS 11: (Serpens caput) best globular! - Messier 5 (ngc5904) - R.A. 15h 16' / DEC + 02 16 - Magnitude: 6.2 (the very best of all!)
OBJECT 12: (Serpens cauda) diffuse nebular and star cluster - Messier 16 (ngc6611) - R.A. 18h 16' / DEC -13 48 - Mag.: 8.3
OBJECT 13: (Ophiuchus) large open cluster - IC 6633 - R.A. 18h 25' / DEC + 06 32 - about 20' arc across, very nice, 65 stars
OBJECT 14: (Serpens cauda   galactic cluster - IC 4756 - R.A. 18h 37' / DEC + 05 26 - Bright (m = 5.1) with lots of stars at low x
OBJECT 15: (Ophiuchus) dark nebula! the best! - B72 (Barnard's nebula) - R.A. 17h 21' / DEC -23 35 - Very interesting, S-shaped

A VISUAL GUIDE TO OUR OBJECTS IN THIS CONSTELLATION SELECTION –

<u>Object 1</u> - Bright Star RAS ALHAGUE (Alpha Ophiuchii)
From the Arabic, this magnitude 2.1 star denotes the "head of the serpent-charmer," and we can well imagine from this middle-Eastern origin the mighty King Cobra as it is wooed by the skills and undaunting courage of the mystical "serpentarius."
It is likely that Ras Alhague is tied with the seven brighter members of the Ursa Major "association" which all appear to comprise a very loose yet close star group some 70 light years distant. Now that you are locked onto the brightest star in this seemingly star-poor region of our sky, we will begin our GO TO tour.

Object 2 - Very beautiful double star 70 Ophiuchi
This double star - at a separation of 2.0" arc has two stars of very nice color contrast. It is unmistakable at higher magnifications, but does require at least a 4-inch on the very best nights. About 200x should split these two 4.2 and 5.9 stars with just a thread of darkness between in that scope. It is an outstanding sight in a 6-inch at 250x, with the fainter of the two stars very, very red and the brighter component a vivid deep yellow; this is one of my all-time favorite color-contrasting doubles. Look for the fainter companion star almost exactly due west of the brighter, yellow star. At a distance of only 60 light years, this star system's odd "wobbling" in space suggests that it is likely orbited by a dark object, perhaps a large exoplanet, that is some 10 times the mass of our own Jupiter.

Object 3 - Variable Star X Ophiuchi....good for all scopes, low power and even binoculars!
Varying from magnitude 6 to 9 at a period of just less than one year (334 days), X Ophiuchi really is a great "beginning" variable star and is ideal for the smaller telescopes with their wide fields of view. The chart link listed below will take you directly to a chart from the American Association of Variable Star Observers (AAVSO) in Massachusetts. Merely right click on the chart and "Save as..." into a folder. RE-SIZE and print and you have a wonderful year's work ahead of you:
https://www.aavso.org/apps/vsp/ . Note for these charts, simply type in the NAME of the variable at top to generate your choice of chart.

Note that there are catadioptic "user friendly" charts in that the field has north at the top and east at right, just the correct orientation for catadioptic telescopes with the right angle mirror or prism in use.

Note the slow and very predictable nature of the light changes in X Ophiuchi as shown in the following light curve derived from about two of the 334-day cycles of this interesting low power variable star:

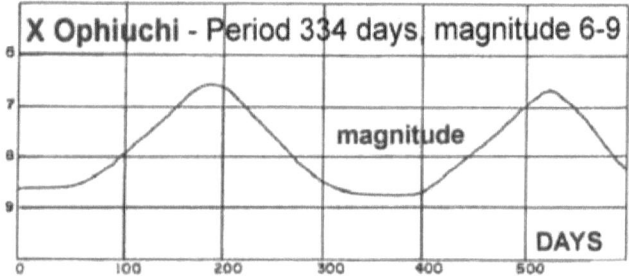

Object 4 - BARNARD'S "*RUNAWAY*" STAR
At first glance this would be a very faint, uninteresting star. At 9th magnitude, the star is visible in 4-inch and larger telescopes and difficult to locate unless you use the star chart provided below. This is a tiny little star, with a total diameter of only ONE-HALF the distance between the Earth and Moon! (140,000 miles across) By comparison, our own sun has a diameter of over 800,000 miles.

"*Barnard's Runaway Star*" has the distinction of have the largest PROPER MOTION (its apparent movement across our skies as we see it relative to more distant stars behind it) of any other star in the Milky Way. It is a magnitude 9.5 red dwarf star (yes, its reddish color CAN be seen in the 6-inch and larger scopes) and moves nearly due north EACH YEAR by an incredible 12" arc - that is about 1/4 the diameter of the planet Jupiter. As can be see from the figure below, this rapid motion can actually be plotted year-to-year.

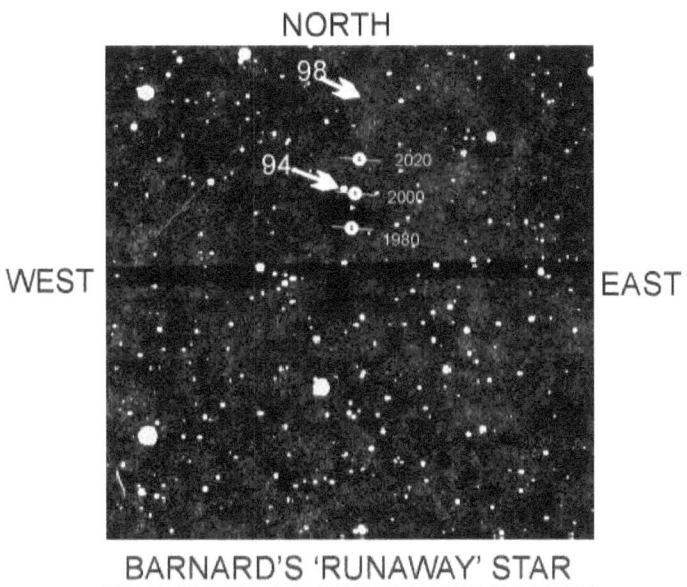

BARNARD'S 'RUNAWAY' STAR
STAR CAN BE COMPARED IN BRIGHTNESS
TO TWO SHOWN

Part of its rapid apparent movement that we see is from the fact that Barnard's Runaway is ONLY 6.0 light years away; it also is physically moving very fast as well. In the accompanying chart, the orientation is REVERSED, in that it matches the field as seen through our telescopes with a right angle prism or mirror, with NORTH at the top and EAST to the right. There are two stars in this chart that I have marked with arrows and numbers, the numbers (decimal points omitted so you won't confuse them for little stars!) are the actual magnitudes of these indicated stars; note that Barnard's Runaway is VERY close in magnitude to each of these stars. At the present time, the star makes a very interesting "optical double" with the fixed star nearby!

Object 5 - One of Many Globular Clusters!
Messier 9
The ninth object cataloged by eighteenth century comet hunter Charles Messier, this is one fine example of the many globular star clusters found in the constellation of Ophiuchus and surrounding areas. For a full in-depth discussion about the globulars and WHY so many of them seem to be concentrated in this general direction, read my ASO Observing Guide: OBSERVING GLOBULAR CLUSTERS, found on the ASO website under the tab ASO / GUIDES/*Observing Guides*.

Be sure and observe ALL the globular clusters on this segment of your constellation TOUR and come back to Messier 9 for comparison! You will immediately realize that this one globular is by far SMALLER and more compact that all the rest! In addition, this globular is more of a crucial test for our telescope size ranges; the 4-inch will not resolve the stars in this globular, but will partially resolve the others on the list. Peripheral stars, a few, can be clearly seen in the 6- and 8-inch scopes begin to break down a few stars toward the center core of this cluster. At a faint magnitude 9.0, it is not at all resolvable in small telescopes, appearing like a bright and very small round "ball" of cotton at medium power.

In your larger two telescopes, particularly the 8", look for two ADDITIONAL CLUSTERS (both very distant globulars that WILL NOT be resolved), the first NGC 6342 (1.5 degree southeast - use a

then go up to about 120X to view) and then NGC 6356 some 1.2 degrees northeast of M-9. The drawing below shows Messier 9 (oriented correctly for catadioptic scopes) as it would appear visually in a moderate-sized telescope.

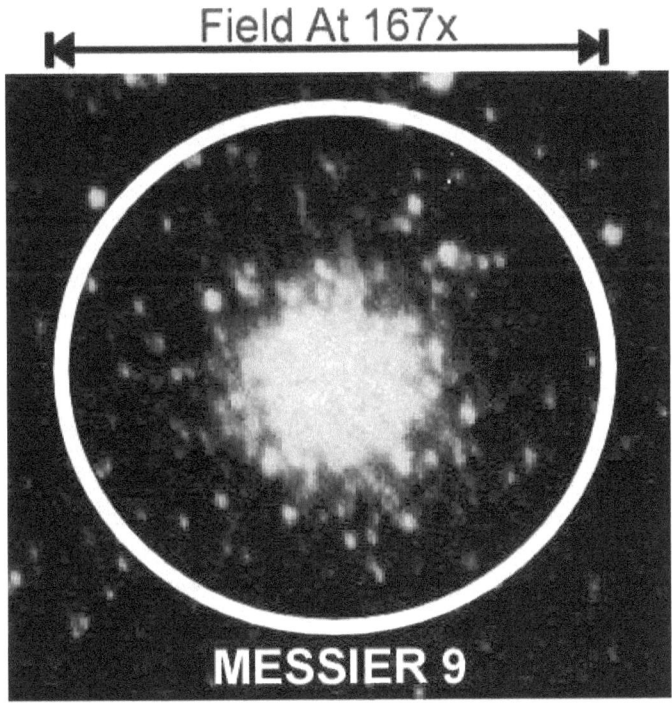

Object 6 - Globular Cluster Messier 10
This is a far better sight in our telescopes than the difficult Messier 9 (above). At magnitude 7.0, this is one of the brightest globulars and presents a wider "scattering" of stars. Unlike the tightly-packed M-9, Messier 10 has its stars spread widely across the field (see figure below, courtesy Lowell Observatory archives), easily resolvable in the 6-inch telescope; at medium power SOME stars around the very edges may be glimpsed with the 4-

inch at 160x and above will show about 70 stars with a very pronounced concentration of brightness right in the center. Look for this central brightening as well with larger telescopes.

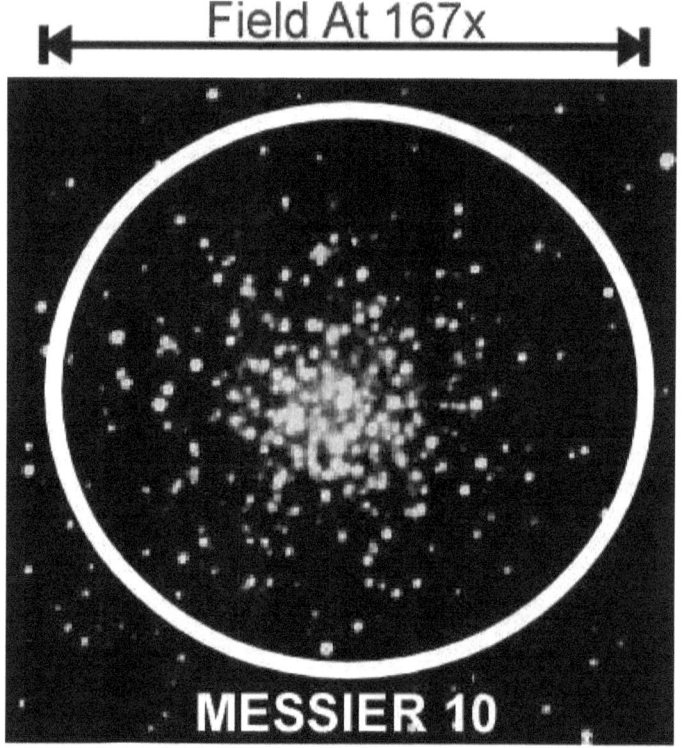

Object 7 - And yet, another Globular Star Cluster! Messier 12

Forming a very nice pair of "faint fuzzies" in the 8 x 50 finderscope and easily in the same lowest power field of views of small telescopes, Messier 12 is only a bit more than 3 degrees northwest of Messier 10. Like Messier 10, M-12 is VERY loose for a globular cluster, almost appearing as a tight open galactic cluster. The star resolution is relatively easy with a 6-inch at about 150x and many peripheral stars can be glimpsed on a very dark night with a 4-inch at around 120x. The 8-inch shows good star images all the way to the center of this object. The very scattering of stars loosely is perhaps attributable to its easy resolution in nearly all of our telescopes. Look for "clumps" of stars, places where many stars appear to be more closely grouped, throughout M-12 in a 4-inch and larger telescopes. Some of these "tight" areas can be clearly identified in the drawing below which has been digitally altered to provide a view that might be expected with the 8-inch at medium-high magnification.

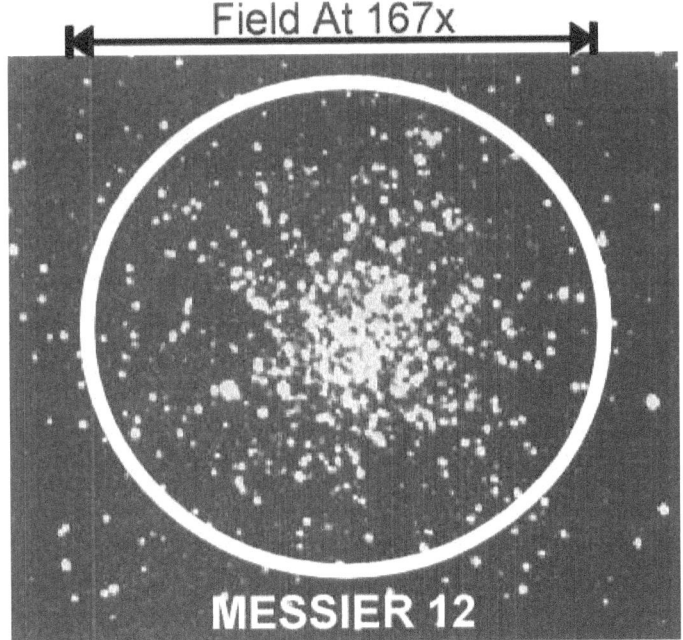

Object 8 - Messier 14 - Can it BE? YES! It's another globular cluster!

I told you that Ophiuchus offered a plethora of globular clusters (if only we knew what exactly a "plethora" comprised). Messier 14 is one of the richest of the springtime globulars, conveniently placed in a part of this constellations where there is little else of interest! This is NOT a bright globular and do not expect to see any star resolution in smaller telescopes. It is VERY tightly packed with a total magnitude of about 9.2 visually, and tiny stars of magnitude 15 to 16 mostly. Because of its packed nature, event the 8" cannot "see" the central structure of this globular and ONLY the peripheral stars may be glimpsed under very dark conditions with both 6 and 8-inch telescopes. On very, very dark sky sites, the larger scopes MAY be able to see "star streamers" than appear as strings emanating from the central part of this cluster. Two such streamers are clearly visible in the following drawing which was digitally altered to render as close a match to VISUAL telescopic appearances in about an eight-inchh telescope. The streamers are at DUE NORTH and at the lower left of the cluster. I have digitally altered this photograph to match both the orientation and the image as seen at about 230x with an 8" scope. Messier 14 is much more distant that M-10 and M-12 which are both "paired" gravitationally.....M-14 is an incredible 70,000 light years distant compared to only 24,000 for M-10 and M-12.

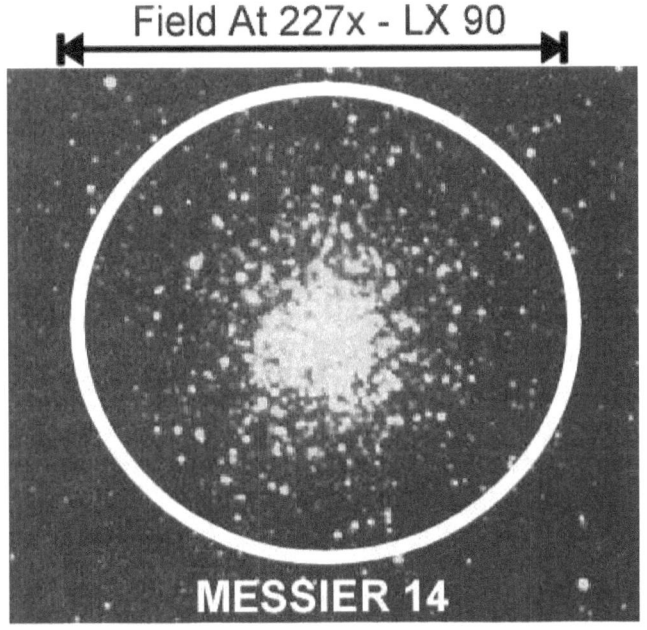

Field At 227x - LX 90

MESSIER 14

Object 9 - And....in case you haven't seen one yet.....a GLOBULAR CLUSTER! Messier 19
Although very, very small by comparison with the more spectacular Messier 5 and others, this is still a very fine globular for small and medium scopes. With individual stars of magnitudes 14th and below, it is not too difficult to begin to realize some resolution with the 6- to 8-inch scopes. However, the true beauty of observing this globular may not be the cluster itself. In all scopes with very low power and wide fields, this provides one of the most striking of all celestial sights, in a very, very rich star field of the fringes of the Milky Way. The cluster in wide field views appears suspended almost 3-D amidst many beautiful stars. As the drawing on the following page shows, the cluster is distinctly oblong, or elliptical; the drawing has been

digitally altered to match the view and orientation of a catadioptic scope and to appear as it might in the 8-inch; the 6-inch actually provides a VERY similar view! This cluster is a very few thousand (but who's counting anyway) light years further than M-10 and M-12, so it is apparently a physically smaller object overall.

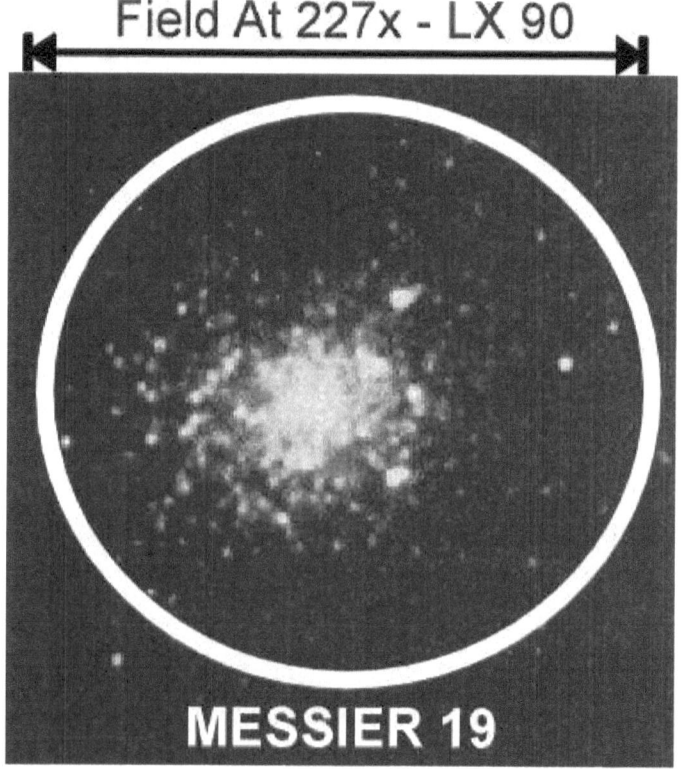

Object 10 - Messier 62 - Yep, another nice!
Right on the Ophiuchus/Scorpius border, you will find one of the most spectacular and large of all the globular clusters. You MUST check this one out, even at low power. Very bright at the center and

surrounded by a faint glow (fairly large) in a small telescope, this cluster can be resolved around its edges with a 3-inch. In the 6-inch there is an absolute blaze of faint stars seen in the very center of this fine object; the view in the 8-inch is breathtaking. After you have taken in this spectacular view at medium-high (about 30x per inch aperture) drop the power down in your telescope and just sit back to look at this fantastic sight with the widest field eyepiece you have. Messier 62 is located amidst one of the *richest star fields* of the Milky Way, and the view is stunning on a very dark night; M-62 stands out like a beacon among many thousands of stars visible in your scope in the same field. With the larger telescopes at very low magnification, it is an awe-inspiring sight, with the faint stars around the edge of M-62 glowing clearly and the bright cluster suspended in the starry field.

Object 11 - Guess what...
We have already explored this wonderful globular in the ASO Constellation Guide installment on Bootes; Messier 5 is the granddaddy of the globulars, huge, bright and easily resolvable. It is a showpiece when you can take your telescope to the darkest site available. I have included this fine object (it is very close to all the others, so a GO TO should not be so difficult) so that you can use it as a reference for all the others you have just viewed. Messier 5 is truly remarkable and serves as a guide for the others you have explored. For a full discussion, please see the Bootes Constellation Guide discussed in Volume One.
Object 12 - Messier 16 (Serpens) Nice Galactic Open Cluster with Nebula (*what? no globular?*)

The "Eagle Nebula" This is a wonderful object for a clear, very dark and moonless sky, often overlooked because of common suburban observing conditions. Messier 16 is a wonderfully rich large open cluster with some fantastic bright nebula surrounding it, but requires a very dark sky site to observe .

This is one object where the low power APO refractor will beat them all! Very low powers and wide field will reveal one of the most spectacular objects of the sky. It is a VERY rich star field, combining a vista of the Milky Way with a very nice galactic cluster and much bright nebulosity. The nebula is NOT conspicuous like the Orion nebula, but requires some real searching and very well dark-adapted eyes. Try moving the telescope slightly back and forth across the star cluster to get the full effect of the bright "glow." A good wide field and very dark skies are more important than aperture on this one, so use your lowest and most favorite wide-field eyepiece, regardless of which telescope you might have.

There are about 100 stars in M-16, all of which can be seen in a 6-inch and larger scope, about one-half of them in a 3-inch and some 17 visible in small scopes. Medium powers on the larger two telescopes MIGHT reveal many intrusions of "dark nebula" inside the lighter diffuse nebula, cone-shaped protrusions that really enhance the appearance. The most obvious and interesting of all is the "Black Pillar" nebula in the SE part of this brighter cloud. When looking at this object, try to image its location....it is in the great Sagittarius Arm of our own spiral galaxy. You are looking across the vastness of space and yet another arm inside our own toward the center of the Milky Way at a great cloud of stars some 8,000 light years distant!

Object 13 - Galactic Cluster IC 6633 (In Ophiuchus)
A nice, often overlooked, galactic cluster. It is quite large (20' arc, almost 2/3 the size of the moon's disk!) and has a total magnitude of about 5.0, so it should be somewhat naked eye on a very dark night. It is an excellent object for the small telescope and nice with the larger scopes with very wide angle eyepieces. Located in the star-rich fields of the nearby Milky Way, you can see about 65 stars in all scopes down to about magnitude 7. If this object were not so large, it undoubtedly would have made Messier's famous list of beautiful deep sky objects.

Object 14 - Another Open Galactic Cluster (in Serpens) - IC 4756
This is a must-see for small telescope users; the rest with more narrow fields of view must rely on a good pair of 10 x 50 binoculars! This is a very large open cluster containing perhaps 80 stars down to magnitude 7 and 8....but it is big: the size of more than two full moons across! So obviously, a field of view of about 1.2 degree will be required, just perfect for the small APO; try the 40mm Plossls on the other scopes or perhaps one of the 32mm Super Wide eyepieces on the 4- to 6-inch scopes; in the 8-inch, you will only be able to get the entire cluster in with a good Super Wide eyepiece....you might try the finderscope, which provides and outstanding view with all stars clearly distinguishable!
Object 15 - A Dark Nebula - "**Barnard's Nebula**" / B72

Well, Mr. Barnard had a star in this Guide and he also has a famous nebula....but not the bright kind. **Barnard's Nebula** is a very difficult (you must use averted vision where you scan around the object but never look directly at it) object, showing as a dark intrusion over a brighter star-filled field of sky. There are so many stars in this bright area of the Milky Way that it appears almost as a background "glow" in the sky on a very dark night. This dark cloud is overriding the bright stars behind it and is clearly "S" shaped, about 30' arc diameter, or about the size of the full moon's disk. Thus, you must use your low power eyepiece in the 4- to 6-inch scopes to get it all in, a 32mm or 40mm in the 8-inch and actually about a 15mm in a small APO for best views. Find this cloud ONLY about 1.5 degree north-north-east of Theta Ophiuchi, which should make it very easy to find. What you should expect to see at very low wide field power is a place "....*where stars aren't*." In other words, in this area just FILLED with tiny star images, attempt to locate this spot by a conspicuous ABSENCE of stars just NNE of Theta Oph.

WANDERING ABOUT....YOUR NEW "USER OBJECT" OPHIUCHUS

This brief GO TO Tour of Ophiuchus and Serpens has revealed a big chunk of our visible Milky Way Galaxy, many globulars and your first "**dark nebula**" and star of large proper motion. But please do not stop here. Go ahead and locate many of the hundreds of other ngc galaxies that are present in this constellation.....there are several more
interesting double and multiple stars that are within the reach of your telescope. In addition, use this

opportunity to actually TWO MORE (!!) USER OBJECTS onto your user objects in the sky program.

LET'S LOAD THE "UNCANNY" FOR USER OBJECTS THIS TIME!

We will load something we cannot see....and also a runaway celestial object.

Take your sky program or keypad and let's load up **"Barnard's Runaway"** stars and ALSO "Barnard's Dark Nebula". These are highly unusual objects, and great for "astro conversations" so you could not find any more appropriate and interesting subjects to add to your growing "User Object" library.

On AutoStar, go to: "Select/Object [enter] and scroll down to "user object" [enter]. Now enter the coordinates listed above for both the "Runway Star" and then as ANOTHER object for "Barnard's Dark Nebula"; under "description" put in a nice title for these two so that it will pique your curiosity every time you see them as you scroll through your list.. You will now have as your FOURTH AND FIFTH TOUR USER OBJECTS two of the most unusual objects that you could possibly select!

Remember to check your user guides and HELP for your particular star program for the correct steps in loading user information.

* * *

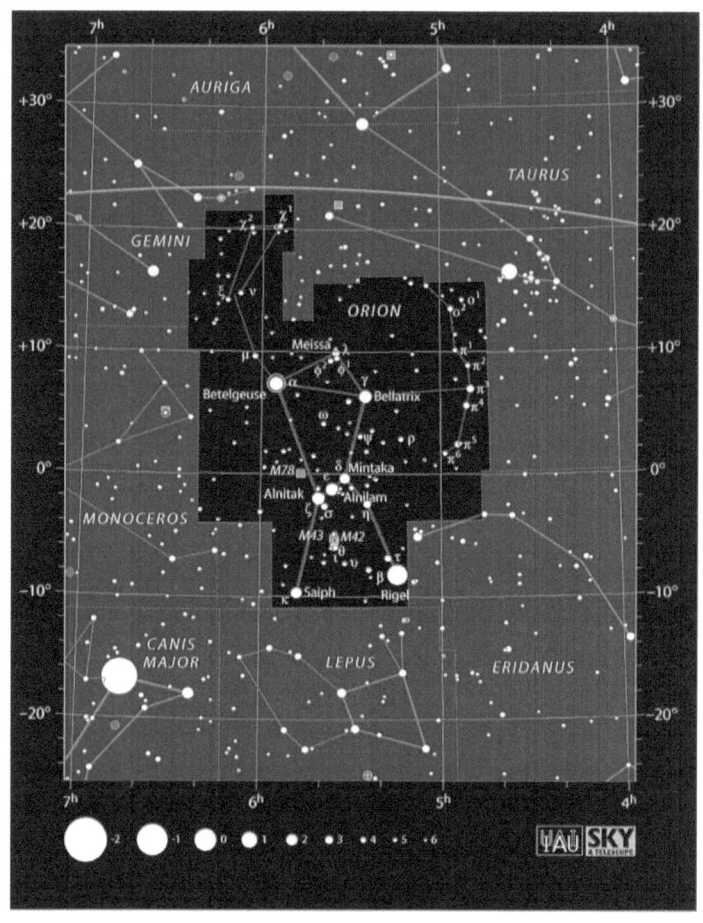

The International Astromical Union
Sky & Telescope
Constellation Chart for Orion

Chapter 23

ORION –
Part One
Making a Mark in the Sky
Legends for Mankind
..the most unforgettable constellation of the cosmos

Presented in three parts as the next three installments of *The Constellations*, the magnificent splendor of Orion justifies that it be presented as:

PART I - ORION: "Making a Mark in the Sky – Legends for Mankind"
PART II - "Observing and Understanding Messier 42 - the Orion Nebula"
PART III - "Revealing the Treasures Within"

Anyone who has ever studied this wonderful constellation realizes that there is simply too much that must be discussed for one presentation. Thus, PART I will feature an overview of the constellation, its stars, mythology, nomenclature and many, many fascinating facts that will make your time with "the Hunter" not only more enjoyable, but more meaningful and educational as well.

PART II will be comprised of a very in-depth version of our typical "GO TO" tour Constellation guides, but with added focus on the true splendor of the Orion Nebula (Messier 42).

PART III discusses the remaining (but certainly not less interesting nor exciting objects, including the **"Horsehead Nebula"** (IC 434), many other reflection nebulae, clusters and several of the hundreds of spectacular double and multiple stars held within the Hunter's haven.

So......join in with me as we explore a world of mythology, cosmology, astronomy and anthropology. Join me as we stalk the mighty Hunter!

THE STARS OF ORION –
Few people who have ever shared a part of their lives with the richness of the deep and dark skies of winter will ever forget the splendor and grace of magnificent ORION. The *"association"* of stars - stars with common distances and common proper motions and thus seemingly "associated" - is brilliant against the cold winter skies, and perhaps equal to any in our celestial sphere outline very accurately the mythological figure they were intended to represent. From its shear size, we can appreciate the strength of the great Hunter; from the brightness of its stars we can witness the energy of fortitude to stalker the celestial fauna; the dark skies behind him provide the mystery of so great a myth that only the name "Orion" evokes when mentioned.

Indeed, unknown to many skywatchers is that this magnificent constellation is actually the "brightest constellation" of the entire sky, having NINE stars of 3rd magnitude or brighter, TWENTY-FOUR nebulae of 10th magnitude or greater, over 180 double and multiple stars visible in amateur telescope, SIX wonderful galactic star clusters, and FIVE stars of the first magnitude, two of which - Rigel and Betelgeuse - are "0" magnitude and among the top bright stars. Betelgeuse (the "Alpha star" of Orion) at magnitude 0.7 is the 11th brightest and much brighter (magnitude 0.2) Rigel is the 7th brightest of the sky.

Many of the stars in Orion are very young, energetic stars of O and B spectral type, members of very close groups located in the spiral arms of our Milky Way galaxy, unlike the "galactic star clusters" which are more commonly associated with the actual central "disk" of the galaxy and the globular clusters which seem to hover over the hub or central nucleus of the Milky Way (as they do commonly with all spiral galaxy. These "association stars", the best example of which are the stars of Orion's "sword" (see figure below) are tens of parsecs apart, yet all formed commonly; over time, the very rotation of the galaxy has resulted in spatial separation of the stars....but so little that it is possible for us to realize that the stars have all moved away from one-another, measurable even within the last century.

The famous Orion nebula contains hundreds of tiny point-like stars (see Part II) that are all part of such an association, with new stars being formed from accretion of the dust and gases of the nebula itself; the brighter stars that we see presently, such as the TRAPEZIUM (theta Orionis - see the Hubble Space Telescope photo below) are the result of formation form this "stellar nursery."

These stars are so young and so common throughout the areas of Perseus and Orion, that this outward motion suggests that the expansion began ONLY 1.5 million years ago!

There are about 85 such associations within our Milky Way galaxy now cataloged with anywhere from 10 to 100 bright, blue stars contained within them. Contrast this with the galactic clusters which contain between 50 and 1,000 RED or BLUE stars in each cluster.....or the globular clusters which possess an incredible 10,000 to 1,000,000 very RED stars in each spherical cluster!

The very core of the nebulous stellar nursery at the heart of the Orion Nebula. Photo by ASO

* * *

For naked eye observers, Orion is embedded deep within the winter skies with brighter star all around, in addition to the brilliant blue-white stars (and one very red one!) that it contains within its expansive borders. To the WEST of Orion are the constellations TAURUS, the Bull (with whom Orion is in many legends engaged in a fierce battle) and ERIDANUS the Celestial River that gave rise to the magnificent constellation of Cygnus, the Swan (summer skies). The bright red star "Aldebaran" of Taurus signals west of Orion, while due EAST of the constellation we find the equally bright, but distinctly more yellow star "Procyon" in the constellation CANIS MINOR, or the Lesser Dog. Likewise, directly SOUTHEAST of our hunter is the brilliant star "Sirius" in CANIS MAJOR, while NORTHEAST of his brightest star (Betelgeuse) is the constellation of GEMINI, the Twins, with its two bright and equal stars "Castor" and "Pollux."

In this wide field photograph you can clearly see the beauty of the constellation of Orion, and the outline of "the Great Hunter" as well as much of the nebulosity associated with the very young stars of this constellation

Orion (pronounced "o-RYE-un") from mid-northern latitudes rises at 9:00 p.m. on about November 1, transits south of directly overhead about 3 a.m. the following morning and sets after daybreak. It is highest in the sky at midnight (midnight culmination) on about December 10th each year.

Within this constellation are *24 diffuse or reflection nebulae*, most of which are so large and so dim that only photographs will reveal them; on the other hand there are some that are visible to the naked eye! Messier 42, the famous Orion Nebula is one such brighter member. On the other hand, there are "finicky" nebulae which elude observation except by the keen-eyed and only under extremely good conditions. An example of such is the equally-famous *"Horsehead Nebula"*, IC 434, which can actually be glimpsed with richest field telescopes under very, very dark conditions, but is all but invisible unless your observing site is perfectly dark and deep clear.

Also in Orion are a bit more than *180 double and multiple stars* that offer fascinating viewing for a range of telescope apertures from 2" to 12" in diameter. Part III of this "GO TO" TOUR Constellation Guide for Orion provides just a few of the wonderful multiple stars that are visible here.....I urge you to consult *"BURNHAM'S CELESTIAL HANDBOOK"*, Volume II - Orion, for a complete listing of these stars, including a brief abbreviated description of magnitudes, separations and position angles, coordinates and a brief description of each of these stars. In addition, I highly recommend loading the "Orion Doubles" tour by **Jose Chinchilla** of Madrid, Spain, tour onto your GO TO scope for some splendid wintertime viewing. This added dimension of your GO TO telescope to allow you to search out the 32 most colorful and interesting doubles/multiple stars in Orion is well worth uploading into your Tour library!

Continuing on deep sky objects in the boundaries of Orion, we find that there are two planetary nebulae suitable for larger amateur telescopes (8" and larger) and at least six galactic clusters that can be logged with all telescopes from the binoculars....to the smal Questar.....to the largest telescopes. For a complete listing and free downloadable star charts for the hundreds of variable stars in the constellation (and scores of them actually inside the Orion Nebula!), consult the listing of the American Association of Variable Star Observers (**AAVSO**) at www.aavso.org which can access all charts and very informative observing information for this and all of the other 87 constellations.

My **Orion Sky Chart** following demonstrates the overall constellation and the main objects of interest for this "GO TO" TOUR; there are hundreds of other objects that simply cannot be listed. To best use this and all Arkansas Sky Observatory charts, merely photo copy the chart, laminate and carry with you and your PC or computerized telescope into the starry night.

Let's talk about some of the brighter stars of this magnificent outline before discussing their pertinence to the mythology and history of Orion the Hunter.

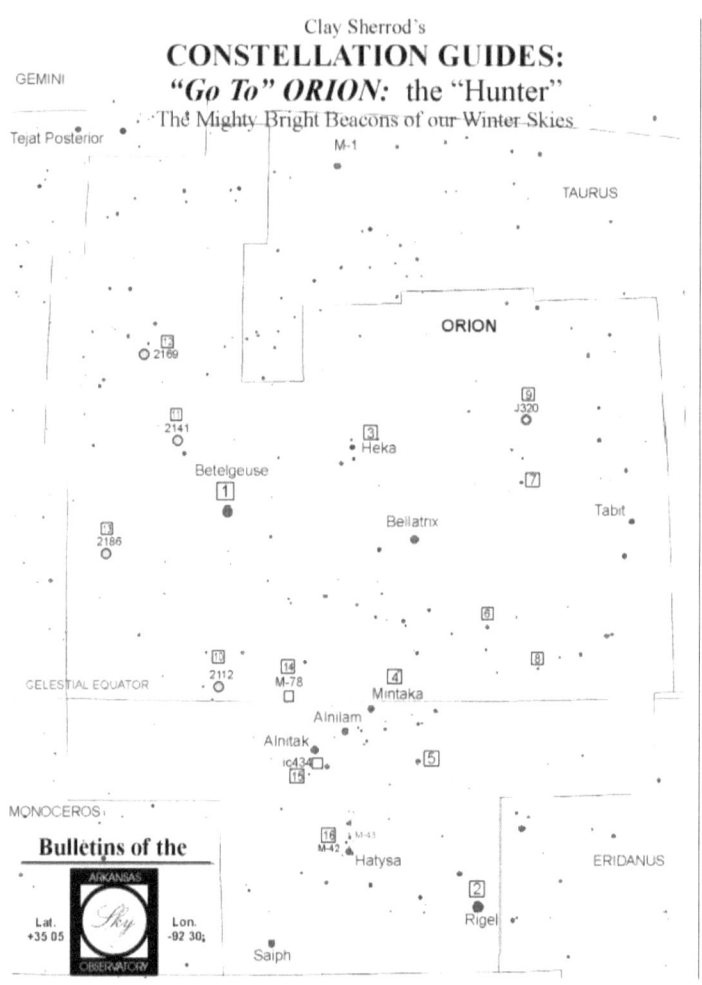

Once you rather "get the hang" of the major configurations of these stars, you begin to clearly see the outline of a great hunter, club in one hand, the hide of a lion in the other, in rapid confrontation with the charging Bull, Taurus, to his west.

First, we look at the "bright four cornerstone stars" that comprise his two feet and two shoulders. In the diagram below (in my way of seeing the constellation....you will see a historical difference to that when we discuss the mythology of the constellation!), the Hunter is chest forward and FACING WEST, eyes directed toward Taurus. The brightest and reddest star of the constellation, *BETELGEUSE* (pronounced: "BAY-tel-gertz," and NOT "beetle-juice"!) comprises Orion's RIGHT SHOULDER; his left shoulder is the bright star *BELLATRIX* (pronounced: "BALE-a-tricks", to the west of Betelgeuse. Much farther south we locate Orion's two feet....*SAIPH* (pronounced: "SAFE") is his right foot while the brilliant white star *RIGEL* (pronounced: "RYE-gel") marks his left foot.

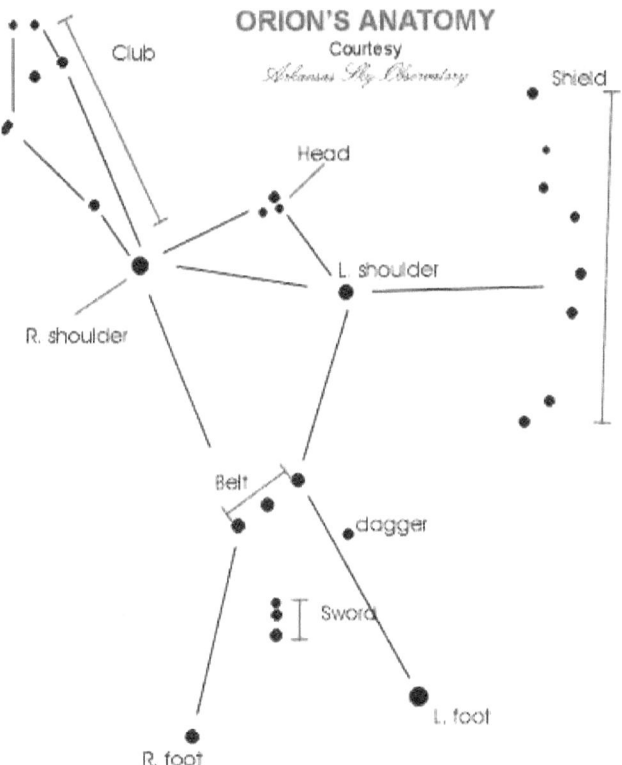

But we are not done with just such a basic outline as with many other constellations! Let us continue this wonderful "dot-to-dot" from star to star. Orion's HEAD can be marked by the triplet of stars midway and slight north of a line extended between Betelgeuse and Bellatrix, this noted by the not-so-bright star *HEKA* (pronounced: "HECK-a"). All three stars of this small triangle should be visible to the naked eye on a reasonably dark night. Now let's talk about his ARMS and his instruments of the hunt.

His LEFT ARM is outstretched to protect himself from the charging bull. There are NO distinct stars that describe this arm, but he is holding the skin of a slain Lion in his left hand - pushed forward to ward off the attack of the great horns of the bull. This shield is marked by eight (8) stars: The northern two stars are "omicron 1" and "omicron 2"; the line of stars distinctly sweeps from north to south in succession with a series of "Pi stars" - numbers Pi-1 through Pi-6 as noted in the drawing above!

Orion's arm is extended way above his head, club in hand, ready to strike the fatal blow to the bull. Rising straight UP from Betelgeuse, we can follow this club (and arm) through the stars (south to north): mu, xi, upsilon, and chi1/chi2 which make the enlarge end of the club.

Orion's waist is distinctly marked from east to west by the "belt stars of Orion," *ALNITAK* (pronounced: "ALL-knee-tak"), *ALNILAM* (pronounced: "ALL-knee-lamb") and *MINTAKA* (pronounced: "men-TA-kuh"). A deadly dagger hangs from the front (west) of his belt, marked by the bright star "eta

Ori" (which oddly was NOT named by the ancient Persian/Arabic namers of the stars).

From his waist hangs a mighty sword marked by the stars (north to south): sigma, theta-1 and theta-2 (also the Orion Nebula!) and finally *HATYSA* (pronounced: "hot-EE-suh").

I urge you to take the time to explore this wonderful field of stars and familiarize yourself with the magnificent arrangement as it applies to the mythology that we are about to explore. The darker the skies, the more incredible the journey through Orion. If you can plan some evening WITHOUT the telescope you will be equally rewarded from the views of magnificent fields of bright stars and expansive nebulae that eludes telescopic sighting but will reveal their secrets to the low power and wide fields of ordinary binoculars.

For a wonderful evening of object-by-object touring of Orion through binoculars and the naked eye, consult the guide you will find at: http://www.dibonsmith.com/ori.htm.

THE LEGENDS OF THE GREAT HUNTER –

The legends surrounding the giant Orion are many and varying....however, the one aspect of the story that time nor culture can change is that ORION has always been associated with a great "Hunter." This is one of the many intriguing aspects of the constellations, the stars and their lore. Throughout the ages, with humanity separated by great oceans, ice caps, mountains and fierce natural predators, mankind has somehow always managed to "spread the word" when it comes to associating certain star patterns in our night skies with lessons of humanity. Orion is no exception.

There can be some interesting comparisons of Orion with the other "hero" of the sky: *Hercules*. The great Hercules, given tests to prove and feats of strength and fortitude to accomplish, is a very well respected figure in both legend and sky, though not quite as prominently displayed by its starry selection. On the other hand, you will see that Orion is a rather self-serving, skirt-chasing ego-maniac (sorry Orion fans.....) whose biggest claim to fame - and his ultimate death - was ridding the world of all animals except mankind!

Truly, this star pattern as we have just seen can be unmistakably associated with hardly other mental images than that previously described. Indeed, once you have "seen Orion" in this fashion, you likely will never imagine any other form.....the hand raised with club to strike the bull....the arm outstretched in protection from the charge of the wild animal.....his mighty sword dangling from the waistband of a loincloth....

Even the great Greek historian and writer **Homer** wrote that Orion was "....*the tallest and most BEAUTIFUL of all men...*"

Although we typically rely on the beauty and ancient mythological stories of the earliest Arabian astronomers for our wonderful star names, even they have been in disagreement as to the proper name of this great human figure....but DO agree that the stars ARE representative of a large man of some kind. The most common association from antiquity is that of ancient Syria, "Ga-bara," and from the Jewish "Gibbor," both of which easily name this figure as simply *"the Giant."* This was later adopted by the Arabian skywatchers as "al-Jabbara" (also meaning "*giant man*") and even later evolving into "al-ba-adur", or *"the strong Giant."*

From ancient Babylon to more recently during the Punic Wars, the rising of the constellation of Orion signified a period of great storms, wind and widespread destruction.

The Egyptians considered the star pattern - *"Sahu"* - as perhaps the most significant of all in the sky. Sahu was the keeper of dead souls, the gatekeeper of life after death. Of course the significance of the afterlife was of great importance to those who could afford the luxury....the priests, princesses, pharaohs and queens, as attested through their nearly-obsessed compulsion at tomb building, the burial of their riches and the embalming and subsequent mummification of the high caste of society upon death. This complex world of life-after-death to the Egyptians was given the name "OSIRIS" and this moniker has become a mainstay throughout ancient Egyptian tomb hieroglyphics. Regarding this important constellation, one of the earliest recorded documents - the *"Book of the Dead"* - reveals the Egyptian belief of some 3,200 years ago that persons properly prepared and of the proper social standing will ascend into the soul and spirit of Orion.

Indeed, Orion does not stand alone to signify his stately standing of the sky: he is accompanied by his two hunting hounds Canis Major and Canis Minor (the large and small dogs, respectively), and immediately beneath his feat crouches the tiny bunny rabbit, **LEPUS** (isn't that a great name for a rabbit?!), obviously in fear of Orion's bragging of ridding the Earth of all its creatures but man.

But just WHO was this guy named "Orion" and how did HE get to be such a bigshot?

To the Celts, Orion was a "*King in Arms*," a "celestial warrior" to the earliest Greek legends, and even the "*Star King*" to the very mysterious and elusively private Saxons of old. Perhaps it is only the Babylonians who did NOT see a great figure of a robust and daring "man" in this pattern as they noted this star group as "Heaven's Light" or the "*Lighted doorway to Heaven*."

Curiously and morbidly, the legends of Orion all seem more focused on how he DIED than how he lived. Interesting, isn't it....the Egyptians considered this great starry sky figure as the very reincarnated SOUL of the god of Osiris, ruler of the afterlife, and most of the stories told of Orion have forever never regarded his expertise of the hunt or the bravery in stalking his prey, nor the magnificence of his life, as was done with his "big-man-on-campus" competitor Hercules. Orion is only remembered by the way he might have died.

Note in this curious 1687 drawing of Orion by celestial cartographer **J. Hevelius** that the figure is drawn exactly OPPOSITE that which is commonly associated with this family star pattern!

For the Hunter-legend we can go back as far as 3,000 years ago into the land and fables of the ancient Hittites. Orion to them, like it appears to be in so many cultures, was a very charming and captivating hunter, admired and cherished by all who knew him. To these Hittites, this great figure was known as "*Aqhat*," young and dashing....quite the ladies' man (another association that seemed to be fairly global). In their legends, there was a goddess who resided over "battles of men," her name being Anat.

But in addition to being infatuated with Orion himself, she was also in great envy of his mighty BOW, which she repeated begged Orion to lend her. As he continued to refuse to relinquish his weapon to her, she ultimately sent a hit man to steal it from him. But rather than retrieving the bow, the thief was forced to kill Orion and accidentally lost the bow into the waters of the sea during the process. In this tale, rather that the RIGHT arm holding a shielding hide from a lion, the Hunter is holding his mighty bow through the same stars....this, too, can be clearly imagined when looking at this constellation.

Still thought of as a "hunter" by the Greeks, the stories - like so many of them did - varied as to the mythological background of Orion....but yet all still dwelled on his DEATH rather than his LIFE. In the earliest "hunter-gatherer" times of ancient Greece perhaps, one story emerged in that Orion was considered "the mountain dweller," a fabulous (and of course "charming") hunter who was SO good at his trade that he vowed - bragged actually - that he would hunt the entire world until he had completely abolished all of the wild animals that walked upon it. This was a bad idea.

Orion was such an adorable young and handsome man that all who saw him - including the goddess who watched over the planet Earth - fell in love with him and his charm. Even though she was madly in love with the hunter, the words of extinction eventually found their way to her ears and she dispatched a great defender of the Earth - a giant SCORPION (the constellation **Scorpius**) to kill Orion with its deadly sting. Thus, for his dastardly scheme to rid the Earth of its wildlife, Orion was put to death and imprisoned in the sky, perpetually chased by the great Scorpion which is always located opposite from the hunter as the sky seemingly rotates above the Earth.....with all its animals well intact.

Then there is the story, also of a charming and "girl-magnet" great hunter, that was from a slightly different Greek influence, one more typical of mythological influences. It seems that Orion, quite the ladies' man, was attracted to none other that the god **Apollo's** sister, **Artemis**, who also was a very excellent and renowned marksman. Very protective of his sibling, Apollo requested the keeper of the Earth to dispatch the mighty Scorpion to eliminate Orion and thus his amorous ways toward Artemis.

Now, in this version and many others, it is not really clear if the Scorpion actually kills Orion, or just engages him in a mighty battle. A possible outcome that is similarly recorded is that Orion did, indeed, do battle with the huge Scorpion, but realized too quickly that this was one "wild animal" that he could not injure nor defeat because of its hard-jointed armor. As the Scorpion increasingly managed to get the upper hand in the duel, Orion was forced to jump into the sea and swim far, far away from the dangerous Scorpion. But Apollo was not done with him....

.....it seems that **Artemis** - hearing the commotion from the struggle - came to her brother's side to see what was happening and was told that there was a horrible enemy swimming far into the ocean....a "black dot" that only she could with her marksmanship could hit and kill so far away. Apollo essentially challenged his sister to demonstrate that she was as good a shot as the world thought....and she struck Orion in the head with her very first attempt, killing him dead. Finished. Vamoose. Kaput... goner. To prove she had hit her mark, Artemis swam far into the ocean to retrieve her kill and realized that she had killed the man who loved her.

Her cries to **Zeus** and the other gods watching over the mortal Earth fell on deaf ears from the pleadings of Apollo and they refused to restore Orion's life. With what little "god-like" power that Artemis possessed, she settled to ultimately raise the spirit of Orion into the firmament and there it circles until today, still constantly pursued by the Scorpion.

Thus, the mighty Hunter was not killed by the Scorpion that was sent to do the job....he was killed by the woman he loved....running away from the Scorpion.

Orion seen in a more humorous light
From a 1460 European Woodcut

Just for the fun of it.....now that we have seen the many faces of Orion.....I have included the actual artwork from a 1460 medieval European woodcut that was intended to be a serious representation of this "heroic hunter and lover." Perhaps after reading the brief mythological references to our great hunter this drawing will add a little levity to an otherwise serious plight of our beloved hunter. In spite of his untimely and very mis-directed death and suspected philandering among the beautiful princesses of the gods, nothing can take away from his splendor in the sky. Personally, I will always remember and think of Orion as a magnificent "larger-than-life" heroic and brave hunter who perpetually battles the great bull with his eye of fire.

NEXT.....PART II: *"Observing and Understanding Messier 42 - The Orion Nebula"*

We will peer deep into the constellation and study perhaps the most famous of all deep sky objects and certainly one of the most observed: *The Great Orion Nebula*. This wonderful stellar nursery, where stars are "born" and pass through their infancy is a veritable education in the early lives of young stars. No matter if you observe this fantastic object with the naked eye, binoculars or with a great telescope, the result is always the same: beauty and amazement at a celestial gift that draws us into the sky like no other.

.....until then....Good Observing and may the stars serve as your sentries as you explore the frontiers of space!

Chapter 24

ORION NEBULA

PART II

Observing and Understanding the Incredible Messier 42 - the Orion Nebula / The Belt and Sword Stars of Orion

Anyone who has ever studied this wonderful constellation realizes that there is simply too much that must be discussed for one presentation. Thus, PART I featured an overview of the constellation, its stars, mythology, nomenclature and many, many fascinating facts that will make your time with "the Hunter" not only more enjoyable, but more meaningful and educational as well. If that overview did not get you excited enough to pursue further the wonderful world of the Hunter, then your DC power source needs a fresh set of batteries. Perhaps this current installment will charge you up:

This Guide - PART II - is comprised of a very in-depth version of our typical "GO TO" tour *Constellations* guides, but with added focus on the true splendor of the **Orion Nebula** (Messier 42), the immediate surrounding area and the fascinating variety of nebulae and variable stars through the "sword" of Orion. There is not a more eye-pleasing, research-generating, and thought-provoking object anywhere in the sky that can take the place of this magnificent emission and reflection nebula, the birth place of hundreds of stars and the bearer of yet stars to emerge into the beautiful heavens.

The remaining installment, PART III, will be our

typical concise GO TO tour and discuss the remaining (but certainly not less interesting nor exciting) objects, including the *"Horsehead Nebula"* (IC 434), many other reflection nebulae, clusters and several of the hundreds of spectacular double and multiple stars held within the Hunter's haven.

So......join in with me as we continue to explore this constellation of mythology, cosmology, astronomy and anthropology. Here we go deep into the stellar nursery of the famous and beautiful ORION NEBULA, an object that continues to amaze even seasoned observers in telescopes, binoculars and to the naked eye!

TOOLS OF HIS TRADE....THE "BELT" AND "SWORD" OF ORION (with some wanderings into "Dark Adapted Eyes")

The closeup star chart featured below shows details within the remarkably symmetrical star field known as Orion's "belt" and his "sword." This "asterism," or group of familiar stars, is often mistaken by novice sky watchers as "the big dipper" because....well, because that's what it LOOKS like! It is a striking view with the naked eye, and even more dramatic in good binoculars on a very dark moonless night when very elusive nebulae such as the "Horsehead" (IC 434 and discussed in detail in Part III) can be sometimes detected by the keen eyed with eyes "dark adapted" over a long period of time.

<u>A Brief Word Concerning "Dark Adapted" Eyes</u> -

"*Dark adapted*?" Perhaps more in Orion's borders than anywhere else because of the wealth of very extended and faint gaseous nebulae, but important for ALL deep sky observing, is the aspect of allowing your eyes to become dark adapted. All this means is allowing your eyes an opportunity to adjust themselves as the very slow-to-respond low light receptors become increasingly active under dark conditions. We are all familiar with two types of "night blindness." This can happen: 1) when going from a bright room or moving away from a bright flashlight into an inky black environment...for many minutes the eyes cannot see the subtle detail that is actually faintly illuminated; and, 2) going from a very dark condition into an extremely bright one, such as driving down a dark highway and suddenly encountering bright headlights that "temporarily blind" the eyes that gaze directly at it.

In astronomical observing, it is very important - particularly for the deep sky observer - to allow much time after coming from a lighted room for the eyes to become dark adapted. The minimum time is 15 minutes with NO white light....optimum time before the very faintest objects and stars can be seen is a full 45 minutes. It is, indeed, a long time to wait....but the results in what your eyes can see by comparison is phenomenal. Full dark adaptation can result in TWO FULL MAGNITUDES gain on your limiting magnitude, both naked eye and telescopic for stars, and up to THREE magnitudes on very faint and small nebulae, galaxies, comets and other subtle deep sky objects.

Once your eyes are fully capable of seeing the

faintest objects possible it is important to keep them that way. Many of us are forced to routinely observe with interruptions, particularly when observing at home. If you are forced to go the lighted indoors for a phone call or "family crisis" and attempt to return to observing once done, your eyes are no longer dark adapted and the entire process (for the entire length of time) must start anew.

Also, the use of a white-light flashlight - no matter how large or small - will destroy the night vision that prolonged darkness has provided. I highly recommend NOT buying an expensive "astronomer's flashlight", but take one you already own and remove the front lens; at your local camera store you can by or special order WRATTEN 25 or 25-A red mylar-like filters in sheet form. One sheet is enough to do about 4-6 common flashlights by cutting out a slightly oversized circle to insert under the glass flashlight lens. This is the same filtration that is used in darkroom lighting and provides a "color" of light that your eyes are not sensitive to, thereby preserving night vision.

A complete discussion of "The Eye in Observational Astronomy" is found in my posting on the Arkansas Sky Observatories' website www.arksky.org .

Orion's Belt Stars

Nearly everyone who has looked more than casually at the winter night skies has stopped momentarily and marveled at the perfect little "diagonal line" that is outlined by the three nearly-equally-bright stars that comprise Orion's Belt (see description in

Part I). A complete description of Orion's "anatomy" as outlined by the brilliant stars of the constellation is give in Part I. This area truly invites exploration with a good pair of binoculars, and this discussion of the Orion Nebula would not be complete without exploring the area around it in brief. A complete look at many of the objects and double/multiple stars within the "belt" and "sword" of Orion is found in Part III of this three-part series.

The belt of Orion is one of the most identifiable and most unforgettable of all asterisms of the night sky. The belt is marked by the bright stars (in order from East to West):

- *ALNITAK* (zeta Orion, pronounced "ALL-knee-tack") - RA 05 41 / DEC -01 57 - from the Arabic "al Nitak" or the "girdle star of the Giant," this is a 1.8 magnitude very "young O-type" white star. It is a triple star, with the closest two being 1.9 and 5.5 magnitude, with the brighter star being very white or yellow-white and the fainter component clearly a lavender or green-blue star. Although separated by 2.6", this is not a particularly easy double star because of the brilliant glare of the primary star. For the 4" and larger telescopes, there is a 10th magnitude third star, although not actually linked gravitationally, nearly due north of the close pair. Like all three of the belt stars, Alnitak is about 1500 light years distant, the three of them actually comprising a VERY loose triple star of sorts! It is nearly due south of Alnitak that we find the famous "Horsehead Nebula," so photogenic, but nearly impossible to glimpse visually.

- *ALNILAM* (epsilon Orionis, pronounced "ALL-knee-lamb") - RA 05 36 / DEC -01 12 - from the Arabic "al Ni-lam", or "belt of pearls" which is a moniker that is thought to have represented originally the entire span of all three stars. This is the central of the three belt stars and the brightest, magnitude 1.7

- *MINTAKA* (delta Orionis, pronounced "men-TAH-kuh") - RA 05 32 / DEC -00 18 - from the Arabic "al Mintakah" or "the Giant's belt." This is the westernmost and the faintest of the three belt stars, yet varies almost 0.2 magnitude in the course of nearly six days. Look north of the bright star for a very lavender-colored 6th magnitude companion to Mintaka. An interesting aspect of this companion is that it is spaced almost exactly a half light year from the primary star, about the same distance that the famous "Oort Comet Cloud" is located from our own sun.

Orion's Sword Stars

Referring to the chart above, the "sword of Orion" is not only an interesting naked eye and binocular object, but it is also the home of perhaps the most famous of all deep sky objects, the Great Nebula in Orion, Messier 42 and much additional diffuse gases that actually glow for us to see them! As can be seen in the chart, the "sword" dangles just as it should from Orion's "belt" stars. There are three brighter stars that comprise the sword and are described following, from north to south: (try hitting these targets using your R.A. and DEC "entered coordinates" by pressing the "mode" key and holding for about 2-3 seconds and entering the correct coordinates given below via the keypad or in the "Coordinates" tab of your sky program).

- 42 Orionis - RA 05 33 / DEC -04 50 - A fairly faint naked eye star of only magnitude 4.5, but a fine double star in 4" and larger telescopes, although the 3" scope may show both components. Look for the secondary star, about 8th magnitude nearly due south and a bit west of the brighter star. This star clearly is distinguishable as a member of the "sword trio" in binoculars, and represents the "top" or northernmost extent of Orion's sword. As with EVERY star discussed here as members of either the "belt" or "sword", 42 Orionis is surround by nebulosity - NGC 1977, a very faint (actually magnitude 4.7, but spread over a 40' x 25' area!)

moon-sized cloud centered on the 4th magnitude star. ALSO....look immediately NORTH of 42 Orionis for the very loose star cluster NGC 1981 which contains 10 stars magnitude 8 to 10. This cluster is clearly visible in nearly all telescopes, but very low power wide field instruments provide an incredibly striking view and actually are MORE likely to reveal NGC 1977 than "higher power" instruments!

- Theta Orionis - RA 05 35 / DEC - 05 23 - The famous "Trapezium," or the heart of the fabulous Orion Nebula (NGC 1976) which will be discussed in much length following. It is located about ½ degree, or about one lunar diameter, SOUTH of 42 Orionis. Theta - which curiously has remained unnamed throughout antiquity - is a fine multiple star, with four brighter components (about magnitude 6.5 each) easily seen in a small telescope at moderately high magnification. However, the 8" telescope will reveal a total of SIX (6) stars in this group, and a 16" scope can show EIGHT on a very dark and steady night. The positions of these incredibly challenging stars relative to the four brightest members of the Theta multiple group are clearly shown a chart I have prepared in the following detailed description. Indeed, modern estimates suggest that the eight stars are merely the brightest members of newly-forming galactic star cluster! Most of the 300-plus stars that have been identified within only 5 arc minutes of Theta and thought to be a part of this most unusual cluster are well beyond amateur observation. Thought to be ONLY 300,000 years old, this could likely be the youngest of all known star clusters in the Milky Way galaxy! The image of the hundreds of stars centered on the Trapezium shown below was drawn by the author using astrometric grids on the 24" f/4 reflector some three decades ago.

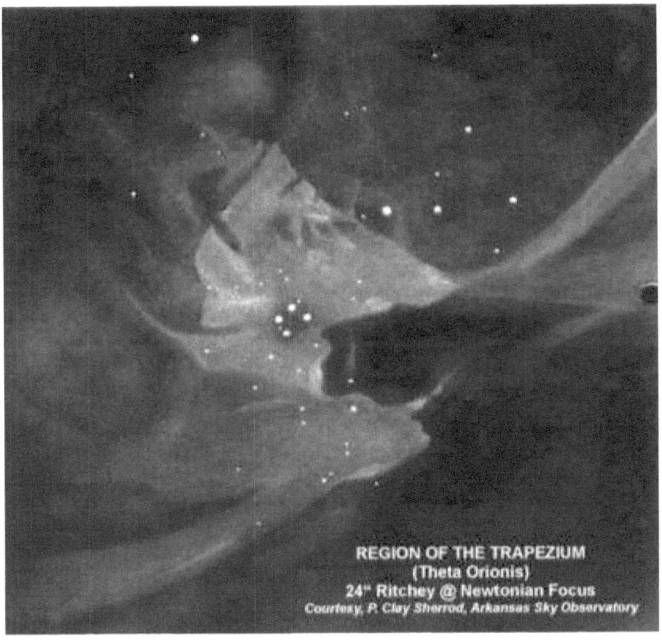

REGION OF THE TRAPEZIUM
(Theta Orionis)
24" Ritchey @ Newtonian Focus
Courtesy, P. Clay Sherrod, Arkansas Sky Observatory

- *HATYSA* - (iota Orionis, pronounced "hot-E-suh") - RA 05 35 / DEC -05 55 - This southernmost star of Orion's sword is also the brightest at magnitude 2.8. It is also the only named star of the grouping. This star also carries another Arabic name, "Nair al-Saiph," or the "jewel of the sword." Hatysa is located about ½ degree south of Theta and the Great Orion Nebula. This is a great triple star for most telescopes. The primary, 2nd magnitude, star has a secondary star about 10 arc seconds southeast; it is an easy object at 6th magnitude and distinctly blue in color. Nearly due east of the bright Hatysa is a faint 11th magnitude star some four times more distant in your telescope than the 6th magnitude companion. Another double very nice "pure white" double star - Struve 747 - is found only 8' to the southwest of Iota, and with is a very easy double for all telescopes with magnitudes of 5.4 for the primary and 6.5 for the secondary star about one "Jupiter diameter" southwest. The 14' round and very faint nebula NGC 1980 surrounds Hatysa, but is an extremely difficult object visually.

As mentioned, every star of the BELT and every star of the SWORD is surrounded by glowing diffuse nebulosity! Indeed, it is likely that every one of these stars - all nearly at the same distance, from 1,400 to 2,000 light years from us - and many, many other very faint ones that are not so obvious, were actually formed out of the primordial gas cloud that is now nearly all "accreted" or gravitationally compacted into masses that we call "stars." The great Orion Nebula is one such example of a very rich cloud of this gas that is still forming "protostars" today; even the smallest telescope can probe deep into the Orion Nebula and see very faint stars which appear embedded in the nebulosity.....stars which were effectively "born" from the very gas cloud through which you view.

THE GREAT ORION NEBULA – PHYSICAL ASPECTS

Before discussing the thrill of *"what you can see"* within and around the Great Nebula in Orion, let us explore some of the facts, details and speculations about this magnificent deep sky object. With the incredible view of the Orion Nebula that is immediately overwhelming when seen in even a pair of common binoculars, it is sometimes easy to overlook the physical attributes that this "classroom of stellar life" affords us.

Distance and Dimensions

For the record books.....the Orion Nebula was the FIRST deep sky object photographed, this in a 50-minute exposure by pioneer astrophotographer **Henry Draper** in 1880. Years before, **William Bond** at Harvard College Observatory had recorded the first photograph of ANY celestial object - the bright star *Sirius* - utilizing the great 15-inch refractor in Massachusetts, the largest of its kind in the world at that time.

The nebula and most of its associated stars which have formed from it (see "variable star" discussion below), is located some 1,300 light years distant which places it somewhat closer than the "belt stars" (1,500 to 2,000 light years). The star association and gas cloud is HUGE....the overall width being some 25 THOUSAND TIMES larger than the entire span of our own solar system, or a total of 32 light years. The very dense central portion (which the majority of what is seen visually in amateur instruments) comprises a portion of space the equivalent of the distance of nearly 5 light years: the space between our sun and its closest neighbor....Alpha Centauri!

The Nature of the Nebula

In celestial terms, this nebula is one of the most recent creation in all of the heavens. Indeed, the nebula and its associated stars, are thought to have begun its formation only 23,000 years ago.....in the recent midst of the Great Ice Age here on Earth. It is in that immediate past that the Trapezium stars actually began to glow while early hunter-gatherers on our planet searched frozen plains for traces of the life-sustaining Mastodon....perhaps there were some inquisitive people among them who glanced skyward and wondered whether that tiny string of star-like objects we now know as the "sword" was indeed becoming brighter night to night in the frozen winter sky.

As discussed in our *AURIGA* Constellation Guide (see the Constellations, Volume One), the expansion of the nebula within this 23,000 year period has resulted in the "expulsion" of three brighter stars (the "runaway stars"): AE Auriga, 54

Arietis and Mu Columbae which for whatever reason were ejected at high speeds (the central part of the Orion Nebula is expanding from its center at an incredible rate of 6 miles per second!) at very near the very earliest years of this remarkable stellar nursery.

Unlike the Merope Nebula and other "reflection" nebula associate with the nearby *Pleiades* star cluster (Messier 45), we see the Orion Nebula and much of the other nebula filling this constellation because it ACTUALLY GLOWS through fluorescence, much as you see the desk in your office from the overhead lighting provided. The wide field photograph on the preceding page from the **ASO Wide Field 130mm astrograph** reveals the tremendous extension of this energetic gases of the nebula. The energy of the hot and energetic young stars (particularly the Theta Orionis group) produces strong ultraviolet radiation, just like electrical current is producing energy within the glowing gases of your fluorescent light bulb. Although - like so much of our universe - the nebula is nearly completely comprise of hydrogen and helium, its "greenish color" that we see visually in telescopes is due to the much smaller percentage of "energized oxygen" in the nebula. On the other hand, nearly all familiar photographs of Messier 42, including spectacular amateur photographs rivaling those of the professional astrophotographer, display the distinctive color of the nebula - DEEP RED - the color associated with the more common and film-sensitive Hydrogen atoms so dominant in the Orion Nebula.

The Variable Stars Within –

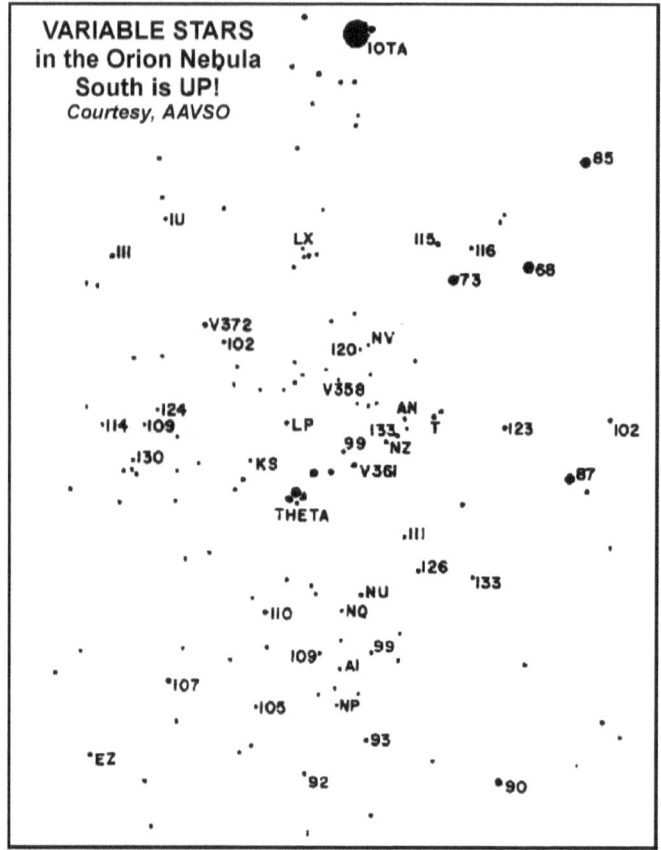

The chart above from the American Association of Variable Star Observers (AAVSO at www.aavso.org) provides the locations of the dozens of variable stars that are within "walking distance" of Theta Orionis and all embedded in the clouds of the Orion Nebula. Without going into considerable detail about these "nebular variables", it will serve to say that each of these stars, like the "runaway stars" and the bright Theta Orionis grouping, was "born" from the nebula. For a review on methods and techniques of observation of such stars, and a discussion on nebular variables, see my in-depth guide at ASO GUIDES/*Variable Star Observing*. The brightest of these stars - V372 - varies from magnitudes 7.4 to 8.6 while the faintest shown here - AI Ori - starts at a very faint maximum of 12.1 and dims to 14.2 with no regular period known. All of the "Theta Orionis Variables" are irregular "T-Tauri" type variables with no distinct period nor clearly defined range of maximum and minimum light. For charts of specific stars, use the AAVSO chart creator at:

https://www.aavso.org/apps/vsp/ . Note for these charts, simply type in the NAME of the variable at top to generate your choice of chart.

THE GREAT ORION NEBULA – VISUAL ASPECTS –
The Thrill of the Hunt.....the Thrill of "The Hunter"

There is truly "something for everybody" in this wonderful winter object. The naked eye can clearly detect at first glance that this is NOT a stellar object....it is diffuse and the eye struggles as if to focus it into a point of light like all other bright stars surrounding it. One look with binoculars and you clearly detect why: in addition to the fabulous cloud of gas, the field of view in a wide field glass is remarkable, with stars filling the field of view like sugar spilled on black satin. On a dark night, the gaseous clouds can be traced in all directions away from the central portion of this nebula, and the nearby fainter one, Messier 43. Averted vision (not looking or staring directly at an object but attempting to see fleeting glimpses of subtle detail as the eye moves and "sees" the object with the eye's perimeter) will reveal countless streaks and globs of nebulae not first seen when merely looking casually at this area of sky.

Perhaps what surprises me so much about this remarkable object is that Charles Messier - from whom the famous "Catalog" originated with no particular order season-to-season - decided to observe and record 41 objects BEFORE including the Orion Nebula!

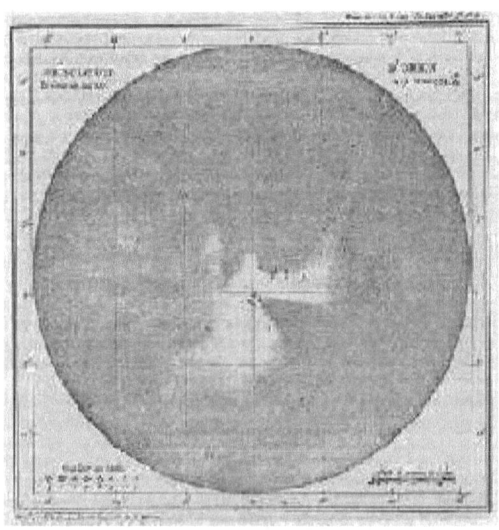

MESSIER'S ORION NEBULA
The Drawing made for his 42nd Entry

The Many Stars of the Orion Nebula

We will start our "GO TO" tour of the Great Orion Nebula at its very heart.....the multiple star known as Theta Orionis, or *"the Trapezium,"* and subsequently the many hundreds of stars that are known members of the "family" propagated by the Orion Nebula.

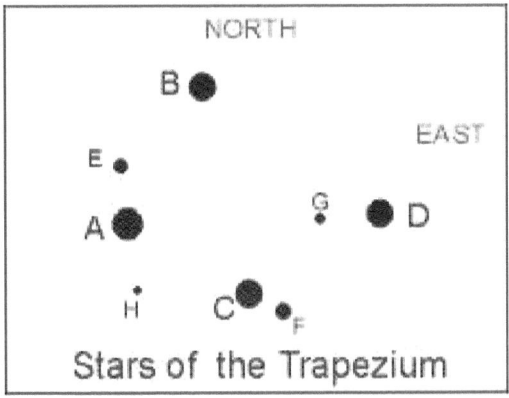

Stars of the Trapezium

My diagram above shows a "reversed" image of the Trapezium stars as might be seen at very high magnification with a large Maksutov or Schmidt-Cassegrain telescope with the diagonal fitted (north at top and east at right). The stars "A", "B", "C" and "D" are of course the brighter members and form the "trapezoid" shape from whence the name originates; all four of these stars are visible in small telescopes using fairly high magnifications of about 40 to 50x per inch aperture. Note that star "C" at magnitude 5.4 is the brightest of the quadrangle and is actually visible to the naked eye, as is the combined brightness of ALL four stars. Magnitudes of the remaining stars are: "A" = 6.8; "B" = (variable, 8 to 8.7); and "D" = 6.3. On a very dark and steady night with Orion very high in the sky, a good 4" scope at about 200x might reveal the 11.3 magnitude star "E", immediately southeast of bright "C". Another faint target star for a 5" and larger scope is "F" at magnitude 11.6 an fainter than the

"E". These fainter members are NOT as easy as one might expect....they appear fainter than the magnitudes provided, but have been confirmed at that brightness. The stars "G" and "H" are visible in a 16" telescope under incredibly steady conditions and can be held steady in a 24" telescope. None but the primary four stars can be photographed by amateur means since the bright clouds of nebulosity accumulate their light on film or CCD, thus masking the fainter stars in exposures long enough to reach a limiting magnitude of less than 8.0.

But these eight stars of Theta Orionis - as spectacular a multiple star as they might be - are NOT the only stars of note within the cloud we know as the Orion Nebula. From 1976 through 1982 I developed and maintained an extensive study which both mapped and monitored (visually and photoelectrically) the seemingly countless stars associated with the Orion Nebula.

An interesting aspect of observing this nebula is "tunnel vision" for nearly all observers, myself included. Even seasoned observers are so mesmerized by the beauty of the gas cloud and the striking contrast of the Trapezium, that little attention is addressed toward the hundreds of very tiny stars that are visible in virtually every telescope....the larger your scope, the more stars that are revealed, as shown in the detail drawing made from astrometric measurements of these Orion Nebula stars as seen under dark mountaintop skies with the 24" telescope operating at about 170x. Nearly ALL of these stars, using standard good quality eyepieces can be seen in a quality 8" or 10" telescope and more than half in telescopes of 4", 5" and 6" diameter!

STARS ASSOCIATED
with the
ORION NEBULA
Arkansas Sky Observatory 1982

If one zooms in even closer to the Trapezium area, even fainter stars might be seen as seen in the next star field positional chart from 1982 using the 24" at about 400x. This is one of the interesting aspect of the Orion Nebula AND its field stars: both taken higher magnifications very well. The nebula, as discussed following, delivers considerable fine "mottling", like "lambs' wool" clouds of a March evening with higher magnifications in very dark skies in any telescope; higher magnification reveals additional stars as well, since the added contrast will allow the very faint points of light to be more easily differentiated from the very bright surrounding nebula.

I encourage every observer to take the time to explore the wonderfully rich star field associated with the Orion Nebula. Start, using the two charts above, in the "center," at Theta Orionis and use the charts as a guide to determine how many of these stars you might visually detect....I can guarantee you that, on a very dark night, you will see the majority of them in most good telescopes of any size! One note, however, is that the drawings made from my 1982 measurements are "astronomically correct"; Maksutov and Schmidt-Cassegrain modern telescopes using the diagonal prism have NORTH at top and EAST to the right, due to a "mirror-imaging" by the prism or mirror diagonal.

Exploring Messier 42 - Touring the Orion Nebula

As mentioned, the French comet hunter and "faint-fuzzy-cataloger" **Charles Messier** took his time getting around to including the Great Orion Nebula as his 42nd object. It certainly was known to him far earlier, and is a conspicuously none-stellar object even to the naked eye.....no doubt this object has been looked upon with wonder by all humankind during the last 23,000 years. Curiously, however, we have mentioned that this region of Theta Orionis was never named by the Arabian stargazers who named everything around it. Surely they would have noted the unusual naked eye appearance of this area.....a "star", but not quite. **Galileo** likewise missed the Orion Nebula in his first telescope around 1610, not surprising as the semi-permanent stars of the sky apparently were of little interest to his earliest explorations. It was not until the following year that a fellow Italian observer "discovered" the unusual nature of Theta Orionis. The first detailed record was by mathematician/ optician/astronomer **Christian Huygens** of Saturn notoriety when he actually sketched the stars of Theta while describing them immersed in the nebulous cloud.

Today - even in ordinary binoculars - the Orion Nebula is simply something that cannot be overlooked so easily. Indeed, I have always encouraged each and every astronomy/science teacher who is attempting to encourage an interest in astronomy to beginners, merely show the student the Orion Nebula on a dark and moonless night. That will do the trick, for Messier 42 is an overwhelming, almost hypnotizing, sight even to the uninspired and the uniformed.

The famed double star observer at Yerkes Observatory, **E.E. Barnard**, noted that the nebula always appear to him as a "ghostly bat" in the field of view of the remarkable 40" Clark refractor. Indeed, this shape - likened more in modern times to friendlier creatures - reminds me of a sea gull gliding through the sky, wings outstretched and arched gracefully from one edge of a wide angle eyepiece to the other.

Every telescope of every size will show remarkable detail of the Orion Nebula. As mentioned, the view in each telescope will likewise change when inserting medium and high magnification eyepieces.

Thus, one rather different approach of this "GO TO" TOUR guide to the Orion Nebula will be a discussion of relative images based on MAGNIFICATION more so than telescope aperture. Where appropriate I will, of course, interject limitations and expectations of various telescope apertures.

Messier 42 - Wide Angle and Very Low Magnification (telescopes and binoculars) - Stretching the Limits:

In binocular and very wide angle richest field telescopic views, the Orion Nebula is spectacular when viewed on the proper night, including:

1) no moonlight to interfere;

2) very clear and crisp night;

3) Orion on the celestial meridian, as high in the sky as it will get from your observing location;

4) eyes totally dark adapter with no extraneous light interference.

Perhaps the most awe-inspiring view of this region is with richest field instruments with about a three degree or more field of view; for with that area covered, all three of the "sword stars" can be included in one very dramatic field of view. In addition the combination of good optics and wide field will ALSO allow the observer the begin to detect the very faint nebulosity that surrounds nearby Iota and 42 Orionis in addition the grandeur of M-42 and M-43 to its immediate north.

In wide field low power instruments, look for the "gull wings" that stretch far outward from the central region of the nebula. These are very conspicuous filaments of gases that stream outward from the dark "gap" in the central dense region of the nebula known as "the fish's mouth" (see my drawings following....the fish's mouth is the dark gap that appears to intrude into the brightest area of the core of M-42). One such filament streaks to the west from the fish's mouth for nearly 2 degrees. Larger telescopes will reveal further extensions of this streamer, but with the sacrifice of field of view. A second and even more interesting filament that contains much lumps and dense clouds and streaking nebulosity arcs gracefully in an opposite direction, southeastward toward Iota Orionis, also extending well over one degree.

In low power instruments, the bright central region near Theta Orionis will be clearly seen in stark contrast to the rest of the faint nebula; also, increasing power just a bit will reveal the fish's mouth in much better detail. The brighter stars of the Orion Nebula (refer to the chart above) will be seen sparkling within the gas cloud, with occasional hints of very faint stars that are just at the limit of sight at this very low magnification.

<u>Messier 42 - Medium Magnification</u> (telescopic only views) - Unlocking the Fine Details:

The drawing immediately above was made by the author about 40 years ago with a 6" *Unitron* refractor at f/16.

THE ORION NEBULA
6" Unitron Refractor @ 38x -Arkansas Sky Observatory 1982
Drawing by P. Clay Sherrod

Note the intense center of nebulosity and the gull wings that are present all the way to the edge of this field of view.

In medium magnifications (about 15x per inch aperture) the observer will begin to see more and more faint stars; within the nebula, expect this limiting magnitude for stars:

3"	-	10.2
4"	-	10.9
5"	-	11.4
8"	-	12.5
10"	-	13.4
12"	-	14.1
16"	-	14.9

Note that these limiting magnitudes are NOT the faintest star your telescope is capable of revealing under ideal circumstances; but observing faint stars that are embedded in bright nebulosity will not allow you to reach that faint limit, and thereby these limitations are empirical limits based on actual observing sessions within the Orion Nebula.

At this medium magnification, examine the full extent....a bit at a time.....of each of the "gull wings" for some very fine filamentary detail. You will be absolutely amazed at the wealth of fine detail once your eye becomes fully dark adapted. Of course, larger telescopes will reveal more difficult detail and fainter extended filaments, but there is more to see in all telescopes by increasing magnification.

Also, center the "fish's mouth" and examine the remarkable mottled detail that becomes clearly visible within the bright central area that is just south of the dark feature. You will note in 5" and larger instruments that this bright concentration of nebulosity describes a bit of a "box" shape with considerable texture within it. Once again, let you eyes wander throughout this area in search of the VERY faint stars that suddenly pop into view with this increased magnification. You will find that if you scan, the peripheral view of your eyes - "averted vision" - will reveal far more faint stars than if you peered directly at the area in which they are located. At this magnification, you might begin to locate some of the very few RED stars that are found in the nebula...search and discover!

Messier 42 - Higher Magnifications - Exploring the Heart of the Great Nebula:At magnifications of about 20x to 25x per inch, you will begin to reveal stars near the limit of light grasp of your telescope...thus the faintest stars will likely be seen at this power. I suggest that you will see TWICE the number of stars at this magnification as with medium magnification. Below is the latest drawing made through a 12" Meade f/10 with the diagonal mirror removed to match orientation with that done with the 6" refractor; you will note that the very fine and delicate detail suddenly becomes richer, when at first glance there might appear to be more detail in the smaller aperture instrument.

THE ORION NEBULA
12" Meade Schmidt @ 86x - Arkansas Sky Observatory 2001
Drawing by P. Clay Sherrod

Also holding up very well under high powers is the central region of the nebula; there will be countless knots and tiny filaments and dark globules that suddenly become visible. At higher magnifications it might require some time for your eye to adapt to the lowered contrast between dark sky and faint nebula, but this wait is well worth it. Also showing up remarkably well under very good conditions are the places "where stuff ain't" as I like to say. These are areas that are totally absent of nebulosity, scattered stars or other objects of note....these are places that stand out in inky stark contrast to nearby areas illuminated by the fascinating nebula. These dark areas are fascinating and add yet another dimension to the incredible and seemingly never-ending discoveries that you can make with the Great Orion Nebula.

And a Word About Messier 43

Messier 43 (ngc1982 at coordinates [2000 epoch] R.A. 05h 36m / DEC -05d 16m) is immediately north of the Orion Nebula and actually appears much as an "appendage" of the great nebula....but it is not. Messier 43 is also an emission nebula with many embedded stars that have formed from the dust and gas of this cloud, just as has happened in the Orion Nebula. It surrounds an 8th magnitude star and is a vague but clearly-definable cloud, misty-appearing due north of M-42 and within the same low power field of view as the "fish's mouth" (above). A beautiful image of this nebula (north up) is seen below as captured in the wide field camera of the Anglo-Australian Observatory.

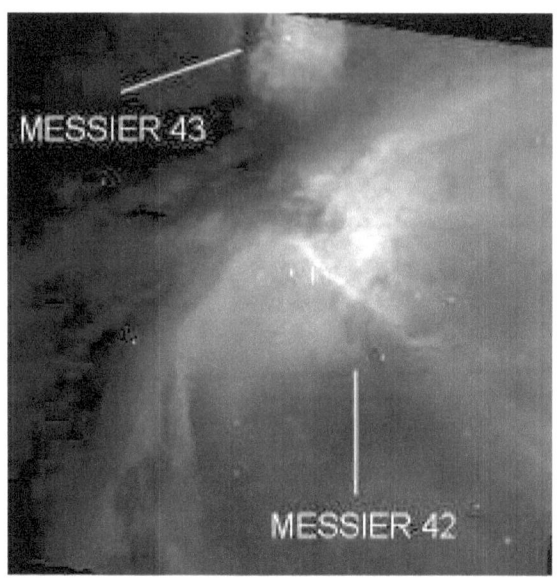

NEXT INSTALLMENT OF THE ORION TOUR
PART III - "Revealing the Treasures Within"

With our next "GO TO" TOUR, part three of this three part series, we resume the "normal" Concise Directory of interesting and unusual deep sky objects to locate in your computerized telescope. Not ignoring Messier 42, but many observers overlook countless of other fantastically rich and exciting objects in Orion....from planetary nebulae and wonderful galactic star clusters....to beautifully colored and challenging double and multiple stars.

Like the Great Galaxy in Andromeda, it is easy for one magnificent object to overshadow and shortchange other objects that are not only worthy of observation, but further away from such showcase deep sky splendors they would be showcase objects in their own rights. So we will explore such wonderful objects in the grand constellation of Orion in Part III.

Yes, there is life in Orion after the Great Nebula....

"He lay on his back in his blankets and looked our where the quartermoon lay cocked over the heel of the mountains. In the false blue dawn the Pleiades seemed to be rising up into the darkness above the world and dragging all the stars away, the great diamond of Orion and Cepella and the signature of Cassiopeia all rising up through the phosphorous dark like a sea-net. He lay a long time listening to the others breathing in their sleep while he contemplated the wildness about him, the wildness within........"
Cormac McCarthy, All the Pretty Horses

The Orion Nebula
As photographed through the 130mm APO
astrograph
At Arkansas Sky Observatories.
From the full color original by the author

Chapter 25

ORION
PART III - YOUR "GO TO" TOUR for Orion
Revealing the Treasures Within.....

In this third part of our three-part, *"GO TO ORION" - Revealing the Treasures Within*, we conclude the Constellation Guide for this magnificent winter grouping. We began with a complete historic and general-interest overview of Orion in Part I followed by a detailed look at the environs in and around Messier 42, the Great Orion Nebula. Here we once again enter the ranks of our "GO TO" concise guides of deep sky objects and interesting double/multiple stars for computerized telescope.

This three part series has been a wonderful test of my concentration, as Orion presents so much fascinating information, objects, history and incredible sights, that focusing in on pertinent objects and information has been quite difficult. There is so much within Orion's borders that unfortunately for space restrictions had to be left out. However, I certainly urge all observers to obtain a very good handbook and star atlas (or computer Sky program) and study the wealth of fine objects found here. As for printed handbooks, no one should be without the tremendous reference books *"Burnham's Celestial Handbook,"* a 3-volume library that is packed with much information of all 88 constellations and the objects contained within them. Look for Orion as the final "chapter" in Book Two of Burnham's.

This guide - PART III - will comprise the in-depth version of our typical "GO TO" TOUR Constellation Guide's Concise Directory of objects for your computerized telescope and discusses the remaining (but certainly not less interesting nor exciting) objects within Orion, including the "Horsehead Nebula" (IC 434), many other reflection nebulae, a great selection of galactic clusters visible in telescopes of all sizes and several of the hundreds of spectacular double and multiple stars held within the Hunter's haven.

For the full listing of all brighter stars (to naked eye limit) which includes: 1) Bayer; 2) Flamsteed; 3) SAO #; 4) R.A. & DEC; 5) brightness; 6) most notable double and multiple stars; 7) variable star designations; I encourage you to visit the wonderful Constellation reference site at http://www.deepskywatch.com/deepsky-guide.html as well as all other constellation Star Tables that he has provided through this wonderful web site. This is a great reference to print and put into a binder for cross referencing star designations, locating double and variable stars, and current coordinates in the sky.

As with every "GO TO" guide, each GO TO object in Orion is discussed for your telescope regarding the type of conditions necessary for you to view it optimally for discern the very faintest details.........magnifications and aperture necessary for most objects, and much, much more. This is YOUR complete guide to get you on your way to exploring the best (and few!) objects in these two constellations. The following listing of "BEST" objects contains the finest or most interesting from my own observing experience and preference.

Use the attached star chart and the following Guide as an excellent reference for your next star party itinerary, or a beginning for further study into the thousands of objects visible in this part of the sky. Truly these extensive *Constellation Study Guides* will most definitely put your sky program to work for you in the most efficient and enjoyable way possible! As a matter of fact, MANY telescope users are now programming their own "Tours" based on these guides, using each constellation as a separate GO TO Tour for the sky library that can be added in or deleted through the main edit screen on your PC or MAC computer.

For a VERY complete listing of NGC objects that are found within Orion, you may access the website link noted above which will provide a complete listing of all suitable NGC objects, their coordinates, magnitudes and various reference and cross-reference material for each object.

We hope you enjoy these comprehensive GUIDES to touring the constellations via your AutoStar and its computer-driven telescope. Each new installment is complete with diagrams, charts and illustrations that you will find nowhere else. Please let us hear YOUR feedback and your observations of each and every constellation after YOU have toured its vast reaches of our skies!

YOUR ORION CONCISE DIRECTORY OF INTERESTING OBJECTS –
In addition to our regular listing of a few selected objects, I have included the complete abstract listing

for the ten (10) nice galactic star clusters that are viewable in this constellation via modest instruments; that listing will be found at the end of this concise description of our TOUR.

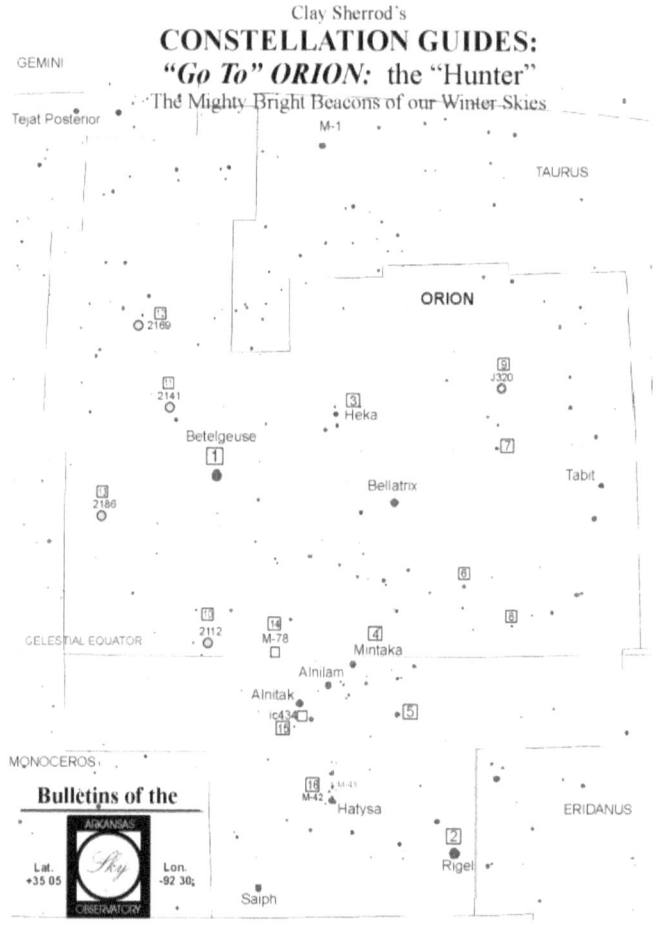

Clay Sherrod's
CONSTELLATION GUIDES:
"Go To" ORION: the "Hunter"
The Mighty Bright Beacons of our Winter Skies

For our visit into lair of the great Hunter, I have chosen the finest (or most interesting) 11 objects in this ORION "GO TO" tour (as with all guides), all objects listed will be visible in most telescopes (some naked eye) from the small Questar through the common 8-inch telescope; of course larger apertures may "show" an object a bit closer and "better," but frequently a wide field and low power view is more desirable than aperture for FINDING the objects initially. Indeed, I strongly encourage you first FIND the target object, or its approximate location through your GO TO function with your lowest power and then - once IDENTIFIED positively - move up slowly in steps with magnification if necessary. Remember, not all objects "like" magnification. Sometimes better "field of view" (such as the wonderful wide fields provided by small telescopes) is desired over light gathering (like the larger 8") and magnification.

The rule for determining "optimum magnification" is that: 1) too low power results in sky background glow detracting or diminishing the contrast against the deep sky object; 2) too high magnification darkens BOTH the sky background AND the object; 3) medium magnification can be achieved at which you have MAXIMUM contrast between the object and its darkened background sky. I have found through three decades of direct observing that about 15x per inch aperture (36x for the 60mm; 55x for the 90mm; 75x for a 150mm; and, 125x for the 200mm).for deep sky observing is PERFECT for most objects. That being said, always remember that DOUBLE or multiple stars require whatever power you can crank out....the seeing conditions are the limiting factor here.

For my complete and comprehensive discussion regarding magnification, seeing conditions and sky transparency, see the ASO GUIDES at the Arkansas Sky Observatories website www.arksky.org .

With all deep sky objects, avoid attempting to observe when the moon is in the sky, even a very thin crescent, as its brightness in the sky will overshadow the very dim contrast afforded by even the brightest deep sky object; if you see the object at all against moonlight, you will NOT see the subtle outlying areas or the full detail of what is presented. Be sure to refer back to *Part II - "Orion Nebula"* of this three-part GO TO series regarding the dark adapted eye and its importance when viewing the great expanse of diffuse emission nebulae within this constellation.

Also, as I always suggest with all of the "GO TO" TOUR constellation lists, a good star atlas and/or chart which will list all the finest objects, constellation-by-constellation. One very handy reference guide is the PETERSON FIELD GUIDE TO THE STARS AND PLANETS, which features complete lists with declinations, right ascensions, magnitudes, and all pertinent information for you to expand your observing horizons beyond this brief GUIDE. For multiple stars and many listings of the finest deep sky objects, the classic work, *Burnham's Celestial Handbook*, Vol. Two is highly recommended.

Note that your computer program or GO TO telescope will NOT have every object listed on every constellation GO TO tour....this is intentional. You can access some of the most interesting objects of the sky directly from their coordinates. It is quite simple for the Meade Autostar as you merely enter these coordinates as follows in the 10-step process (other PC programs are similar):

NOTE: The procedure is specified here because nearly all control software for all computerized telescopes and sky program operate off of basic primary and sub-menus. Although the Autostar key strokes are unique for that device, the process logic is the same and all software operates similarly.

1) Press the "MODE" key and hold down for 3 seconds and release;
2) Displayed will be the current Right Ascension and Declination of the center of field of view of where your telescope is presently pointed (assuming that you have properly aligned from "home position");
3) [NOTE: if you have the Meade electric focuser attached to any of the ETX or LX telescopes, holding down the "MODE" key will bring up the "Focus" command first....merely scroll (lower right scroll key) down one step to access the RA and DEC to enter your desired coordinates]
4) Press the "GO TO" button on AutoStar;
5) This will change the display and you will note the cursor blinking over the first digit of RIGHT ASCENSION (R.A.); merely use the number keys and dial in the R.A. of the object you are searching for;
6) When done, press "Enter;"
7) This moves the blinking cursor over the "DEC" coordinates;
8) [NOTE: the declination, unlike R.A., can be either positive or negative and you will see the "+" or "-" sign displayed depending on where your telescope is aimed at that time; if it is NOT the desired setting (plus or minus), merely use your arrow key to move the blinking cursor OVER the

"+" or "-" sign and change by using either of your lower corner SCROLL KEYS;
9) Proceed to enter the DEC using number keys;
10) Press either "Enter" or "Go To" when finished and the telescope begins slewing to your desired object!!

The constellation tour Star Chart provided on a previous page will get you started on your journey for this constellation.

Following is the concise object list for your "GO TO" tour of ORION; you may wish to find the majority of the objects from the Sky Library (for example, you can easily go to the beautiful ORION NEBULA (Messier 42) if you pull up "Object/Deep Sky/Messier/..then type in '42'...." and then press "Enter", followed by "GO TO" to access this very delicately star-laced gem. On the other hand, if you want to experiment and become a "better computer user" try entering the exact R.A. and DEC coordinates (given in EPOCH 2000 coordinates in the listing below) of that same object as described by holding down the MODE key. You will find the accuracy of entered GO TO's to be somewhat less than those stored in the sky library, but the capability of acquiring unlisted objects is fantastic!

Of course, another method of Autostar acquisition of the Great Nebula is to go to the deep sky named objects by: SETUP / OBJECT / DEEP SKY / NAMED...and scroll until you see "Orion Nebula." At that point press Enter which will bring up much data and information about this glorious object if you keep scrolling....however to access the object via automatic GO TO, simply press "GO TO" on the key pad at which time the telescope will take off for Messier 42.

Yet again, you can use the handy "NGC OBJECT" component of your computer or handbox library. In addition to the designation "M42", this object has the label NGC 1976. On Autostar, key in SELECT / OBJECT / DEEP SKY / [scroll to: "NGC" and press "enter"]; now merely key in the number "1976" and press enter to activate the Autostar on this object...continuing to press "Enter" will allow handy information about this object to be displayed on your keypad....merely press "GO TO" to move the telescope to the Orion Nebula.

You will access your FIRST GO TO target - (usually the alpha star in each constellation) - via the command "SETUP / OBJECT / STAR / NAMED....and scroll to "Betelgeuse" then press "Enter" and subsequently "GO TO" to move your this bright red supergiant star. Remember also that many distinctive objects are sometimes listed among the "named" objects. So, likewise for that object you might merely go to SETUP/OBJECT/DEEP SKY/NAMED....and then scroll alphabetically to the "common" name of the object if you are not already there; press "enter" and then GO TO and your scope is off and running!

You may also access the constellation by: SETUP/OBJECT/CONSTELLATION/"Orion".....Enter....GO TO, which will take you close to the central position of the constellation's boundaries OR to the brightest star of that constellation, depending on program..

The Concise List

(featuring Epoch 2000 coordinates)

OBJECT 1: bright star - *BETELGEUSE* (alpha Orionis) - R.A. 05h 55' / DEC +07 24 - Mag. 0.7 - Brilliant red supergiant

OBJECT 2: bright star/double - RIGEL (beta Orionis) - R.A. 05h 15' / DEC -08 12 - Mags: 0.3 & 6.7, wonderful double star!

OBJECT 3: quad star - HEKA (lambda Ori) - R.A. 05h 35' / DEC +09 56 - Mags: 3.4, 5.6, 11 & 11.5 - Can you see all four??

OBJECT 4: a double/triple? - MINTAKA (delta Ori) - R.A. 05h 32' / DEC -00 18 - Mags: 2, 6.5 & 14 - #1 and 2 easy...#3 tough!

OBJECT 5: test double - eta Orionis - R.A. 05h 25' / DEC -02 23 - Mags: 4 & 5 - test for 3", only 1.5" space - nice star!

OBJECT 6: nice easy double - rho Orionis - R.A. 05h 20' / DEC + 02 51 - Mags: 4.5 & 8.5 - good double for 3" and up!

OBJECT 7: test double - 14 Ori - R.A. 05h 08' / DEC + 08 30 - very tough for 5"; not easy in 8" scope!

OBJECT 8: variable star - W Ori - R.A. 05h 05' / DEC +01 11 - easy semi-regular, ranges from 6.5 to >10 in about 210 days

OBJECT 9: planetary nebula - ngc2002 - R.A. 05h 39' / DEC + 09 04 - Mag: 12.0 w/ 14th mag. star - very tough, 8" or above

OBJECT 10: galactic cluster - ngc2112 - R.A. 05h 54' / DEC + 00 24 - Mag: 8.6, about 90 faint stars - fairly large

OBJECT 11: galactic cluster - ngc2141 - R.A. 06h 03' / DEC + 10 26 - Mag. 10.8, 100 stars (fine in 5" and up!)

OBJECT 12: galactic clusters - ngc2169 & ngc2194 - R.A. 06h 09' / DEC + 13 57 - Mag. 6.4 / 9.2 - tiny clusters with about 18 / 100 stars...cool!

OBJECT 13: galactic cluster - ngc2186 - R.A. 06h 12' / DEC + 05 26 - Mag. 9.3 - very small w/30 stars. Good in 5" +

OBJECT 14: diffuse nebula - Messier 78 (ngc2068) - R.A. 05h 47' / DEC + 00 03 - Mag. 10.3 - looks like small comet!

OBJECT 15: *ORION NEBULA* - Messier 42 & 43 (ngc1976 & 1982) - R.A. 05h 35' / DEC -05 23 - details *in Part II*

OBJECT 16: "*horsehead nebula*" - IC434 - R.A. 05h 41' / DEC -02 25 - very faint, seen only in wide field optics and dark skies

A VISUAL GUIDE TO OUR DEEP SKY OBJECTS IN ORION

Object 1 - Our "Starting" Bright Star –
"*BETELGEUSE*" (alpha Orionis - see Part 1 for pronunciation!)

Throughout the "GO TO" tour *Constellation* guides we have discussed the beauty and many times symbolic meanings of the Arabic names which most of our bright stars still retain today. We have "Aldebaran" as the "eye of the bull," "Deneb" signifying the "hen's tail," and now we have the difficult-to-pronounce BETELGEUSE....the "armpit of the giant," from the romantic Arabic "Beit al-geuze." Not the most alluring nor creative name, but nonetheless straight and to the point. Betelgeuse is NOT always the brightest star in Orion, that distinction belonging to bright white Rigel to its southwest. As a red supergiant however, the star is quite variable and is the brightest of all known variable stars. Thus, many times Betelgeuse can reach "0" magnitude, or slightly brighter than Rigel....other times it fades to depths fainter than magnitude 1.5. This incredible star - located only 540 light years distant has a volume exceeding - are you ready for this? - 160 MILLION TIMES our own sun. Because of its tremendous size and proximity, Betelgeuse can actually show in the world's largest telescopes, a "disk", rather than a point of light for its surface rather than a point of light like all other stars. This is an incredibly interesting star for further reading and I strongly encourage those not familiar with the evolution of stars to take the time to learn more. Recommended reading is the classic work, "Red Giants and White Dwarfs" by Robert Jastrow.

Object 2 - Another "Brightest Star" - the beautiful double star RIGEL (beta Orionis)
Because of Betelgeuse's variability, I suppose we would slight the equally-bright RIGEL, the beta star in Orion, if we did not give it top billing along with the giant's armpit star. Not much more creative, the name is also from the Arabic "Rijl Jauzah al Usra" with only the "Rijl" part remaining in recent times. Nonetheless the long form of the name tells us that this star is "the left leg of the giant man." Rigel is nearly twice as far as Betelgeuse, and like it red counterpart, is also a supergiant star....only a brilliant white one, with an incredible luminosity 57,000 times the output of the sun. During your next winter or early spring star party, let your visitors thrill at just how bright that actually is: if we could move Rigel to the distance of many of our closest stars - like our brightest star in the sky, Sirius - it would be only slightly dimmer than the

full moon! Other than shining like a sparkling diamond in very low power views near the winter Milky Way, Rigel is best known as one of the prettiest double stars for amateur telescopes. Almost due south of the brilliant magnitude 0.3 star is a much fainter 6.7 magnitude star that "should" be easily seen in the smallest of telescopes...IF the brightness and distance of 9" arc were the only factors. However, I have found the companion VERY difficult in a good 4" refracting telescope because of the glare of Rigel itself. The fainter star contrasts wonderfully in larger telescopes. In an 8" or 10" glass the brilliant white coloration of Rigel makes an outstanding comparison for the true blue color if its companion star. In Part I the Orion association of stars was discussed, noting that many of the brightest stars of the constellation were actually a part of a very large "clustered grouping"; Rigel is certainly though to be one of these many gravitationally-bound stars, although it appears to be slightly closer to us than the 1,200 to 1,500 light years for the remaining Orion association stars.

Object 3 - Beautiful Double Star - HEKA (lambda Orionis) - that you can turn into a Quadruple star! Pronounced "hee-KAH", this is a well known double star and beautiful double star, the primary being magnitude 4.2 and the secondary, exactly northeast from the brighter on (Position Angle 44 degrees), is magnitude 6.1. Separated by 4.5" arc, telescopes. There should be no mistaking this star when you " GO TO" it....it is the brightest of three stars that comprise the "head of the giant", the other two being phi-1 and phi-2 both of which are about 4th magnitude.

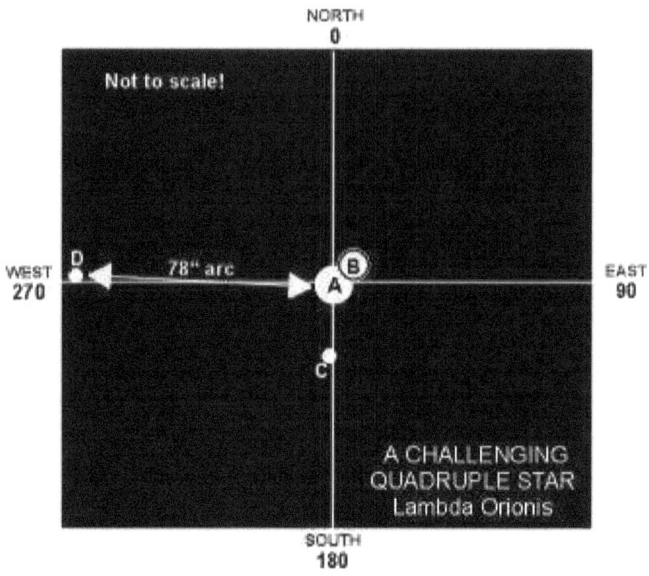

However, observers with a 4" telescope and more ideally and 8" and larger will realize double the fun on this star as there are also two ADDITIONAL stars with respect to Heka, both faint in the magnitude-11 range. The "C" star (see diagram above) is almost exactly due south of Heka by nearly 30" arc and is slightly brighter than its other twin, magnitude 11.5 and located slightly more than TWICE that distance from Heka and almost exactly due WEST (Position Angle 271 degrees). Use medium-high magnification (about 20x per inch aperture) or more to reveal these fainter stars. The 4" and 5" telescopes will be pushing the limit to glimpse these very faint objects, but under good dark and steady skies they should be "reachable." This quad makes a very pretty sight in an 8" scope at about 160x.

Object 4 - Another Good Double - *MINTAKA* (delta

Orionis) - Here is one for everyone!....easy to find and just as easy to see!

Here is one that does not require a lot of grunting and groaning to locate. Delta Orionis, "Mintaka" at magnitude 2.2 is the fourth brightest star in Orion and forms the westernmost "belt star" (See Part II) of the Giant's garments. In addition to being an easy double star target for almost every telescope, this is also a variable star for the keen eyed observer who wants to watch it brighten and dim every 5.7435 days. Its magnitude variation is about 0.2 from maximum to minimum so it is just detectable to the human eye without error as to when it is brightest and dimmest....you can use the other nearby stars (see the list in Part II) as a judge of its brightness night-to-night.

However, *Mintaka* is best known as a beautiful and easy double star. Being part of the "Orion Association" (Part I), Delta Orionis is - like all the others - part of a group of very young energetic white stars, some 1,500 light years distant. Its companion is also a very white star so do not expect any color contrast in this pair....the brightness contrast will be enough to get you excited. Look for Mintaka's fainter 6.7 magnitude companion almost due north of the bright star, almost 1 minute arc, or slightly greater than the planet Jupiter would extend in your eyepiece. There will be no mistaking the companion as it will be the brightest of the many stars that you are likely to see in this beautiful low power field of view. Use about 10x per inch maximum to appreciate the star-rich field.

Object 5 - Another double star - A good test star for the 3" and 4" scopes! Eta Orionis

With so many other bright stars nearby it is a wonder that Eta Orionis, magnitude 3.3 did not get named. For it is the one star that actually "boxes in" the belt stars at a right angle giving the grouping a confusing "dipper shape." Nonetheless, this is the fairly bright whitish star that is directly south of *Mintaka* (see above) and west of the "sword grouping" which includes the Orion Nebula. This star is really not as easy as its 1.4" separation indicates. The following listing:

2.5"	inches	=	1.81"	arc
3.0"		=		1.51
3.5"		=		1.30
4.0"		=		1.13

Shows what separations should be attained in good telescopes of smaller aperture. Thus, it appears that a 3-inch should resolve it (barely) and it should be a cake walk with a five inch; however, this is not the case; on a very unsteady night recently I noted that - in an 8" Schmidt-Cassegrain - it could easily have been overlooked as a double from the blurring of the two Airy disks from bad seeing. Use pretty high magnification (about 20x per inch minimum) to get the most out of your telescope on this star. On a very steady night I encourage you to increase the power to about 35x or 40x and note the distinct "purplish" color of the fainter 5.2 magnitude star which is now nearly due north of the Eta Orionis.

Object 6 - Now....here is one for the 3" scopes!
Double Star Burnham 1053
Once again, you will access this star either by using the R.A. and DEC manually entered via the hand controller, or by the "old fashioned" setting circles

on your scope (they actually work very well....and it's fun!). Burnham 1053 is a very close star for the 3" to 5" size range, separated by 1.4" arc, right at the Dawe's limit for the 3" scope. The primary "A" star is magnitude 7.5, so it is not the brightest of our Auriga objects by a long shot. Center that yellowish star and increase the magnification to about 200x and look for a faint 9.6 magnitude star near DUE NORTH (a bit to the west) and very close. You will require a very steady night and very high magnification (take it on up if the air will hold it!) in the smaller of the two scopes.

Object 7 - Another Good Double for Smaller Telescopes - Rho Orionis
This is yet another double star that will be a great target for smaller telescopes, but is an excellent double for all. Rho Orionis, magnitude 4.6 is imagined to be part of the left arm of Orion supporting the shield of animal skin or bow, whichever you chose to see as the Giant's weapon of choice. It is easy to locate in the telescope and once found, look for its companion to be just north of due east in the medium power field of view; I recommend about 15x per inch to adequately show the magnitude 8.6 secondary star....be sure and use a bit more magnification once found and look for the distinct blood-red colors of these two stars! With a separation of 7" arc, this should be an easy target with a 3" and a beautiful sight for its color in 5" and larger telescopes.

Object 8 - The Semi-regular Variable Star "W ORIONIS" - Excellent for all telescopes!
This is an EXCELLENT variable star for all telescopes, but exceptionally so for smaller

instruments, since rarely are there "semi-regular variables" with a magnitude range through their entire maximum-to-minimum cycles that remain bright enough for continuous observations with smaller instruments. For a discussion on variables, and the importance of regular monitoring of unpredictable ones, see my complete discussion of variables and observing techniques at the website of the American Association of Variable Star Observers, AAVSO.

W Orionis varies somewhat periodically at a rate of "maybe" 210 days; but the cycle is marked and interrupted occasionally by small "jitters", jumps and declines in magnitude and periods when no change takes place at all. Semi-regular stars, although thought to essentially be pulsating Mira-type stars, are not completely understood as to why their variations are not entirely predictable. Thus, YOUR observations of them are vitally needed over a long period of time. Indeed, with this star ranging in magnitude from a bright 6.4 to fainter than 10.5, you can compile your own light curve after only less than a year, giving details and a visual history of the erratic variations - as well as its regular cycle.

As with all variable star observing, the ideal source for information, comparison star charts, and to report YOUR valuable observations (badly needed on these irregular stars, by the way!), is the American Association of Variable Star Observers (AAVSO - www.aavso.org) in Cambridge, Mass. The charts for nearly every variable star - as well as "new stars" (such as novae, and stars not previously known to be variable) that appear from time to time - can be downloaded off the internet into a file in your computer; save the file and bring it up.....the image will be huge. Resize this image to fit your page, resave, and then print for a good chart to use at the telescope!

To make estimates on this semi-regular variable, use the AAVSO "a/G" chartS found at:
https://www.aavso.org/apps/vsp/ . Note for these charts, simply type in the NAME of the variable at top to generate your choice of chart.

This is the wide field finder-type chart which features brighter stars over a larger area of sky from which to pick given comparison stars. Also, note that this is the AAVSO "standard" chart and not the "reversed chart"; the reversed editions of the charts are newer and are matched to the reversed field of view of the Maksutov and Schmidt Cassegrain telescopes, which result from the diagonal prism or mirror.

Object 9 - A Faint Planetary Nebula - NGC 2022
Here is an object that is definitely for "the faint at heart." Located just east of the "head of Orion" star HEKA, this rather large (28" arc...about the diameter of Saturn's globe in your eyepiece!) gaseous remains of a stellar explosion is ONLY magnitude 12.4....remember that is not "stellar" magnitude but rather the magnitude of such a star if spread out over this rather large 28" arc. Thus: very faint indeed. However, based on my actual observations of this object, it IS visible in the 8" aperture telescope and becoming increasingly more interesting in larger instruments. On a very dark night a really well-collimated 8" or 10" telescope might reveal the faint 14.6 magnitude central star to this very elusive planetary nebula. NGC 2022 is one

of only two - and is the brightest of - planetary nebulae in the entire constellation of Orion...a constellation virtually FILLED with other nebulae: reflection and emission, but only two faint planetary nebulae!

Object 10 - A Nice, Often Overlooked Galactic Star Cluster - NGC 2112
NGC 2112, at magnitude 8.6 and containing about 90 to 100 very faint (10th to 12th magnitude) stars, is an often ignored object in Orion in lieu of the many fascinating nebulae that pack this constellation. This galactic star cluster, spread over a relatively small 12' arc circular area, is located almost due east of *Mintaka*, and makes a fairly even triangle with that star in the western end of Orion's "belt" and *Alnitak*, the easternmost belt star which is southwest of NGC 2112. Very low power instruments may be able to actually get Messier 78 (discussed below) and this cluster in the same field of view, since NGC 2112 is only a few degrees due east of M-78. This cluster is clearly visible as a faint circular patch of light in smaller instruments at medium magnifications (about 15x to 20x per inch aperture). However, with the 4" and larger telescopes more and more stars are visible; look for a glowing circle with about 25 or 30 stars in a good 4" scope on a dark night, while the 8" should reveal some 80 of the brightest stars.

Object 11 - Another Nice Galactic Cluster, NGC 2141
Although fainter (magnitude 10.6) than ngc2112 discussed above, this is a cluster that should be visible in most telescopes. In smaller instruments this object appears similar to NGC 2112, with no

particular detail other than a faint smudge of light, though dimmer, of about similar size to that of ngc2112. The 4" telescope will likely not be able to see any individual stars in this cluster as the brightest are about magnitude 11.8 at best. However a 5" or 6" telescope will begin to reveal some "glittering" among the nebulous appearing background, while the 8" should reveal perhaps half of the nearly 100 faint stars packed into this cluster.

OBJECT 12 - A Brighter (albeit smaller and less impressive!) Galactic Cluster - NGC 2169 (and NGC 2194)

This tiny star cluster (only 5' arc!) is located in the same wide field view as the brighter star "xi Orionis" which marks the base star of Orion's "club". Since this location is amidst the wonderful winter Milky Way, rich with faint stars throughout nearly every telescopic field of view, this is a place to let your telescope - and your imagination - wander through the star-drenched blackness of winter's night. NGC 2169 contains only 18 stars, but all are bright enough to discern in the 3" scope, and this smaller aperture is perhaps providing the best view of this spectacular area....not so much because of this fairly sparse cluster...but more so because of the proximity to xi (magnitude 4.48) and the many stars that are visible in all apertures throughout this region. It perhaps more than anything else will serve as a benchmark for your further random explorations throughout this very beautiful area of sky, sweeping southeastward into the rich Milky Way and nebulae of Monoceros. In larger telescopes (5" and up), observers should sweep from ngc2169 only about 2 degrees southeast to find another nice cluster, and one more rewarding to larger apertures: NGC 2194. This is a very star-rich cluster of magnitude 9.4, containing at least 100 stars packed into a VERY small circular area only 8' arc across! At first this resembles a very sparse globular cluster, but the 8" and larger telescopes will begin to show the even scattering of this galactic star cluster.

OBJECT 13 - Yet Another Galactic Cluster – NGC 2186
Located smack-dab in the middle of Orion, very close to its eastern border with Monoceros is an isolated galactic star cluster, ngc2182. It is about 5 degrees east and a bit to the south of bright red Betelgeuse. This cluster is right on the western edge of the winter Milky Way, and hence provides some spectacular viewing of very faint stars limited in the field only by the size of the telescope you are using. NGC 2186 is tiny (5' arc) and has only 30 stars, magnitude 10-11, but is a brighter 9.3 magnitude overall and clearly distinguishable in the 3" and 4" telescopes, appearing like a very small unresolved faint planetary nebula or globular. The 5" and 8" scopes will begin to reveal many of the stars, and the larger the telescope, the better the view. With all telescopes use medium-high (20x per inch aperture) to view this faint and small object.

OBJECT 14 - A Nice Diffuse Nebula - Messier 78 - (this is the kind of object that kept confusing Messier as "comets!")
Messier 78 is an interesting object in the 4" telescope and up, and certainly can be viewed with a 3" but with lesser detail. Its overall brightness is very faint (10.3) when you consider that it is spread out over an extended 8' x 6' field of sky. This is a

very difficult object visually, but does have its merits.

This object has been more confused as a "comet with a double nucleus" than any other object in the sky! Like the Great Orion Nebula, M-78 has gases that are visible to us because they actually glow....not reflect. There are two very active white stars embedded within this diffuse nebula, both about magnitude 10.2 and thus better seen in telescopes 4" and up. These two "equal" stars are arranged in a nearly north-south orientation some 53" arc apart. They become quite clear in the 5" scope and a bit of nebular detail becomes visible as you move up in aperture to the 8" and larger telescopes. With larger telescopes, look for Messier 78 to be somewhat "comet-shaped," in that one end of it (the northwest end) is very blunt like a small comet's head, and the other (southeast) end appearing very diffuse and simply "disappearing into darkness" much like an outstretched comet's tail.

This is a very interesting object to study under medium (about 20x per inch aperture) magnifications in telescopes of any size; the longer you look at it under very dark skies, the more you begin to actually see of it! Like most of the stars and most of the nebulae in Orion, all of this is somewhat interconnected at a distance of some 1,500 light years, give or take a few parsecs. The two embedded stars and the spectacular nature of this "little brother" to the Orion Nebula is well captured in the following photograph as captured in the 2MASS sky project cameras.

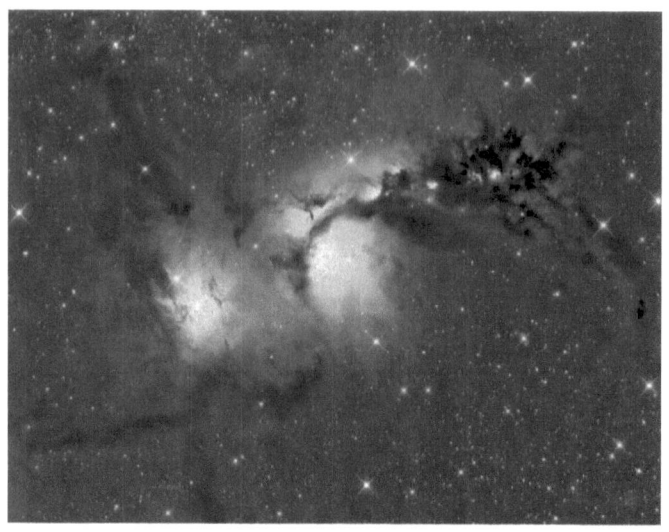

OBJECTS 15 - Messiers 42 and 43 - The Great Orion Nebula (and his sidekick)

For a complete discussion of the Great Orion Nebula and the fantastic area surrounding it in the constellation of Orion, refer to my Part II of this three part Orion "GO TO" TOUR. Messier 42 and its associated stars of theta Orionis (the "trapezium") comprise the central star of Orion's "sword," while Messier 43 is a faint path of displaced nebulosity and bright stars immediately NORTH of the Orion Nebula. The two comprise a spectacular sight on a dark winter's night at the lowest power you can muster in any telescope. Seen also in Part II, the classic beauty of the 100" Lick Observatory's portrait is reprinted here so that the smaller area of M-43 is NOT confused for NGC 1977, which is the brighter diffuse cloud seen at the bottom (North) of the photo. Messier 43 is the small cloud immediately below the "fish's mouth" in this photo and almost touching the huge M-42. Note the bright star Iota (HATYSA....see Part II) at the top (South) of the Orion Nebula. Below this photograph is the Lowell Observatory's 42-inch telescope's image of a close-up of Messier 43 (this image - of M-43 - is reversed from that of Lick Observatory's, with NORTH up and EAST to the right, as seen in a Maksutov or Schmidt-Cassegrain telescope).

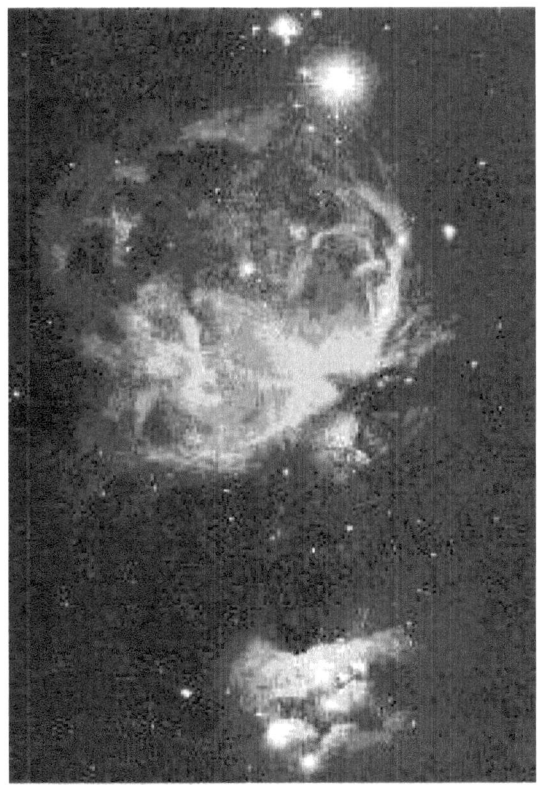

MESSIER 42 (CENTER)
and Messier 43 (bottom)

OBJECT 16 - The Famous HORSEHEAD NEBULA - IC 434
I remember talking with a young man many years ago at a major dark sky public observing event

following one of the local astronomy club member's comments. He had just gotten a new 8" Newtonian telescope from Criterion, a "Discovery" I think it was. This young and energetic man was full of the energy that all of us feel when getting such a new and - in 1975 a BIG - telescope. It was a very cold early November event to coincide with the Leonid Meteor shower which would hit shortly after midnight when Orion would be high in the sky. Never having used the telescope in so dark a sky as we had on Petit Jean Mountain, one would assume that one of his first major targets would be the Great Orion Nebula. Not so.

"*I'm going to see the Horsehead!*" he proudly proclaimed, patting the less-than-shiny white tube of the new telescope. "Finally going to see it!", stroking it like the fur on the back of a cat's neck. He, of course was referring to the famous "Horsehead Nebula" just south of Zeta Orionis, an object astronomers refer to as "IC434." Actually, this object is as large as two moon diameters and is magnitude 1.9! Why SHOULDN'T he expect to see this most famous of dark and bright nebulae combined in the shape of a horse's head?

"*In that?*" I asked with reservation. He looked proudly at the Criterion.

"*Yep...it's an 8-INCH*!" Much exclamation put on that point, and I did not have the heart to tell him he could have had a 20" right then - actually more like 3 a.m. - and still would be unable to see the object. But sometimes practical experience is the best teacher, albeit most of the time a disappointing lesson. But his reasoning was quite good. He was moving up from a 4" refractor with considerably less light gathering and from a darker-imaging f/15 focal ratio to a fast f/6 in the Newtonian and a brand new 25mm Kellner eyepiece. The object is BIG and it is BRIGHT. The following photograph demonstrates how distinct the Horsehead is photographically, with Zeta brightly shining above (north) of it.

The Famous Horsehead Nebula - IC 434
35-minute exposure, 24" Reflector @ f/4 - SO 410 Kodak

However, much to his tremendous disappointment he search and searched the area, using a very carefully plotted "map" of the field stars I have prepared for the group ahead of time. Using these stars from any good star program or atlas, you can easily plot where it "should be" just like extrapolating where you are on a map in relation to two towns on any highway.

"I THINK I might see it..." he would continue to shout over the rest of the group, now huddled around a small 5" Jaegers lens f/5 refractor using about 15x. In this small, short and very much NOT impressive little scope here was the Horsehead, very faintly but clearly distinguishable. No mistake about it in the nearly 3 degree field of view, right where it should be due south of Zeta and a bit of a right angle to Sigma Orionis to its west. The intrusion of "Barnard 33", a very dark nebula that light cannot pass through over IC 434 (the bright nebula behind it) was an impressive yet very ghostly and vague image. Then the gentleman with his homemade octagonal wooden-tube 4-1/4" f/4 hand held richest field telescope likewise spotted the Horsehead.

Meanwhile, back at the 8" it continued. *"Is it RIGHT or LEFT of that star?"* we could hear him ask, still not convinced that he was seeing it....I could sense disappointment setting in quickly. He was not going to see the Horsehead Nebula in that telescope.

Very low and wide field instruments are required to visually detect IC 434 and Bernard's dark nebula 33. Indeed, it can even be glimpsed in the large 11x80 and 20x80 binocular if one can hold them steady enough; I have seen it this fall in early morning skies from the mountain in a 2.5 inch scope with a 26mm Plossl eyepiece! But I cannot see it in the ETX 125, nor in an 8-inch and certainly

NOT in the 12" and larger telescopes. For the Horsehead, one needs:

1) the very darkest skies;
2) good low power optics with excellent dark sky contrast;
3) the Horsehead to be as high in the sky as possible;
4) totally dark-adapted eyes;
5) wide field of view; and,
6) a good finder chart printed from one of the PX sky programs to identify exactly where it should be located relative to stars that will be seen in wide angle views.

The Hubble Space Telescope has no problem "seeing" this object, and it is relatively easy to photograph, even by piggybacking a camera with a telephoto lens and exposures of 10 minutes or greater on fairly fast (ASA 400+) film. (see my complete discussion on getting started in piggyback astrophotography at the Arkansas Sky Observatories' website: www.arksky.org . The image of the Horsehead Nebula from the Hubble below clearly shows the dark Barnard 33 nebula as it overrides the bright nebula behind it, offering a somewhat "3-D" impression of this remarkable object.

There are actually two (2) components to the Horsehead Nebula. First is the bright emission nebulae that fills the sky between Zeta and Sigma Orionis, running almost perfectly north-to-south.

This nebula is known as "IC 434" and is an incredible 60' arc (one degree!) long north to south (that is the length of two full moons edge-to-edge) and 10' east-west. It is believed that the illumination of IC 434 is from the energy of the star Zeta, much like Theta Orionis activates the glowing of the Orion Nebula (see Part II). Although it appears as if the nebulosity just "stops" abruptly at its eastern edge in the photographs above, it actually is still there, merely covered up by the dark nebulae that also makes up the famous Horsehead, or dark nebula B-33. Indeed, you will note, even if you do NOT see the Horsehead, that there is a sudden "absence of stars" to the east of a one degree imaginary line running south from Zeta Orionis....the stars are there....they are just covered up by light absorbing dark interstellar matter, the very stuff that covers up the bright nebula behind it to create the effect we know as "the Horsehead." Look at the photograph above and imagine it in 3-D....the dark nebulae is closest to you, while the bright cloud of gas is behind it. Imagine it as fog filling in rapidly on a cold spot on an English moor.

My friend would not be too disappointed if he knew that the Horsehead is NOT visible in the world's largest refractor telescope visually either! Indeed, the larger the aperture and more "powerful" the telescope, the less likely one is to actually glimpse this object with the eye. Using the criteria listed above, your chances are best for spotting the nebula, but also knowing where to look is a much benefit. I mentioned the small charts that I had circulated to assist observers in locating the Horsehead. A sample of that is shown below,and you can click and print this chart for use at the telescope. NOTE: for this chart, I have reversed the field so that it will appear as in a Maksutov or Schmidt-Cassegrain, with North at the top and EAST at the right, a mirror image of the way the sky actually would appear!

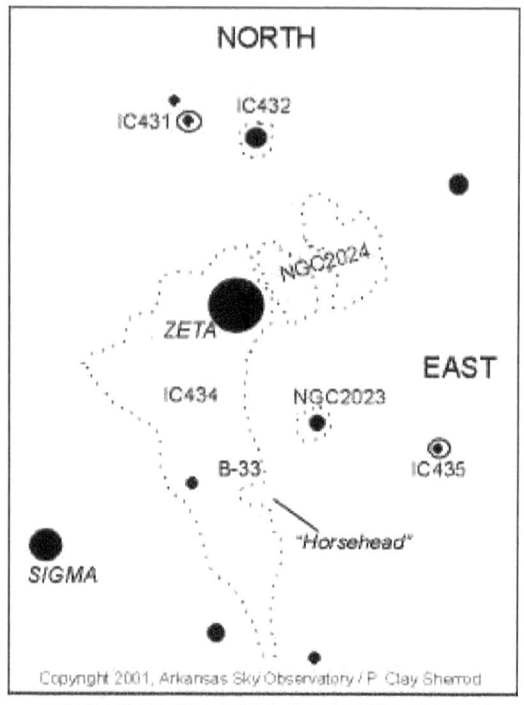

HORSEHEAD NEBULA FINDER CHART

Note the two faint stars embedded within the brighter IC 434 just to the west of the Horsehead and also the brighter star SIGMA ORI (discussed as our "User Object" for Orion, below). Drawing an imaginary line eastward from bright Sigma through the faint star marked with the ARROW, you will find the Horsehead an equal distance on the other side of that fainter star than it is from Sigma. Also note another method of finding the Horsehead using these stars. The nebula makes a right angle almost perfectly between Zeta and Sigma, due south of Zeta and nearly due east of Sigma.

While in this area, note the two emission nebulae NGC 2023 and IC 432, both of which are far easier objects than is the Horsehead. A good 4" scope at very low magnification will show both of these as a medium-brightness star surrounded by a "halo" of light....very nice objects to add to your list of exotic targets! For larger telescopes and equally rewarding object is IC 431, just to the west and slightly north of IC 432. Merely "star-step" your way through this exciting region of Zeta Orionis and see how full of wonderful objects and star groupings it truly is!

Just as so much within the mighty borders of Orion the Hunter, the Horsehead (B-33 and IC434 combined) is some 1600 light years distant, and thus is part of the huge association of bright white energetic stars and emission nebulae that comprise this wonderful region of our visible universe.

WANDERING ABOUT....YOUR NEW "USER OBJECT" IN ORION

Just southwest of Zeta Orionis and nearly due west of the Horsehead (see the finder chart above) is a wonderful multiple star for moderate-sized amateur telescopes, SIGMA ORIONIS. This star has a combined magnitude of 3.7 and a "total" of five stars in a very complex multiple star system. The brightest of these is the 4th magnitude "A" star, like nearly all other stars of the Orion Association a very hot energetic white star in the 1,400 to 1,500 light year distance range. Using the chart below, let us look at this complicated orrery of stars. The star "B" is a 5.9 magnitude white star as well, but this star will likely not be seen in amateur telescopes. At only about 0.3" distance away in Position Angle 200 degrees (just west of due south from the brightest star), it is theoretically visible in a good

16" under good conditions, but typically cannot be split into two stars. Only on extremely steady nights and magnifications approaching 800x to 1,000x have I ever seen this star in the 24" Cassegrain. But.....there's more!

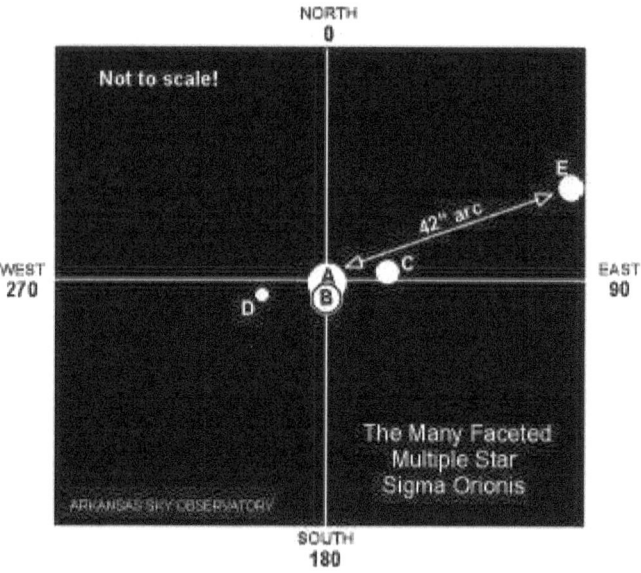

Look for the 7.3 magnitude "C" star in Position Angle 84 degrees, or nearly due WEST of the bright unresolvable pair A-B; it is bright enough and far enough away (13" arc) that it should be an easy object in all telescopes at medium power (about 15x per inch in most cases). About the same distance away, but nearly opposite the brighter Sigma is star "D", which is much fainter at magnitude 10.2 and 11" arc distant. A fifth star (making FOUR that can be seen with amateur telescopes) is much brighter and easy to spot in the smallest of telescopes, this "E" component being a 6.4 magnitude star at Position Angle 61 degrees, or rather northeast of the bright primary star(s).

This is a wonderful object to add to your growing "User Object" library for the Autostar and is a fine multiple star that is not included on the Autostar or Nexstar list of double/multiple stars. Its epoch 2000 coordinates can be entered as: R.A. 05h 38.7m / DEC (-) 02 36.

On AutoStar, go to: "Select/Object [enter]...." scroll down to "User Object" [enter]. Now enter the coordinates given above for "Sig Ori", using the number keys on AutoStar. After entering the coordinates and pressing "Enter" yet again, scroll down one and you can list the magnitude of the object as "3"[Enter].

For different sky programs, the logic is the same; use the HELP tab to learn the correct steps to enter coordinates for all of your User Objects.

With the addition of this star, you now have a wonderful and relatively complicated multiple star in the collection of celestial curiosities for your next star party or family outing with the telescope!

*   *   *

"I turn my telescope to Barnards Loop and M42, glowing in Orions sword. Stars are fires that burn for thousands of years. Some of them burn slow and long, like red dwarfs. Others- blue giants-burn their due so fast they shine across great distances, and are easy to see.As they Starr to run out of fuel,they burn helium, grow even hotter, and explode in a supernova. Supernovas, they're brighter than the brightest galaxies. They die, but everyone watches them go."

Author: Jodi Picoul

Modern computer-controlled telescopes give
today's observers the Universe at their fingertips
Image courtesy Meade Instruments

Chapter 26

PEGASUS
An Upside-Down Mighty Winged Steed Born from the Blood of a Bad Hair Day

In this *Constellation* guide, "GO TO PEGASUS" of the series we begin to get a taste of impending fall....a season that brings with it not only very clearly defined and wonder-filled bright starry constellations, but also the precursor of those crisp and deep clear nights of September and October in the Northern Hemisphere.

Although containing only ONE deep sky object - **Messier 15** - Pegasus is a wonderful constellation for close study, with many faint galaxies, scores of wonderful double and multiple stars and some fascinating early lore and history associated with the constellation itself as well as many of the stars within it.

Pegasus is located high north of the Celestial Equator (the "0" demarcation line in declination) and hence all objects for our "GO TO" tour will have positive ("+") declinations. The most identifiable "sky mark" of this very large constellation is the famous "**Great Square**," its namesake derived from the nearly perfectly square asterism (pattern) formed by the bright stars *Alpheratz* (actually in Andromeda to its east), *Algenib*, *Scheat* and *Markab*, a large quadrilateral that forms a square almost 12 degrees wide in declination and stretches some 15 degrees (almost a perfect "hour" in Right Ascension) across the sky.

At dark on August 15 each year, from about 35 degrees north latitude, the square of Pegasus will be seen rising in the ENE sky about 9 p.m. The star *Enif*, far to the square's west will rise significantly earlier - about 6 p.m.....that will give you some idea of the great expanse of this constellation! Enif passes (culminates) nearly overhead from that latitude on Midnight of August, while the Great Square sees midnight culmination nearly ONE MONTH LATER! The constellation is very favorably placed for all-night viewing during August and September each year.

One Horse....two wings.....many tales –

There perhaps is no other mythological fable that has invoked so many variations of a theme throughout the early Arabic and Grecian golden period of gods, heroes and tall tales that Pegasus....the beloved horse (although it came to be through evil means) has been the central figure in a wide variety of exploits and fireside stories of old.

There has never been much dissention in that this constellation represents the fabled "Winged Horse" or "Flying Horse". In fact, so important is the Greek legend of this flighty animal that it was immortalized in one of the most symbolic of all American institutions: the oil industry. We all have seen the muscular and mighty flying steed which graces the red, white and blue emblazoned signs of the *Mobile Oil Company*.

That is a fitting tribute that could hardly be rivaled by any other association.

In Greek mythology, and reinforced by the Arabian namers-of-the-stars through the beautiful star names given to the brighter members of this constellation, the large star pattern represents a beautiful, peaceful and dedicated (yet most unusual) airborne horse that sprang from the spilled blood of the horrid snake-headed troll of a woman, *MEDUSSA* after she was slain by the Greek hero *Perseus* (who, by the way, follows the horse quite closely to the East in the skies). But the flying horse concept is only one of many variations and evolutions of the original association of the constellation:

1) <u>Association with fresh spring waters</u> - the earliest Greek derivative of the name "Pegasus" comes from the translated words for "Ocean Springs"; later versions tweaked the words slightly to denote "the strong one," for his ability to buck so forcefully on the mountains of Cornith that his hooves cracked open the Earth and released the pure spring waters of Peirene, still flowing freely today thanks to the mighty bucking bronco. In addition, Pegasus appears to have mythologically tapped (hoofed?) the spring waters of Mount Helicon nearby. Although obviously important to the writers of the legends, the particular significance of a horse providing spring water to a land not particularly drought-stricken is puzzling.

2) <u>Heroic feats</u> - As if springing from the bloods of snake-headed Medusa was not enough, Pegasus was well know for some very superhuman (oh...."superhorse") deeds. First, and most obvious, was the fact that he could fly. Several Greek interpretations of the origin of the Pegasus name suggest that it signifies NOT an association with the

ability to crack open the earth and provide spring water, but that of STRENGTH instead. Pegasus, although tamed into submission by Athena, was still will and heart and brave by ancestry. It was this horse that carried all the lightening and thunder bolts that were so skillfully tossed around by the great Greek god Zeus. It was Pegasus who carried the hero *Bellerophon* en route to slay the horrible *Chimaera* (an odd creature with the heads of a lion, a snake and a goat). As legend has it, however, the winged horse and Bellerophon had a "falling out..." Literally. Bellerophon decided that, because of his great deeds and heroics, he was "godlike" and could ascend via the winged horse up to the house of the gods at Olympus, much to the ire of Zeus. The horse grew weary of all the heroic escapades - and assisted by the sting on the rump by a mighty insect sent by Zeus - Pegaus simply bucked him and threw him to earth, blinding and crippling him forever. As a just reward to the steed, Zeus allowed the horse to ascend to Olympus in the sky forever sparkling in the constellation that we see today, to never carry a rider again.

3) <u>Biblical and religious associations</u> - Pegasus has also been linked as the horse which carried Nimrod in the earliest Jewish legends as well as the actual image of GABRIELE the archangel.

Perhaps one of the most curious aspects of just how we have assigned a horse - much less a "winged horse" - to this star pattern is the fact that, in order to make the stars "work" even close to resembling a winged horse, you must look at the pattern upside down...this is similar to the curious situation of Hercules, another Greek hero, who also is denoted upside down....something very unbecoming of such significantly-respected heroes of old!

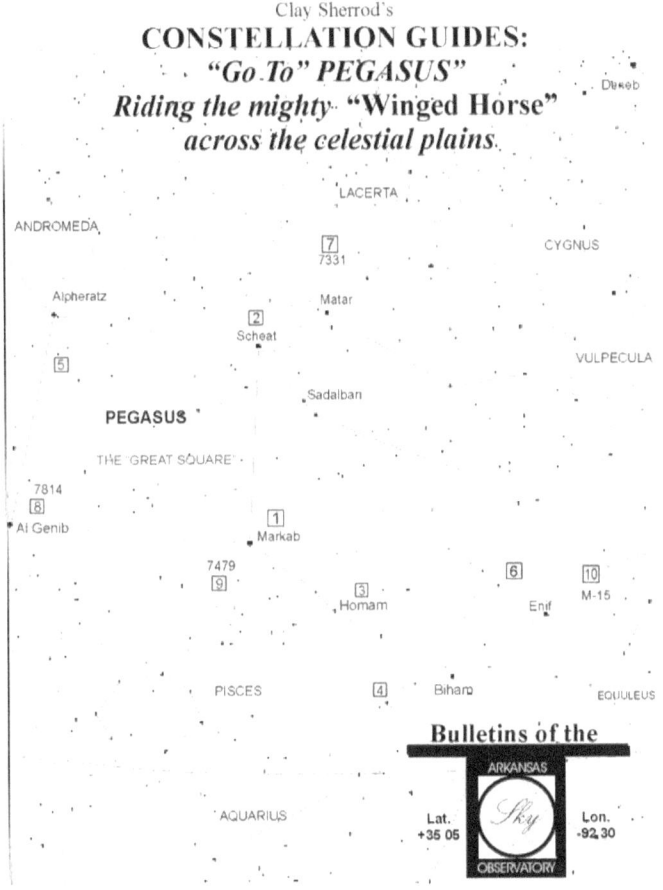

Regarding the interesting objects in the huge constellation of Pegasus, refer both to the detailed "GO TO" tour map shown above and to the constellation listing on your computer or keypad program.

Note from the sky chart included here that the CELESTIAL EQUATOR passes directly south of

Pegasus. This is the reading "0" degrees on your properly adjusted declination setting circle. All angles NORTH of this equatorial line are positive ("+") and all angular measures (declinations) south of the celestial equator are negative ("-"); hence you will see references in this "GO TO" GUIDE to only "+" declinations when referring to Pegasus objects since all are NORTH the "0" declination equator.

In addition to all the GO TO objects for your Pegasus Tour, you will note an abbreviated listing of many "ngc galaxies" that are good, albeit difficult, objects for the medium sized scopes at the end of the Concise Listing which follows.

Each GO TO object in Pegasus is discussed for your telescope regarding the type of conditions necessary for you to view it optimally for discern the very faintest details.........magnifications and aperture necessary for most objects, and much, much more. This is YOUR complete guide to get you on your way to exploring the best objects in Pegasus with your computerized telescope and its GO TO function. The following listing of "BEST" objects contains the finest or most interesting....as with our "GO TO" tour of Coma Berenices and Virgo (both of which are in the *Constellation* guides) there is ALSO a bonus comprehensive listing of ALL NGC GALAXIES (they are on your handbox and all PC sky programs) that can be seen in at least one or more of our telescope size ranges *(faint ngc objects that are NOT visible in at least an 12-inch are NOT included in these "GO TO" TOURS)*.

Use this attached star chart and the following Guide as an excellent reference for your next star party itinerary, or a beginning for further study into the thousands of objects visible in this part of the sky. Truly these extensive Constellation Study Guides will most definitely put your AutoStar to work for you in the most efficient and enjoyable way possible! As a matter of fact, MANY telescope users are now programming their own "Tours" based on these guides, using each constellation as a separate GO TO Tour for the AutoStar or PC sky library that can be added in or deleted through the main edit screen on your PC or MAC computer.

We hope you enjoy these comprehensive GUIDES to touring the constellations via your AutoStar or PC sky program and its computer-driven telescope. Each new installment is complete with diagrams, charts and illustrations that you will find nowhere else. Please let us hear YOUR feedback and your observations of each and every constellation after YOU have toured its vast reaches of our skies!

YOUR PEGASUS CONCISE DIRECTORY OF INTERESTING OBJECTS –

While there are literally thousands of faint galaxies within the borders of Pegasus, only a handful are visible to modest telescopes and even more are beyond reach of all but telescopes 12 inches and larger. Thus, only the brightest (and THEY are still quite faint!) and in some cases, most challenging, objects are chosen for this brief tour. Two of those galaxies listed CAN be glimpsed adequately in a telescope under 5 inches.

I have chosen the finest (or most interesting) 17 objects in this PEGASUS "GO TO" TOUR; as with all GUIDES, all objects listed below will be visible in most telescopes (some naked eye) from 3-inch to 8-inch; of course larger apertures may "show" an object a bit closer and "better," but frequently a wide field and low power view is more desirable than aperture for FINDING the objects initially. Indeed, I strongly encourage you first FIND the target object, or its approximate location through your GO TO function with your lowest power and then - once IDENTIFIED positively - move up slowly in steps with magnification if necessary. Remember, not all objects "like" magnification. Sometimes better "field of view" (such as the wonderful wide fields provided by small refractors) is desired over light gathering (like the larger 8-10 inch) and magnification.

The rule for determining "optimum magnification" is that: 1) too low power results in sky background glow detracting or diminishing the contrast against the deep sky object; 2) too high magnification darkens BOTH the sky background AND the object; 3) medium magnification can be achieved at which you have MAXIMUM contrast between the object and its darkened background sky. I have found through three decades of direct observing that about 15x per inch aperture, for deep sky observing is PERFECT for most objects. That being said, always remember that DOUBLE or multiple stars require whatever power you can crank out....the seeing conditions are the limiting factor here.

For my complete and comprehensive discussion regarding seeing conditions and sky transparency, see my discussion of this topic in *GUIDES - Frequent*, on the ASO website, www.arksky.org

With all deep sky objects, avoid attempting to observe when the moon is in the sky, even a very thin crescent, as its brightness in the sky will overshadow the very dim contrast afforded by even the brightest deep sky object; if you see the object at all against moonlight, you will NOT see the subtle outlying areas or the full detail of what is presented.

Pegasus is dominant in late summer and early fall skies; riding mid-way on the Celestial Equator, it is suited for long-period observing for all telescope users both north and south of the equator. All deep sky objects and difficult double stars are ALWAYS best observed when they are located nearly overhead (or as high in the sky as possible), thus requiring the observer to look through the thinnest portion of the Earth's "lens" of atmosphere and haze.

As with all of the "GO TO" tour constellation lists, I recommend a good star atlas and/or chart which will list all the finest objects, constellation-by-constellation. One very handy reference guide is the *PETERSON FIELD GUIDE TO THE STARS AND PLANETS*, which features complete lists with declinations, right ascensions, magnitudes, and all pertinent information for you to expand your observing horizons beyond this brief GUIDE.

For starting instruction on all computerized systems and Apps, please see the introductory discussion begging in Volume One of this series.

The **constellation tour Star Chart** included here

will get you started on your journey for this constellation.

Following is the concise object list for your "GO TO" TOUR of PEGASUS; you may wish to find the majority of the objects from the PC sky library or AutoStar library (for example, you can easily go to "MESSIER 15" if you pull up "Object/Deep Sky/Messier Object/..type in '15'...." and then press "Enter", followed by "GO TO" to access this very nice and bright globular cluster. On the other hand, if you want to experiment and become a "better AutoStar user" try entering the exact R.A. and DEC coordinates of that object as described above after holding down the MODE key. You will find the accuracy of entered GO TO's to be somewhat more than those stored in AutoStar, but the capability of acquiring unlisted objects is fantastic!

Of course, for named objects such as the starter star "**Markab**" (alpha Pegasi), you might choose to merely key in SELECT / OBJECT / STAR / NAMED....and scroll to Markab. Enter and then press " GO TO" and you are off to your first object!

OBJECT 1:   bright star - MARKAB (alpha Pegasi) - R.A. 23h 02' / DEC + 14 56 - Magnitude: 2.5 / NW "corner"
OBJECT 2:   bright red star - SCHEAT (beta Pegasi) - R.A. 23h 01' / DEC + 27 49 - Mag: 2.5 to 3.1, variable reddish!
OBJECT 3:   test star - HOMAN (zeta Pegasi) - R.A. 22h 39' / DEC + 10 34 - See if you can find the 11th mag. star!

OBJECT 4: tough double - 37 Pegasi - R.A. 22h 27' / DEC + 04 11 - Mags: 5.8 & 7.0, really tough **for 6"+**

OBJECT 5: good one for LX 90 - 85 Pegasi - R.A. 24h 00' / DEC + 26 49 - Mags: 5.8 & 8.5 - very close, 0.8"

OBJECT 6: nice variable - AG Pegasi - R.A. 21h 49' / DEC + 12 23 - Mag: 9 to 6; a permanent nova? 6th mag. now!

OBJECT 7: spiral galaxy - ngc7331 - R.A. 22h 35' / DEC + 34 10 - Magnitude: 9.7, very good object for 5" +

OBJECT 8: spiral galaxy - ngc7814 - R.A. 00h 01' / DEC + 15 51 - Magnitude: 12.4, very difficult but distinct!

OBJECT 9: spiral galaxy - ngc7479 - R.A. 23h 02' / DEC + 12 03 - Magnitude: 11.6, interesting shape, very long!

OBJECT 10: globular cluster - Messier 15 (ngc7078) - R.A. 21h 28' / DEC + 11 57 - Magnitude: 6.7, very dense, rich

OBJECTS 11 THROUGH 19 - NGC GALAXIES PEGASUS (other than those listed above) listing of "ngc" galaxies in Pegasus in order of RIGHT ASCENSION (and NGC # order)

NOTE: These NGC galaxies may NOT be detailed in the following "Visual Guide"
as are those Objects 1-10 listed above...use the abbreviated descriptions as they follow the order:
NGC# / R.A. / DEC / MAGNITUDE / SIZE (in minutes arc -'-) / GALAXY TYPE , description

ngc7814 / 00 01 / +15 51 / 12.4 / 3.0 X 0.8 - spiral, very edge-on and large...but dim, 8-inch or larger

ngc7177 / 21 58 / +17 29 / 11.9 / 2.1 X 1.1 - spiral, oval tilted, fairly small and should be seen in 5-inch
ngc7217 / 22 06 / +31 07 / 11.0 / 2.6 X 2.3 - nice large & fairly bright spiral, larger telescopes easy object
ngc7332 / 22 35 / +23 32 / 11.8 / 2.3 X 0.6 - very "squashed" elliptical, starlike and faint - larger telescopes
ngc7448 / 22 58 / +15 43 / 11.2 / 2.0 X 1.0 - face-on spiral, fairly bright and possible in 4-6 inch scope
ngc7479 / 23 02 / +12 03 / 11.6 / 3.4 X 2.6 - very nice but faint barred spiral; use larger scopes for this
ngc7741 / 23 41 / +25 48 / 11.6 / 3.0 X 2.0 - similar to above, at same magnitude and size (triplet with below)
ngc7742 / 23 42 / +10 29 / 11.9 / 0.9 X 0.9 - very tiny and faint elliptical; probably only in 8-inch+; starlike
ngc7743 / 23 42 / +09 39 / 12.0 / 1.6 X 1.4 - # 3 of the above triplet, faint tiny spiral, possible in 6-inch

....SO LET'S GALLOP AWAY ON OUR CELESTIAL STEED!! (refer to the Pegasus Star Chart for all the objects described in detail on the guide)

A VISUAL GUIDE TO OUR DEEP SKY OBJECTS IN PEGASUS –

Object 1 - Bright Star - "*MARKAB*" (alpha Pegasi)
Our starting point for every "GO TO" TOUR is always (or usually!) the brightest star of the constellation or region. At a distance of 110 light years and a magnitude of only 2.5, the Rosette stone of Pegasus is not much of a spectacular object.

However, this is a really good time to discuss the rich history of star names - beautiful and enriching star names - that were placed upon so many of our brighter stars by the ancient Arabian sky watchers. The brighter stars of Pegasus were aptly named for the most part, perpetuating the lore of the winged horse and its importance in early civilized mythology. Let us examine some of the names of the special stars of Pegasus. Remember that the horse is UPSIDE DOWN in out skies....the four stars of the "Great Square (*Markab, Scheat, Al Genib* and *Alpheratz* (this one in Andromeda) comprise the horses' BODY; (see chart above; Pegasus' front legs are marked by the stars Matar and *Sadalbari,* and his head a great distance to the west, set by the AutoStar reference star "*Enif.*" Let us look into some of the colorful horse-related star names for select Pegasus landmarks:

Alpha - "Markab" - this is most definitely a "horse named-star," except in some of the very earliest Arabic associations. Earliest designations associated this star with "Marchab," the ship on the water. The range of names that place this star as part of a horse's anatomy vary throughout Arabic times, including "Markab" (the horse's saddle), "Matn Alfaras" (the horse's shoulder), and finally "Yed Al Pheras", or the horses hoof.

Beta - "Scheat" - Do you remember another "Scheat" name associated with a star in Aquarius (there is, and you can find it at ASO GUIDES/Constellations/Aquarius) . The name is from the Arabic "Al sa'id" , denoting the horse's foreleg.

Gamma - "Al Genib" - Two connotations are associated with gamma Pegasi: "Al Janb" is the Arabic term for the "side" or flank of the horse, while "Al Janah" appears to be the more likely origin of this star's name, meaning "the horse's wing." This is the brighter star located at the SE corner of the Great Square.

Epsilon - "Enif" - this frequently used Autostar alignment star is one of the western most of all stars in Pegasus, and derives its name from the Arabic "Al An'f", or the head (nose) of this upside down horse.

Zeta - "Homan" - Here is a total break in horse lore....the Arabic sky watchers, for whatever reason, chose to NOT associate this star - clearly in the forearm formed by the extension southwestward from Markab - to the horse at all! The earliest places the significance of this star with the fortunate fates of all warriors and heroes, or "Sa'd Al Na'amah." Far to the right of this is "the whispering star," or "Al Hamman", a later designation. Other names assigned to this seemingly popular star were: "Sa'd Al Na'amah," or the "lucky Ostrich Star," and Na'ir Sa'd Al Bahaim, or "the bright and fortunate star of the two beasts."

Eta - "Matar" - although comprising the upper part of Pegasus' front leg, this star is named "Al Sa'd Al Matar, or "earliest rains," also having nothing to do with a winged horse nor its legends. So, we can clearly see that unlike today, where rocks, firebombs and pubic dissent rage over territorial rights of Arab lands.....only thousands of years ago were they fighting over the proper naming of the beautiful stars of the dark Arabian skies.

Object 2 - "Scheat" - (beta Pegasi) - A very Nice Crimson Star - Naked Eye Variable!
This is a nice naked eye variable star and also a good star to examine telescopically at very low powers to afford a wide enough field of view for comparisons to other stars of different colors nearby. Scheat is the "upper right" star (NW) in the "Great Square an shines a magnitude 2.5....but it varies irregularly to as faint as 3.2 with no identifiable period; with so many stars of similar magnitudes in the naked eye field of Pegasus and beyond, this is an excellent star to keep an eye on. It is farther than Markab by 100 light years and is a star similar to Betelgeuse in Orion, growing physically larger when brightest and smaller when dimmest. In the modest telescopes, you should be able to pick out to companions (optical - not true physical - doubles to this star). There is a 9th magnitude star due EAST of beta Peg at about 4' arc distance; in addition, an third optical companion can be glimpsed at 11th magnitude SW of the brighter star at about 1.2' arc distance.

Object 3 - "MATAR" (zeta Pegasi) - For 4-inch scope users: Try out your visual acuity on this optical double!
Actually this is a challenging optical double star for all telescopes, not nearly as easy as its separation might suggest. The larger telescopes should make easy game out of this one, but ONLY if you know where to look! These stars have nothing physical in common and are separated by vast distances in space. Zeta is the closer of the two stars, slowly moving away from us at 210 light years. The faint

11th magnitude star is a full one minute (') arc (62" arc) away, just barely east of due SOUTH from 3.2 magnitude Zeta. This should be quite a challenge in a 3-4 inch telescope and I would be interested from users of that scope who have been able to spot the star. Knowing where to look is 99 percent of the battle.

Object 4 - A very tough double star - 37 Pegasi – Challenge for an 8-inch....test your eyesight with something even smaller!

Right now the double star 37 Pegasi, a 5.5 magnitude star only 4.5 degrees north of the star we just discussed (Zeta), is as far separated as it will be during our lifetimes and as far as it has been since its discovery in 1889. Its orbit is "laid out" to our line of sight, thus it appears that the secondary star is merely moving back and forth relative to the brighter star! At present, the companion, magnitude 6.9 is exactly NE of the brighter star by only 1.0" arc, and their similarity in brightness make this a very challenging star for both a 6 and 8-inch telescope. Even under the very best conditions, this is at the limit of resolution for the 5" scope yet should be cleanly separated in the 8" with powers of 220x or more.

Object 5 - 85 Pegasi - Another Fine Double, and Another Test for a 6-inch telescope

From the positional chart above, you can see that the 11.2 magnitude companion star of 85 Pegasi is just barely west of SOUTH; I have drawn the chart to correspond with the field of view of a Maksutov and Schmidt-Cassegrain, with North at top, and East to the right as it appears with the diagonal mirror in place.

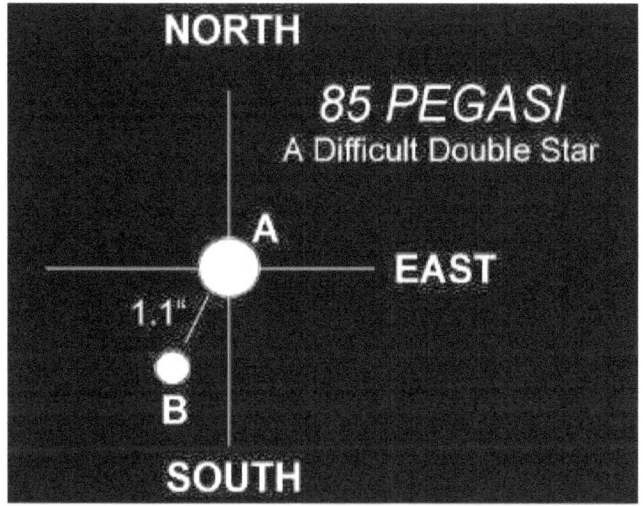

From the positional chart above, you can see that the 11.2 magnitude companion star of 85 Pegasi is just barely west of SOUTH; I have drawn the chart to correspond with the field of view of a Maksutov and Schmidt-Cassegrain, with North at top, and East to the right as it appears with the diagonal mirror in place. The current separation of only 0.8" arc is right at the limit of the LX 90, but I have seen the faint star clearly with the ETX 125 on several occasions this year. Note that in TWO years, the star will have increased its separation to near maximum (1.0") and will be due SOUTH of the primary. This is a short-period double orbiting in a period of only 27 years, and the true double is ONLY 40 light years distant from Earth! Although the main star - magnitude 5.8 - can be spotted easily in the very small telescopes, the faint and close companion star will not be seen.

Object 6 - A Nova In Progress? AG Pegasi - **Robert Burnham's** "Permanent Nova"

A very interesting object to talk about, yet one with little action right now....but you never know. In the mid-1800's this star suddenly brightened from it normal magnitude of around 9.2 to brighter than 6 in only 20 years; during and after this brightening the star began to show very peculiar spectral changes as well, indicating that the star was turning more red, and evolving quickly to a very "late" type star of the Red Giant category. At present the star is at magnitude 8.7 and exhibits very minor light variations as it SEEMS to be once again slowly fading back down to its original magnitude. This star can be easily monitored in all scopes - provided that you can zero-in on the RIGHT star! It does have a reddish hue about it and that might help you to locate and center it.....once you "think" you have the correct star, increase the magnification to about 20x per inch for observation in all scopes.

The American Association of Variable Star Observers has only a limited resource chart available for this curious star which I have revised to include magnitudes of comparison stars and shown below. Using this chart, which has NORTH at top and EAST to the right as you would see in your telescope, you should be able to find this medium-brightness star in the not-so-crowded star field. Note that it is ONLY 3.3 degrees northeast of bright Epsilon Pegasi, so find that star first and zero in on AG Peg from there!

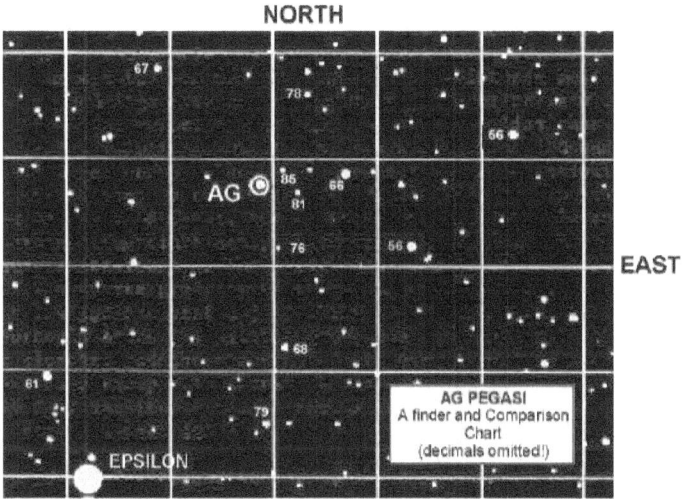

Object 7 - A Beautiful and Relative Easy to See Large Spiral Galaxy - ngc7331

Almost exactly 4 degrees due north of bright Eta Pegasi ("Matar", magnitude 2.9) is a very fine nearly-edge-on spiral galaxy. This has a total magnitude of 9.7 and stretch is a very thin oval shape for 10' x 2' across" This is a very large object, but relative faint since its brightness is spread out over a linear dimension that is nearly one-half the span of the moon's disk! Very similar to the famous Andromeda Galaxy, ngc7331 is much farther than M-31, at a distance of 50 million light years, compared to 2.3 million for Andromeda. As can be seen in the beautiful photograph of ngc7331 below taken through the giant 200" Palomar reflector, the galaxy is oriented almost perfectly N-S in our scopes; in the ETX 60 and 70, you can see a very faint but distinct "smear" of light from this if you look about one field north of Eta Peg. Once found increase magnification in those scopes to about 70x for direct observation; it is clearly seen in a 4-inch and in an 8-inch the object is just fantastic in very dark skies. Averted vision and a power of about 120x in a 6-inch or 8-inch are recommended for best views of this beautiful galaxy!

Object 8 - Another Nice, But Faint, Galaxy - ngc7814 - The "Little Sombreo"to us than it is....ngc 7814 is a miniature "Sombrero Galaxy", oriented edgewise and appearing photographically very much like that object. In the LX 90 the dark dust lane bisecting the oval shape can be clearly seen on a very dark night. However, this galaxy is very faint (less than magnitude 11) and appears ONLY as a distinct ellipsoidal glow in telescopes over 6-inches; I would not expect to see much if anything in smaller telescopes of this object. It is a relatively large (5.1 x 0.9' arc oval) galaxy and what light we see is very highly concentrated near the center "hub" as you can clearly see in the Mt. Wilson 100" telescope photograph shown below.

Object 9 - A Galaxy With an Interesting Shape – ngc7479
Here is a good (but very faint!) galaxy to test your eyesight on. Seen only in telescopes 6 inches and

larger, this spiral galaxy is fairly large at 3.4' x 2.6' across but - as the photograph below taken through the 61-inch telescope of the U.S. Naval Observatory shows - is quite thin and "spindly". Note in the photograph the very distinct S-shape to the galaxy and particularly the "arm" which extends UP (north) in this photograph.

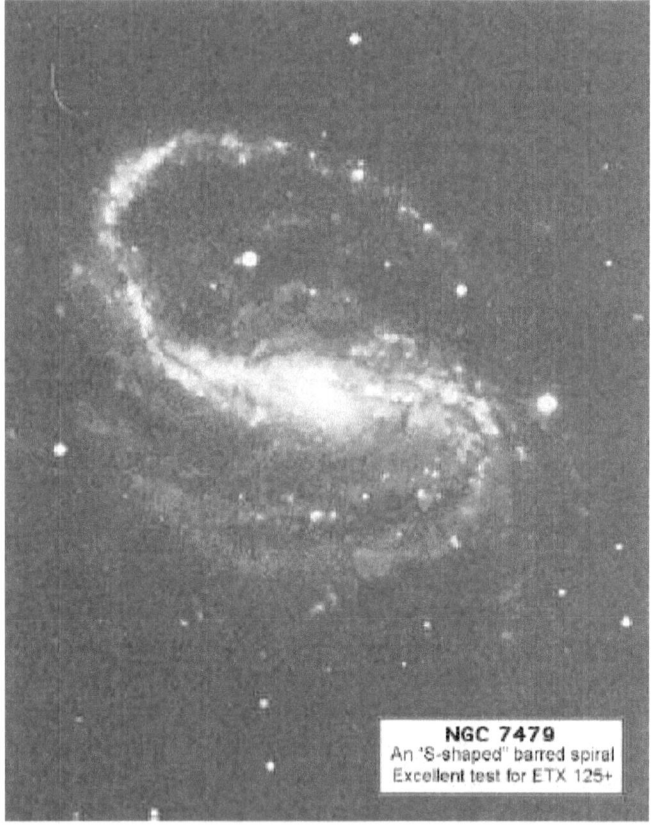

NGC 7479
An 'S-shaped" barred spiral
Excellent test for ETX 125+

This arm, as well as the thin inner portions of the galaxy can be clearly glimpsed in a 6-inch scope even though the total magnitude of the galaxy is a pale 11.7 visually. In addition, I can see a short extension to the right (east) of center in the galaxy as well, but only in 8-inch and larger scopes This galaxy CAN be seen in a 4-inch if the skies are very deep and moonless. Expect little or no detail in small telescopes other than a "pencil-shaped" elongation that appears to run NE to SW.

Object 10 - A Large and Bright Globular Cluster - Messier 15
It's never fair to compare globular clusters to one-another, particularly all the others to the likes of Messier 13 in Hercules or Omega Centauri in our southern skies. Nonetheless, Messier 15 can hold its own in very beautiful and rich clusters, probably among the BEST five in the sky. Having revealed over 140 variable stars in this very distant (39,000 light years!) globular, M-15 is a very bright (5.9) and large (7.4' arc) object for our telescopes. However, it is pale by comparison to the larger Messier 13 which stretches nearly TWICE the span of M-15.

In small telescopes, the object is unmistakable, appearing as a "coarse" and large oval mass, very distinct against the background stars (actually MOST of the stars that you will see in your field of view are CLOSER than the globular cluster!). Moving up to a 4-inch, the outer "rim" stars can be distinctly seen and perhaps in this scope can best be seen a peculiar nature of this cluster...NOT seen in the larger and brighter counterparts. Messier 15 is more elliptical than spherical, and this oval shape is unmistakable in a 3 to 4-inch at medium (about 75x) powers. Because the stars become more and more resolved with increased aperture, larger telescopes do not exhibit this elliptical shape as readily. Expect many stars to be seen in an 8-inch, with a very high concentration (more than with Messier 13) of stars near the center 1/3 of the cluster.

WANDERING ABOUT....YOUR NEW "USER OBJECT" IN PEGASUS

Here is a very interesting QUADRUPLE star for your Autostar library and one that you would likely overlook and never pursue if you could not GO TO it directly. The star is known as "**STRUVE 2879**" and is a wonderfully bright and easy star for our telescopes (consult with my diagram below for the actual positions of these stars in your telescope). Note that ALL of the four stars will be easy (under dark sky conditions) in the 6-inch scope and larger.....THREE of them with a 3-inch and at least TWO with smaller telescopes. We will call the primary (or brightest) star "A" with its magnitude of a relatively bright 6.4 (can be seen in the 8 x 21 finder and above).

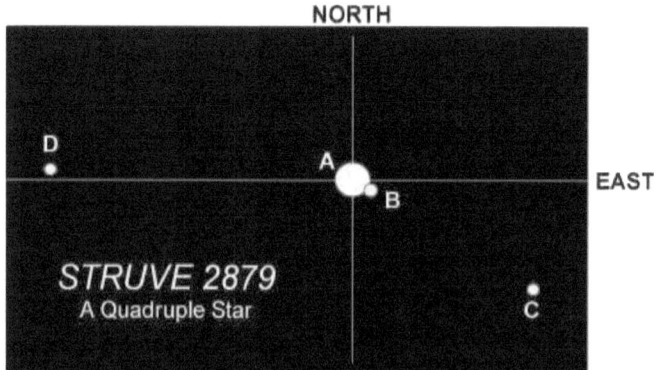

The first companion star "B" is magnitude 8.0 and is the most difficult for all but a 6-inch and larger to locate, right at the limit of resolution for a 3-inch -

1.4" arc. Look for this star using fairly high magnification (about 40x per inch aperture) due SE of "A".

The second star, "C", will be found from "A" at exactly the same angle (SE) except it will be much farther and dimmer than star "B".....about 66" arc, or slightly larger than Jupiter would appear in the same eyepiece, but definitely right in line with stars A & B! At magnitude 9.5, this star will definitely be seen in ALL telescopes, even at medium magnification.

Now look for star "D", magnitude a faint 11.0, about TWICE the distance that "C" is from "A" (124" arc D to A) nearly DUE WEST of "A". This star will NOT be visible in the ETX 60 or 70 and will present somewhat of a test for the ETX 90 as well, although it should be an easy object due west of the primary star in 6-inch and larger telescopes.

On AutoStar, or your computer sky library, go to: "Select/Object [enter]...." scroll down to "User Object" [enter]. Now enter the coordinates given above for "Struve2879", using the number keys on AutoStar. After entering the coordinates and pressing "Enter" yet again, scroll down one and you can list the magnitude of the object as "6"[Enter].

Now, in addition to black holes, quasars, novae, places of historical interest and other curious and weird among the celestial showcase....you have a quadruple star in your USER OBJECT DATABASE! All of these are wonderful for conversation-starters and crowd-stoppers at the big astronomical events!

The sky is always better shared.
Computerized telescopes allow all of us to be experts in not only finding objects quickly in the night sky, but also provides for instant acquisition of information on the objects we observe.

Photo courtesy Celestron

Chapter 27

PERSEUS
A Gallant Hero and Rescuer of Fair Maidens

Perhaps one of the "top ten" most remembered and/or recognized constellations - PERSEUS - is our next Constellation guide for the modern GO TO computerized telescope. To almost all who see this familiar star pattern - just as it was thousands of years ago - it is nearly immediately visually "paired" with the conspicuous "W"-shaped asterism of *Cassiopeia*, the Queen to its west.

Likewise, it is hard to have studied either of them and not remember that - located midway between their brightest stars - the famous "Double Cluster" is positioned as perhaps the finest galactic cluster(s) visible to the naked eye and binoculars. Undoubtedly, there are those that would argue that the *Pleiades* (Messier 45 in Taurus and our second closest star cluster) rivals any such cluster due to its beautiful bright seven-to-nine naked eye stars. However...there is a silent mystery about the double cluster that you do not find with something so conspicuous as the Pleiades: it hints at so much you cannot see to the naked eye. It literally INVITES you to explore further, to use optical aid, to increase light gathering and magnification and hence probe deeper into its core and see fainter stars. The location of the Double Cluster amidst the generously thick groupings of distant stars of the Milky Way adds to the beauty of the star fields throughout this region.

The star chart below for your use in the "GO TO"

TOUR shows the connecting "lines" of the brighter stars; however, since the 1970's (we all DO remember the 1970's do we not?) the star pattern appears to the naked eye to trace out a popular figure that originated in those ancient times known as "the Keep-on-Truckin' Man.", a happy little cartoon fellow walking rapidly to his left, arms outstretched and in full stride.

Even today, you will see the "Keep-on-Truckin' Man" cartoon character emblazoned on 18-wheeler mud flaps and on commercial truck decals throughout the United States and Australia.

The constellation of Perseus is located in a wonderful part of the sky for two reasons. First, as it rises higher and higher in the eastern sky beginning in late summer, it beckons dusty telescopes and the "night-allergic" among us outdoors as crisp, cool fall skies slowly replace the humidity- and mosquito-laden skies of summer. Second, this portion of our night skies - regardless of the season - is one of the most intriguing from the standpoint of such a wonderful array of deep sky objects and fantastic double/multiple stars.

Because of the positioning of our solar system in the Milky Way galaxy you will note that Perseus, much like its nearby neighboring constellations, is void of galaxies bright enough to be observed in common amateur telescopes; look for those in the spring and early summer skies....in the fall, we enter the realm of the GALACTIC CLUSTERS. Hence, as with the constellation of Auriga, I am adding as a bonus at the end of the Concise Listing of Perseus objects for this TOUR an abbreviated directory of all galactic clusters that can be viewed in small to large amateur telescopes.

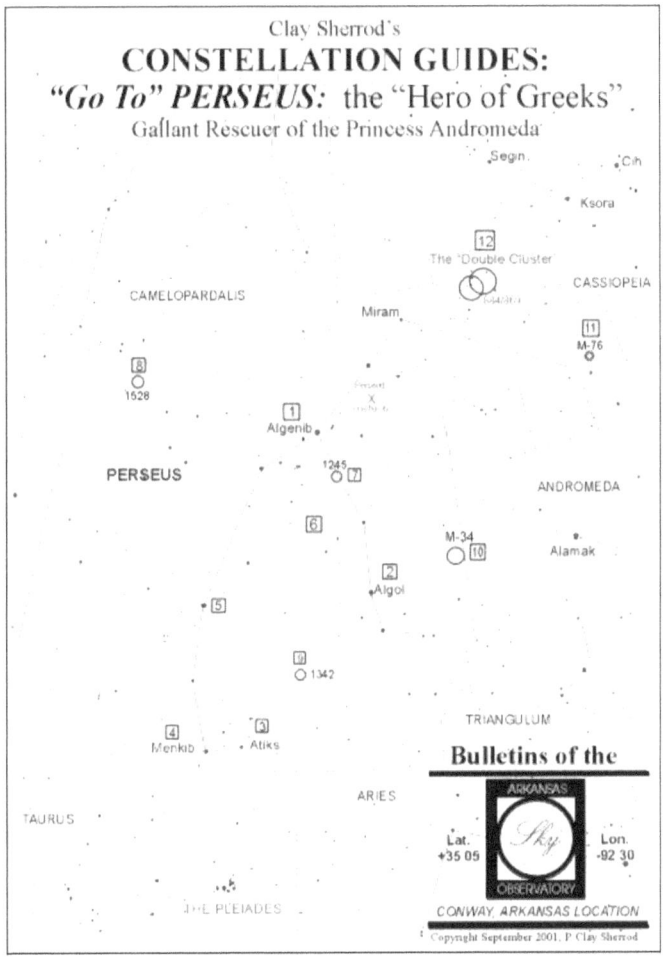

Take the time to explore this wonderful area with the naked eye and with binoculars or a very low power, wide field telescope. From the chart above you note that Perseus is in an enviable position in the sky. To its west are the wonderful object-rich boundaries of Cassiopeia and Andromeda, to its east is the charioteer Auriga, also filled with wonderful galactic clusters and to its south is Taurus, the bull, with its two brightest and closest star clusters, the Hyades and the Pleiades.

Indeed, even scanning around the bright primary star of the constellation, *Algenib*, you will see clearly in binoculars that it, too, is associated with a very large and bright scattering of cluster stars as part of the Perseus Milky Way.....examine this area near Algenib on the darkest of nights with wide field telescopes and/or good 7 x 50 binoculars for a sight you will remember for the rest of your life.

From mid-northern latitudes Perseus just dips below the extreme northern horizon right at midnight on May 15 each year, the only time at midnight that all stars of the constellation are hidden from nighttime view. Conversely, midnight culmination of its brightest star, *Mirfak*, when the star is highest overhead at midnight always towards the end of the first week in November every year.

Within the borders of Perseus are 12 brighter galactic stars clusters that are of interest to amateur astronomers, all of which are visible in the widest range of telescope sizes, and certainly some even in binoculars. Only ONE of these clusters is designated as a Messier Object, although several others could just as easily be so assigned. I have always been amazed that the Double Cluster(s) was not so designated by **Messier**, and that these fantastic twin clusters remained unnamed throughout history except much later as an "NGC" object. That is very much like in horse races.....they apparently "Also Ran..."

In addition to these "fixed" celestial objects which grace Perseus, there is another celestial treat that we experience once each year from within its borders....the **PERSEID METEOR SHOWER**, which emanates only four degrees south of *GAMMA PERSEI* (more on this later!).

Perseus (pronounced EITHER "purr-SEE-us" OR "purr-SEUS", depending on your upbringing) is a mythological giant among tall tales. An explanation for the alternative second pronunciation follows.

MYTHOLOGICAL LORE #1 - The Conception and Eventful Life of Perseus as a Hero The mighty mythological kingdom of Argos (from which came *Jason* and his famed "Argonauts") was ruled by *King Acrisius* who had but one daughter, the most beautiful of all the kingdom. However, being warned by a soothsayer that his own grandson will kill him later in life, Acrisius decided to hide *Danae* away so that the possibilities of her having any children would be impossible. As legend has it, the high god ZEUS himself however, took a liking to the beauty of Danae that he could see through the chamber walls when no mortal man could. Well, one thing led to another and eventually Danae conceived a son by Zeus - PERSEUS (or "per-Zeus"...the latter pronunciation). Alarmed at the sudden realization that he NOW had a grandson, Acrisius sent both Danae and her son far away to *Seriphos*, a beautiful but isolated island ruled by the *King Polydectes* who - you guessed it - fell in love with Danae. To have his undivided attention from her, the king sent Perseus packing away on many death-defying and perilous missions. The most memorable (boy, does this sound like afternoon

TV?) was to go out and return with the head of *Medusa*, the horrible snake-headed Gorgon who turned mighty men to stone by merely glancing at her horrible face and figure. He ultimately tricked Medusa by approaching her with his shield to his eyes so that he could not look and did, indeed, cut off her head! But, unlike daytime drama television, a winged HORSE (Pegasus) sprang from Medusa's headless body which was quickly mounted and tamed by Perseus to return him back to Seriphos. But.....on his way back, he hear the horrible screams that could only be coming from a woman tethered to a bunch of cragged rocks next to a sea shore, (probably a reference to Gibraltar of Cadiz, or if this tale is even older, it is likely a reference to the huge outcropping rocks overlooking the Red Sea from the coast of Israel) a woman who no doubt was about to be eaten alive by the dreaded sea monster Cetus (now a docile whale). Sure enough, he was RIGHT! He and Pegasus swooped down just in time to rescue the fair maiden *ANDROMEDA* and ultimately would become her princely husband.

So did he kill his grandfather? Yep...hit him in the head accidentally while practicing his celebrated discus throw for an upcoming sporting exhibition at a funeral, thus keeping the old soothsayer an honest man.

MYTHOLOGICAL LORE #2 - [short and "G-rated" version] - The Philandering Heroics of Perseus, to keep it short, the part about the rescue of Andromeda who had been chained to the cragged rocks by the sea shore in wait of the horrible Cetus sea monster is the same. It is how Andromeda came to be chained there in the first place, and just what relationship Perseus ALREADY had with the family of Andromeda, notably her father *KING CEPHEUS* and mother *QUEEN CASSIOPEIA* who was suspected of being "romantically attracted" to the young and charming upstart PERSEUS.

Cassiopeia was an arrogant and vain woman, who acclaimed her beauty far better than that of the romantically-toxic sea nymphs, among them the Gorgon Medusa who ask King Neptune (or Poseidon) to punish the Queen for her excessive vanity and arrogance. Hence, Neptune summoned the great sea monster CETUS to ravage the kingdom of Cepheus an order that was quickly learned back home by Cepheus through his soothsayer sage. According the wise sage, the only way to prevent the fall of the kingdom would be for Cepheus to sacrifice his ONLY daughter, Andromeda, to the menacing monster; hence she was agreeably chained to a remote cragged bunch of rocks in await of the monster.

Because, apparently, from her attraction to Perseus and the fact that Perseus was "really" in love with Andromeda, not her mother, Cassiopeia had no objections to this decision (at least there is no mention of any pleas of compassion anywhere....).

Nonetheless, Perseus heard Andromeda's faint whimpers and swooped down just in time to reflect Medusa's ugly image, snakes and all, onto the awaiting eyes of Cetus who immediately turned to stone.

....and they all lived happily every after (except

Cassiopeia, who was still stuck with Cepheus). I told you it was the abbreviated and G-rated version.

*   *   *

As with every "GO TO" tour guide, each GO TO object Perseus is discussed for your telescope regarding the type of conditions necessary for you to view it optimally for discern the very faintest details.........magnifications and aperture necessary for most objects, and much, much more. This is YOUR complete guide to get you on your way to exploring the best (and few!) objects in these two constellations. The following listing of "BEST" objects contains the finest or most interesting from my own observing experience and preference.

Use the attached star chart and the following Guide as an excellent reference for your next star party itinerary, or a beginning for further study into the thousands of objects visible in this part of the sky. Truly these extensive Constellation Study Guides will most definitely put your AutoStar or PC star program to work for you in the most efficient and enjoyable way possible! As a matter of fact, MANY telescope and GO TO users are now programming their own "Tours" based on these guides, using each constellation as a separate GO TO Tour for the sky computer library that can be added in or deleted through the main edit screen on your PC or MAC computer.

We hope you enjoy these comprehensive GUIDES to touring the constellations via yourcomputer-driven telescope. Each new installment is complete with diagrams, charts and illustrations that you will find nowhere else. Please let us hear YOUR feedback and your observations of each and every constellation after YOU have toured its vast reaches of our skies!

YOUR PERSEUS CONCISE DIRECTORY OF INTERESTING OBJECTS –

In addition to our regular listing of a few selected objects, I have included the complete abstract listing for twelve (12) nice galactic star clusters that are viewable in this constellation via modest instruments; that listing will be found at the end of this concise description of our Perseus tour.

As we explore the realm of the home of our Greek hero, I have chosen the finest (or most interesting) **12 objects** in this PERSEUS "GO TO" tour (as with all guide, all objects listed below will be visible in most telescopes (some naked eye) from the smaller apertures; of course larger apertures may "show" an object a bit closer and "better," but frequently a wide field and low power view is more desirable than aperture for FINDING the objects initially. Indeed, I strongly encourage you first FIND the target object, or its approximate location through your GO TO function with your lowest power and then - once IDENTIFIED positively - move up slowly in steps with magnification if necessary. Remember, not all objects "like" magnification. Sometimes better "field of view" (such as given with smaller telescopes) is desired over light gathering (like an 8-inch) and magnification.

The rule for determining "optimum magnification" is that: 1) too low power results in sky background glow detracting or diminishing the contrast against the deep sky object; 2) too high magnification darkens BOTH the sky background AND the object; 3) medium magnification can be achieved at which you have MAXIMUM contrast between the object and its darkened background sky. I have found through three decades of direct observing that about 15x per inch aperture, for deep sky observing is PERFECT for most objects. That being said, always remember that DOUBLE or multiple stars require whatever power you can crank out....the seeing conditions are the limiting factor here. For a complete discussing on magnification and how it applies to YOUR telescope, visit my review under the GUIDES section of this website.

For my complete and comprehensive discussion regarding seeing conditions and sky transparency, see the ASO/Guides tab at www.arksky.org .

With all deep sky objects, avoid attempting to observe when the moon is in the sky, even a very thin crescent, as its brightness in the sky will overshadow the very dim contrast afforded by even the brightest deep sky object; if you see the object at all against moonlight, you will NOT see the subtle outlying areas or the full detail of what is presented.

Also, as I always suggest with all of the "GO TO" tour constellation lists, a good star atlas and/or chart which will list all the finest objects, constellation-by-constellation. One very handy reference guide is the *PETERSON FIELD GUIDE TO THE STARS AND PLANETS*, which features complete lists with declinations, right ascensions, magnitudes, and all pertinent information for you to expand your observing horizons beyond this brief

GUIDE. For multiple stars and many listings of the finest deep sky objects, the classic work, *Burnham's Celestial Handbook*, Vol.. 3 is highly recommended.

The constellation tour Star Chart above (click on and save to a file on your PC; then open it and resize to fit the page and print for a very handy at-the-scope star chart) will get you started on your journey for this constellation.

Following is the concise object list for your "GO TO" tour of Perseus; you may wish to find the majority of the objects from the computer sky library (for example, you can easily go to the beautiful and rich "Double Cluster", NGC #'s 869 & 884, if you pull up "Object/Deep Sky/NGC/..then type in '869'...." and then press "Enter", followed by "GO TO" to access this famous and rewarding wonder of our nighttime skies. On the other hand, if you want to experiment and become a "better computeruser" try entering the exact R.A. and DEC coordinates (given in the listing below) of that same object as described by holding down the MODE key. You will find the accuracy of entered GO TO's to be somewhat less than those stored in AutoStar, but the capability of acquiring unlisted objects is fantastic!

Other sky program and Apps work similarly to the above described; consult your User Guide or HELP for details.

In addition, remember that you can GO TO this object DIRECTLY as well (as with many of the more recognized "named" deep sky wonders.

Merely key in SELECT / OBJECT / DEEP SKY / NAMED....and scroll down alphabetically to "Double Cluster." Press "enter" and then "GO TO" and the telescope will take you there! Pressing "enter" again will continue to provide you with much valuable information about this object and any object that you have selected via the keypad.

You will access your FIRST GOTO target - (usually the brightest star in each constellation) - via the command "SETUP / OBJECT / STAR / NAMED....and scroll to "**Algenib**"", then press "Enter" and subsequently "GO TO" to move your this bright star. NOTE: the star Algenib goes by many names, but is entered into most sky program libraries under that name. Remember also that many distinctive objects are sometimes listed among the "named" objects. So, likewise for that object you might merely go to SETUP /OBJECT/ DEEP SKY/ NAMED and then scroll alphabetically to the "common" name of the object if you are not already there; press "enter" and then GO TO and your scope is off and running!

You may also access the constellation by: SETUP/OBJECT/CONSTELLATION/"Perseus"..... Enter....GO TO, which will take you close to the central position of the constellation's boundaries.

OBJECT 1: bright star - *ALGENIB* (alpha Persei) - R.A. 03h 21' / DEC + 49 41 - Magnitude: 1.8 - nice lavender star!
OBJECT 2: demon star! - *ALGOL* - (beta Persei) - R.A. 03h 05' / DEC + 40 46 - Mag. 2.1 to 3.4 in 2.9 days!

OBJECT 3: double star - *ATIKS* (omicron Per) - R.A. 03h 41' / DEC + 32 08 - Mags: 4 & 8.5 - challenging!
OBJECT 4: try this double - *MENKIB* (zeta Per) - R.A. 03h 51' / DEC + 31 44 - Mags: 2.9 & 9.4 - a toughie!
OBJECT 5: nice easy double - epsilon Per - R.A. 03h 55' / DEC + 39 52 - Mags: 2.9 & 8.1 - wide, fun double
OBJECT 6: Nova Persei - GK Per - R.A. 03h 28' / DEC + 43 44 - Mags: 10.9 to 14....odd variable
OBJECT 7: galactic cluster - ngc1245 - R.A. 03h 11' / DEC + 47 03 - large, good for wide fields, low power
OBJECT 8: galactic cluster - ngc1528 - R.A. 04h 11' / DEC + 51 07 - Mag: 6.2 , 80 stars - NICE, large!
OBJECT 9: galactic cluster - ngc1342 - R.A. 03h 28' / DEC + 37 09 - Mag: 7.1, 40 stars, pretty cluster, low power
OBJECT 10: galactic cluster - Messier 34 (ngc1039) - R.A. 02h 39' / DEC + 42 34 - Mag. 5.5, 80 stars (fine!)
OBJECT 11: "the Little Dumbbell" - Messier 76 (ngc650) - R.A. 01h 39' / DEC + 51 19 - Mag. 11.8 - nice planetary!
OBJECT 12: the "*DOUBLE CLUSTER*" – ngc869/884 - R.A. 02h 17' / DEC + 56 54 - fantastic and memorable sight!

** Plus: <u>GALACTIC CLUSTERS IN PERSEUS</u>:** Following is an abbreviated listing of the twelve (12) brightest and most interesting open, or galactic, clusters in the constellation of Perseus; this area of the sky if VERY rich in wonderful star fields for scanning with the very low power and wide field instruments as well as a standard pair of 7 x 50 or 10 x 50 binoculars, since it is located just on the edge of the dark and rich winter Milky Way skies. The listing following gives the NGC #, the R.A. and DEC of the object, the Magnitude, Size, # of stars and a brief and very concise description and/or notes as necessary:

ngc RA DEC (+) MAG SIZE ('arc) #stars & description

0744 04 47 55 14 9.4 11 25 -
very distant and faint, medium sized, stars faint
0869 05 17 56 55 4.4 36 350 -
"hPer", NW of the two "double clusters"
0884 05 22 56 53 4.7 36 300 -
"xPer", the companion of ngc869....fantastic!
0957 05 25 57 18 7.2 10 40 -
compact and fairly bright, stars about 10th mag.
1039 05 25 42 34 5.5 18 80 -
M34, see description which follows
1220 05 32 53 10 11.8 02 10 -
tiny and VERY faint, only 10 stars, about 12th mag.
1245 05 49 47 03 6.9 30 40 -
very large, sparse but brighter stars, low power!
1342 05 58 37 09 7.1 15 40 -
nice, fairly compact scattering of 9-11 mag. Stars
1444 06 11 52 31 6.4 04 15 –
tiny, but fairly distinct; use medium mag. for this one
1513 06 46 49 23 8.8 12 40 -
not so bright, but many stars about 10-11 mag.
1528 04 11 51 07 6.2 25 80 –
very nice, often overlooked, fairly bright, low power!

1545   04 17   50 05   8.9   18   25 -
faint, and very sparse sprinkling of stars

A VISUAL GUIDE TO OUR DEEP SKY OBJECTS IN PERSEUS

Object 1 - Our "Starting" Bright Star - "ALGENIB" Also known as "Al Genib", "Al Chemb," "Mirfak" (most common), and "Marfak," alpha Persei has gone by many names - all Arabic - over the centuries. Although Autostar and other popular GO TO hand control libraries refer to this star as "ALGENIB", this name is shared with other stars of the sky, notably and most confusingly Gamma Persei and also nearby Gamma Pegasi!! The most common association with this star by far is "Mirfak," which translates as "elbow of the Hero," or "Perseus Elbow." This huge and bright (4,200 more than our own sun!) star is 570 light years distant, making it one of the most remote of all 1st magnitude (it is actually magnitude 1.8) stars. This star is surrounded by an absolutely beautiful field of stars for the casual scanning with binocular or small richest field telescope. Wide field views reveal thousands of stars surrounding Algenib, particularly a group of fairly bright stars immediately surrounding the bright star for as much as 2 degrees. This is a true physical cluster with nearly 110 stars appearing to be gravitationally held along with Algenib; this cluster will be overlooked in the telescope; use either a very wide field, low power richest field scope or binoculars......you will then see this often missed cluster, one of the closest to our solar system!

Object 2 - *THE DEMON STAR* - "*Algol*"

You can be the judge on which version/story led to the actual naming of this star. Algol is one of the most famous of all variable star, and is the ONLY variable to have its own "home" on your Autostar whereas you can actually determine the upcoming minimum brightness of this star by going to SETUP / SELECT / EVENT / [scroll down to "Minimum of Algol and press "enter"]...

The star is so precisely synchronized in its variation (from magnitude 2.13 to 3.40 in exactly 2.86739 days) that you can actually SET YOUR ACCURATE WATCH by this star! However, if you want even MORE accuracy, slew to R.A. 04h 44m / DEC + 36d 38m to the variable "AW Persei" nearby, which varies from magnitude 7.96 to 9.08 is exactly (and I mean EXACTLY!) 2.02873293 days! Both stars are "eclipsing binaries" and are discussed in my article on Algol under the GUIDES tab of this website.

The actual common name in use today comes from the Arabic "Al Ras al Ghul", or the "*head of the demon,*" referring to Medusa's head (see explanation above) as Perseus is holding in his left hand. Does this reference then, elude to the "*Demon Star*" from such a designation? One would think so, unless further ancient mythology is revealed. The Hebrews referred to this star similarly, but having NOTHING to do with Medusa nor Perseus as a hero....it was "Rosh ha Satan" or "*the Head of Satan*" to their scholars. This designation has NOTHING to do with wicked women...only the fact that this star to the naked eye DOES change its appearance, among thousands of stars well known to the ancients that DID NOT vary at all....so something must be "different," or "evil" about this flickering heavenly light!

So, perhaps the name comes from its variability, or from its association with the wicked Gorgon Medusa....or both. It is yet another interesting similarity in ancient star lore by two totally separated cultures and time, yet the ultimate translation results in great similarity. I have discussed this curiosity of the human mind and association with "the Great Bear" of Ursa Major, being so associated by both the Greeks and the Chinese, through wide spans of time, though no contact was apparently made between the two people to share such knowledge.

At three million miles in diameter, this star (the larger primary) is about three times larger than our own sun and is located fairly close to us, at about 100 light years.

Object 3 - Double Star "ATIKS" - Omicron Persei - A challenging object
This is a very difficult double star for the 5" or 6" telescope even though the separation is a full 1.0" arc; the primary star is magnitude 4.0, while the "B" star is much dimmer, only magnitude 8.5. Also known as 38 Persei, the primary ("A") component is a very slight variable although it will not be noticeable to the visual observer. Look for the fainter companion star nearly northeast of the brighter star. This is not as easy as you might think even in an 8" telescope!

Object 4 - Another Good Double - Zeta Persei - MINKIB (also home of the "California Nebula")

This area gives you double for the money. Zeta Persei, a nice and colorful double star for most telescopes, a rapidly expanding group of associated stars and the "California Nebula," ngc1499, a 2.5-degree streak of nebulosity to the immediate north of Zeta. For reference, Zeta marks the "foot" of Perseus. Zeta, along with omicron, xi, chi, 40Per, 42Per and about half-dozen more stars are all moving rapidly AWAY from a common center just north of Zeta which most likely was their place of "birth." Indeed, it is thought that the California nebula is a remnant of the large cloud of primordial gases and dust that eventually led to the formation of these - and perhaps other - stars. This group, and their trajectories, I have plotted on the following Lowell Observatory photograph.

THE ZETA PERSEI STAR ASSOCIATION
Arrows indicate direction of travel from common center (?)

Zeta is a nice double star, magnitude 2.8 for Zeta with a magnitude 9.4 companion a full 13" arc nearly due south of the brighter star. This should be an easy object for 3" and larger scopes with medium magnification, but a bit tougher for smaller apertures due to the great difference in brightness between the two stars. Look for some nice color contrast here, with Zeta appearing almost green while the fainter star - with medium to high magnification - appearing pretty much dull blue.

NGC 1499 is immediately north of Zeta about 5 degrees; this is a huge (nearly 3 degrees E-W) cloud and very difficult to make out visually; piggyback photography reveals it quite easily on red-sensitive film. A good "nebula filter" on a very dark night with the widest possible field your scope can offer "amy" reveal this 4th magnitude (but spread out!) cloud of primordial gases.

Object 5 - A Fine Double Star - Epsilon Persei
Here is an excellent, but not so easy double star with a wide magnitude difference from the primary (mag. 2.9) to the secondary star (mag. 8.1). Look for the fainter star nearly 10" arc away almost due NORTH of the brighter star. Look for the striking color differences as well in these two stars. Most observers see Epsilon (the brighter one) as a blue-yellow color, while the fainter companion is almost "gray" or a steel blue color. Medium, to medium-high magnification is required even with such a great separation, this due to the wide difference in brightness between the two components.

Object 6 - An Unusual Old Nova that is "Still Around!" - Variable Star GK Persei
This is a very interesting star to observe and to at least say "I have seen it...." Observe this area

ONLY in medium magnifications for best views, but to observe the light fluctuations and monitor them carefully of GK itself, you will need to use at least 175x to 200x for accuracy (and probably to even locate the star!). Let's start with the star itself and the Nova of 1901. This now-faint star reached a magnitude that rivaled all but the fifth brightest stars of the sky, magnitude 0.2 on February 23 of that year. Only two days prior it was at magnitude 2 or 3, skyrocketed to brilliance and then fell to below naked eye view just as quickly. However, it was monitored and verified by astronomers around the world, with Perseus high in the winter sky. Using detail sky charts supplied by the American Association of Variable Star Observers (AAVSO), this star can be monitored nightly (and should be!) with users of 5" and 8" telescopes as a very faint 13th magnitude rapidly fluctuating star. Without warning or any hints of activity however, it can suddenly surge to 10th magnitude for a brief period and then subside once more.

To locate and monitor this chart, merely go to: https://www.aavso.org/apps/vsp/ . Note for these charts, simply type in the NAME of the variable at top to generate your choice of chart.

This is both the low and the high power, narrow field "a & g" chart that shows stars to magnitude 13.9 very close to GK Per. It will assist you in finding this faint star among the others and alsogive you actual confirmed magnitudes of others star with which to compare. NOTE that this is the "reversed" chart made specially for users of catadioptic telescopes such as our GO TO scopes, and thus NORTH is at the top and EAST is to the right, just the way it appears in the eyepieces of Schmidt Cassegrain and Maksutov telescopes.

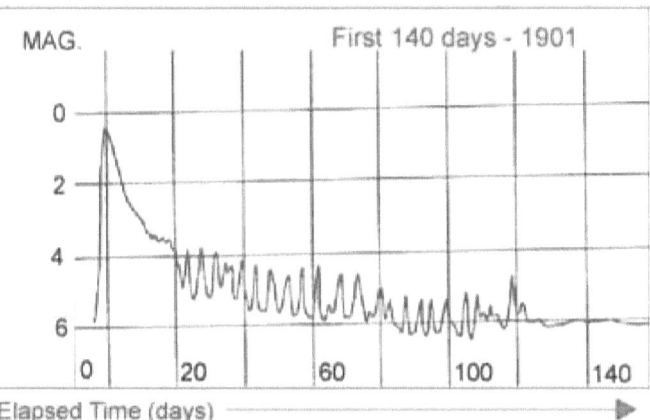

GK PERSEI- Nova Persei 1901 - Still a very unusual and active variable

Note in the light curve from 1901, adapted from that compiled at the Harvard College Observatory in Cambridge, Mass, the star's incredibly rapid rise to maximum, followed by an equally impressive demise. The fluctuations around 6th magnitude continued throughout the remainder of 1901 after which the star slowly faded to its present-day 13th magnitude; HOWEVER....the fluctuations we see today are almost exactly as you see on this chart around the 6th magnitude range, except the star is 7 magnitudes dimmer now, curious RIGHT ABOUT the magnitude the star was PRIOR to this remarkable outburst that placed it 210,000 times more luminous than our own sun! This star deserves to be monitored by observers in telescopes 5-6" inches and larger on a regular basis, with all observations of activity reported to: www.aavso.org. For astrophotographers with large telescopes, there is some nebulae (planetary nebula gas residue) that now surround the remains of this famous nova.

Object 7 - A Nice Galactic Cluster - NGC1245

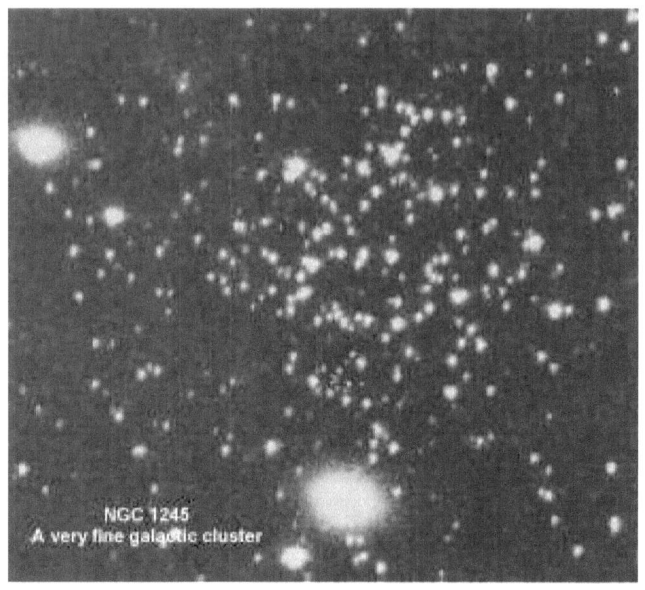

Here is a nice cluster that is easy to locate even small telescopes directly south of Alpha Persei by about 3 degrees. It contains nearly 100 stars of nearly equal brightness as can be seen in the photograph above from the 13" telescope at Lowell Observatory that was used in the discovery of the planet Pluto! It is a very rich cluster and quite large, almost identical in sky area to the disk of the moon (30' arc). The 5" scope will clearly show many of the individual stars, but the 8" will reveal its true beauty. Since the cluster is nearly 1/2 degree, wide field and very low power views are desirable, something that limits the ability to see the faintest stars except in the larger apertures. It appears as a nice round "smudge" in smaller instruments, but is located in a VERY rich and rewarding part of the Perseus Milky Way near Algenib (see above).

Object 8 - Another Nice Fall Cluster - NGC 1528 & 1513

Another "double for your money object", when searching out ngc1528, you can also spend some time exploring the slightly fainter and 1/2 size ngc1513 to its SW by only 2 degrees! NGC 1528 is a very large (25' arc) open cluster with about 80 stars that are fairly bright 8 to 9th magnitude, thus making this an exceptional cluster for small telescopes and even binoculars. When viewing telescopically, very low, wide field views are a must to get the true "cluster effect" realized. This is one of those clusters that reveal more beauty to small, richest field instruments than to larger "big dog" telescopes, so "...this one's for YOU," small scope users! On the other hand, larger scopes will greatly appreciate their increased aperture when it comes to exploring the exciting ngc1513....look for some 44 stars grouped in an area about 12' arc, half the size of ngc1528, and all about an equal magnitude 11. This will appear like sprinkling sands in sunlight with the 8" at about 80x to 120x magnification. The cluster IS resolvable in any 3-inch, though with difficulty and requiring very dark skies.

Object 9 - Yet Another Perseus Galactic Cluster – NGC 1342

Unlike the previous two clusters, ngc 1528 and 1513 which are very symmetrical and uniformly shaped, ngc1342 - located midway from Algol to the "foot" of Perseus (zeta Per) - is an irregular galactic cluster. It is quite large (half the size of the moon's apparent disk, 15' arc) with some 50 stars scattered loosely within this area. Most stars are of 8th and 9th magnitude and this cluster, being in an uncharacteristically star-poor area of central Perseus, is a easy target for even small scopes, with its total integrated magnitude of 7.1 visually.

Object 10 - Messier 34 - A Fine Galactic Cluster in Perseus **Pair of Double Stars as a Bonus!** Don't get me wrong, as I still consider M-34 a fine cluster, with a total magnitude of 5.5 containing over 80 stars packed into a large 18' arc area of sky, right on the western border of Perseus, nearly into eastern Andromeda. However, I still marvel at the fact that this object - which is barely visible in a finderscope and not attainable in today's skies with the naked eye - was warranted a listing by **Charles Messier** in his famous group of objects, yet the much BRIGHTER and LARGER ngc869 and ngc884 (the two clusters of the "double cluster", below) were ignored in his listing! Look a full binocular field northwest from Algol for this very bright and conspicuous cluster of 8th through 13th magnitude stars, certainly NOT a disappointment by any means (but I am still upset with Mr. Messier and have informed him of that fact).

Always use the lowest power and widest field of view possible in your telescope to view this bright scattering of stars located nearly 1500 light years away. Indeed, the true beauty of the cluster is often lost in larger telescopes and the use of higher magnifications leads to disappointment. However, larger telescope will enjoy the BONUS two double stars that are located in this wonderful object! Look carefully at the photo/chart below: first (north, near the center of M-34) is the twin 8.5 magnitude star pair "h1123", a brilliant white grouping of two stars identical in color and brightness! This is a very easy pair, with a full 20" arc of black sky separating the two...they will appear as two of the brightest stars in this cluster. Note that my chart has NORTH at top and EAST at the right, exactly as the image will appear in the Maksutov or Schmidt Cassegrain telescope with the diagonal mirror/prism in place.

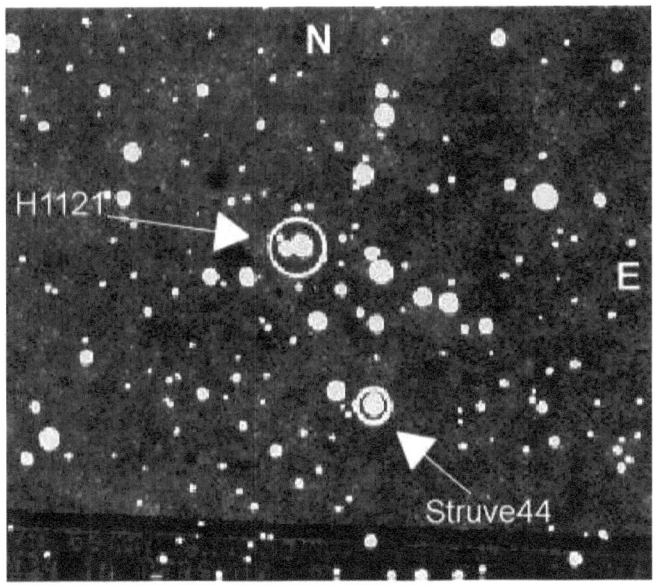

MESSIER 34
and two associated double stars

Look near the cluster's edge, only about 6' SE (lower left in the photo) of the first double for "Struve44", a very close double (1.0" arc) that is a challenge for the ETX 125 and likely beyond the resolution of smaller telescopes. These stars likewise are about 8.2 magnitude, with the fainter one being to the northeast (Position Angle 55 degrees) of the main star.

Objects 11 - The "Little Dumbell" Nebula – Messier 76

This faint planetary nebula is easily located within the same very low power (i.e., a degree field of view) as the brighter star Phi Persei, the westernmost bright star in the constellation and right on the Cassiopeia border. Although this fairly large (157" arc long by 87" arc wide) box-shaped nebula is only assigned a faint magnitude 12.2, I have seen it steadily many times in a 3-inch Questar at 80x. This is a large object, so increasing the power past 30x per inch results in losing it altogether. It is remarkably easy in the 8" scope, with the "dumbbell" shape clearly discernible against a beautifully black star-speckled field of view. I will not argue with the estimate of brightness, but it has always been my experience that this object is brighter than the assigned 12th magnitude. At any rate, it is a challenge and an object certainly worthy of locating.

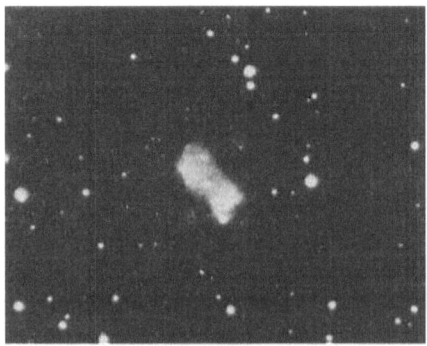

MESSIER 76
The "Little Dumbbell" Planetary

Objects 12 - NGC 869 and NGC 884 - the "DOUBLE CLUSTER"

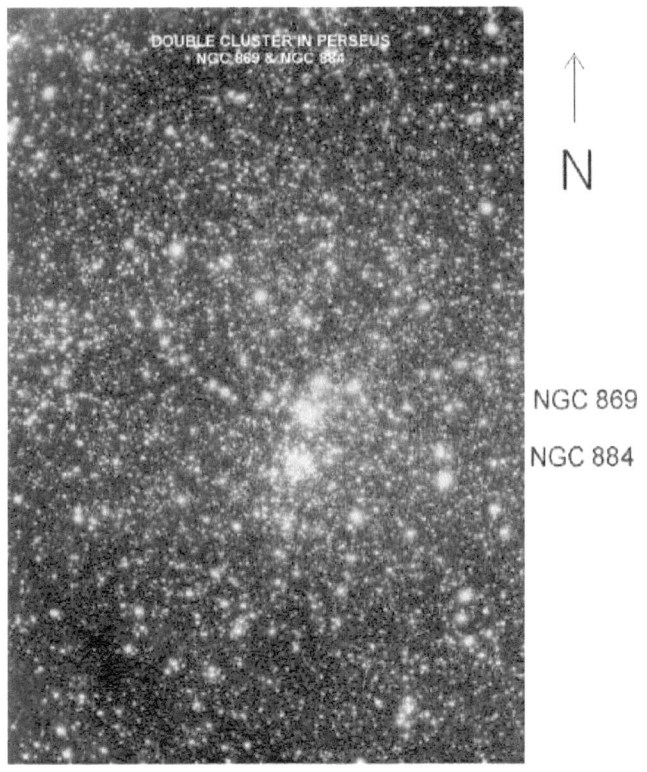

This is undoubtedly the finest star cluster in the sky for small telescopes; it is richer in stars (a total of 650+ in both clusters!) than the beautiful "*Wild Duck*" cluster, Messier 11; it is far more fascinating telescopically than either M-45 (the Pleiades) or the Beehive in Cancer. It is truly a sight to behold in good binoculars on a dark and moonless night when the hero Perseus glides silently overhead with the "Demon Star" carried behind him....

In the photograph above, north is at the top and clearly shows the appearance of this cluster in wide field telescopes with a limiting magnitude of about 11. This photo was taken on red-sensitive film in 1978 through a 200mm f/3 lens guided piggyback for 20 minutes from the observatory on Petit Jean Mountain.

The double cluster has always acted as a "beacon" of sorts in my astronomical pursuits, its position conveniently plotted midway between Gamma Persei and Epsilon Cassiopeia; on the darkest nights the naked eye can perceive of the wonder of these two jewel boxes of the sky, with sparkles hinting of what must lie in wait of telescopic enhancement. In mythology it is these two groups that signify the actual "handle" of Perseus' great sword with which he slew the Gorgon Medusa.

Use the lowest possible magnification - and a wide field of at LEAST one full degree - to take in the entire expanse of these two magnificent clusters; indeed, even outside of the clusters themselves, the star field of the fall Milky Way is overwhelming and one can spend hours scanning slowly through the nearby neighborhood of these two wonderful galactic star clusters. In reality, these two clusters are fairly far away, at a distance of nearly 7,500 light years. Compare that with other famous clusters (the closest to us listed first):

1. The Hyades - Taurus - 130 light years
2. The Pleiades - Taurus - 450 light years
3. The Beehive - Cancer - 525 light years
4. The Wild Duck - Scutum - 5500 light years
5. The Double Cluster - Perseus - 7500 light years.

There are likely as many as 400 stars in ngc869

while almost the same number (320) are thought to be in nearby ngc884. NGC 869 is the cluster more north and west; the 4th magnitude star "10 Per" is embedded in the extreme southeastern edge of ngc884.

WANDERING ABOUT....YOUR NEW "USER OBJECT" PERSEUS

Now here is a totally different GO TO "User Object" for you....we are not actually going to enter an "object" as you might think it, but rather a "position" that will become very important for you at least one or two nights out of every year. You will thank me on those occasions around August 11-12 each year when the moon is absent from the sky.
In August 1971, when I had the world by the tail and thought I had pretty much seen it all in astronomy, I was taking part in a huge - the first of its kind in Arkansas - public star party that brought literally over 10,000 people to a small hilltop near a community park. There were guest speakers and telescopes from all over the south at this event. Mars was at its closest point since 1956 and the moon had already set for the evening.

There was a small Questar 3" telescope mounted far off from the crowd, the scope we later learned was owned by a prominent banker from Hot Springs, Arkansas. His tiny chrome telescope peered diligently toward the constellation Perseus, which was just coming up in the low northeastern skies.

"Whatcha' looking at?" we asked in our attempts to bring the gentleman more into the spirit of things

from his shadowed hide-away. Without even moving from the eyepiece, he said,
"Perseid Meteors."

Perseid meteors?! Well, we were quick to explain that you did NOT need a telescope to observe meteor showers, that the best way is with the naked eye from the comfortable reclines of a soft lawn chair.

"Nope...." he grinned as he replied. "Take a look."

And there, aimed dead center at R.A. 03h 00m and DEC + 58 degrees 30 minutes, his telescope was locked right onto the RADIANT of the famous Perseid meteor shower which was peaking the next evening. And, yes, he WAS looking at meteors.....like lightening bugs headed right for your windshield on a hot Arkansas night, the faint meteors appeared to be heading right for us, just like "warp 1" throws the stars right into the screen of the U.S.S. Enterprise as Kirk gazes "out there."

I had never seen anything like it and would TODAY not believe it was possible had he not shown that phenomenon to me....debris from the once-glorious comet Swift-Tuttle actually streaming out of its path and vaporizing as it encountered the Earth's atmosphere. I have seen this sight many times after that, but have not seen the gentleman who put me on to this wonderful opportunity since.

The Perseid meteor show radiant is located at coordinates: R.A. 03h 04m and DEC + 58 00m, and can be found marked as the red "X" on this "GO TO" guide star chart for Perseus (above).

On AutoStar, (or use the correct tabs/commands for your particular computer sky system) go to: "Select/Object [enter]...." scroll down to "User Object" [enter]. Now enter the coordinates given above for "Perseids", using the number keys on AutoStar. After entering the coordinates and pressing "Enter" yet again, scroll down one and you can list the magnitude of the object as "1"[Enter].

Perhaps with this very unique addition to your GO TO user objects, YOU can impress some unknowing individual or group at your next August star party! It is truly a sight to behold, and must be caught within one or two days either BEFORE or AFTER the actual maximum; for the best results observing this tiny "opening" of the sky within 6 hours of maximum will reveal a spectacular show!

Next Constellation GO TO" TOUR Installment: TAURUS, the might bull of the sky, with its red eye zeroed in on the sky's great hunter, Orion. We will explore the remarkable Hyades and even more fascinating cluster of the Pleiades, or "Seven Sisters." In addition, we will investigate the object that "started it all"....Messier 1, the Crab Nebula, so faint and fuzzy that even Messier was likely shocked that he found the thing!

*   *   *

"Humanity has the stars in its future, and that future is too important to be lost under the burden of juvenile folly and ignorant superstition...."
Isaac Asimov

Perseus carries the head of slain Medusa
With here hair made of snakes
From Johannes Hevelius

Chapter 28

PISCES
....there's something really fishy about this part of the sky....

This is our computerized telescope Cosntellation guide for the large star pattern known as PISCES....a fish in the sea of stars.

This constellation is a bit different than many others - and largely ignored - because of its huge size, few bright stars, no naked eye benchmarks, and a very limited offering of deep sky objects. it covers an enormous chunk of the southern skies, suitable for observers in both hemispheres and spans from 23 hours to over 2 hours in right ascension and incredibly from + 35 degrees to -05 degrees in declination. Yet within this entire constellation there is NOT ONE galactic cluster, globular, planetary nor diffuse nebula which can be seen visually with amateur instruments.

Of galaxies, there are a total of 18 brighter than 14th magnitude, but only EIGHT (8) suitable for amateur observations and all but ONE of those are limited strictly to the 6-inch and larger telescopes! Thus....there is only ONE true "visual deep sky object" that can be found with moderate amateur telescopes in all of PISCES! But hope can be found if using a modern CCD or video telescope camera.

Maybe now do you understand why this large constellation is largely ignored - other that by "astrologers?"

Now - by "deep sky object" I am referring to non-stellar objects and never include the remarkable double and multiple stars that might be seen. This is one category in which Pisces rules. It has no less than 81 fine multiple stars, the majority of which can be detected with some range of amateur instruments. Of these 81 multiple stars, many are the brightest stars in Pisces, including Alpha itself!

Thus, in addition to a somewhat abbreviated description of the "viewable" galaxies in Pisces with this "GO TO" TOUR, we will explore many of these brighter multiple stars and some which are challenging and, simply, beautiful!

Pisces (pronounced "PIE-seez") is just ONE of seven constellations in this region that has some association with "water," and reveals the fascination of water subjects (perhaps due to arid conditions of the "original" Arabian namers of the stars?) to the earliest stargazers. In addition to the "aquatic" fish, we have the Water Bearer of *AQUARIUS, CETUS* (a whale), *PISCES AUSTRINUS* (southern fish), *CAPRICORN* (the "seat goat"), *DELPHINUS* (the dolphin), and *ERIDANUS* (a celestial river). The importance of seafaring to the earliest people and the need for good navigational targets such as stars provide close to the southern horizons, likely gave rise to this "watery association."

A familiar, yet rather faint "asterism" (conspicuous outline or pattern of stars within a constellation) is the CIRCLET of stars which comprise the "*western fish*" of Pisces. The Circlet is shown in our star chart for Pisces and is comprised of the stars:

Gamma, Kappa, Lambda, Iota, Theta, 19 Piscium and 7 Piscium.

Note that 19 Piscium is discussed as a variable star in our list of "GO TO" object in this chapter.

You will note that many of the southern constellations (the far south constellation of *ARGO* - a "ship" - is truly representative of this) were so referred due to the earliest outreaches of civilizations as they explored the vast southern passageways around the world, using the then-unknown southern stars as their benchmarks for celestial navigation. Since so many of the bright southern star groups were NOT named nor recognized, the sailors merely began referring to them by aquatic and navigational names (i.e, "*SEXTANS*" the sextant) which carried over to astronomers back home who subsequently "assigned" the constellations. (see *Constellations Volume Two* for a brief discussion on how our currently-accepted **88 official constellations** came to be!).

Pisces is ALSO fairly unique for such a large constellation in that it has NO star brighter than nearly magnitude 4.0....indeed, "*Alrisha*", the alpha star (in the Bayer designation of the Greek alphabet) is only "scraping" to be a bit brighter at magnitude 3.94!

Because of this absence of brighter stars, this is a very difficult constellation for most observers to find and learn. The trick is to find the "Water Jar" first – dark moonless night required! – and the rest of the patterns will fall into place. While a great winged horse (Pegasus) flies north of the celestial fishes, the great sea monster (Cetus – Vol. One of Constellations) swims to its south.

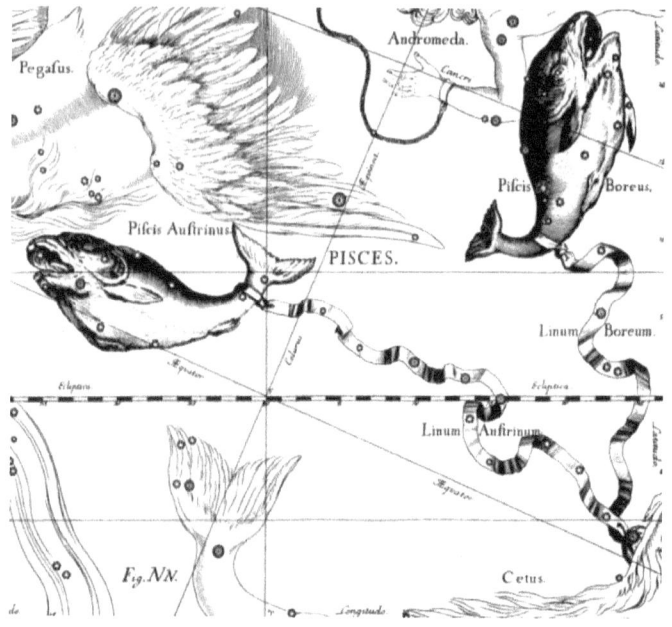

*The fish of the constellation Pisces
As drawn by celestial cartographer J. Hevelius*

Astrologically and to the earliest Arabian stargazers, *Pisces the Fish* was the LAST of the twelve "houses of animals". You will note in our star chart that the Celestial Equator (0 degrees declination) passes almost nearly midway north-south through Pisces and thus objects in this constellation can have either positive or negative declinations. In addition, note that the ECLIPTIC in which the sun, moon and planets appear to move relative to Earth also passes from due east into Pisces and crosses out of the southwest corner. Thus, it is not uncommon for the planets and more frequently the moon to be located in this large constellation for extended periods of time.

Of all the constellation legends, Pisces perhaps has one of the most interesting history.

In the earliest Greek mythology, the constellation represents TWO fish....one of them is a disguised goddess *VENUS* while the other is none-other than the love-providing *CUPID*. Neither, of course, were "fish" in Greek mythology but were forced in legend to assume the fish identity to escape a deadly monster - *TYPHON* - who had brought havoc to the Olympic gods. The great god *Zeus* had rallied the Olympians to defeat their arch enemies the "Titans" who were the sons of the evil goddess *Gaea*. As revenge, Gaea conceived another son, Typhon, who was reared on evil and was taught destruction and sinister ways which he used with supposed supernatural powers and strength to slowly and methodically reduce the power of the Olympians.

To escape, many of the "more important" ("....women and children first!") Olympians learned that if they assumed the characteristics of animals, they would escape notice and destruction by the evil Typhon. Thus, the chosen mode of disguise for Venus and Cupid was....fish - a seemingly odd association with the goddess and the heralder of love....In the beautiful drawing by **Hevelius** shown On the previous you can clearly see the two fish (can anyone tell which is Venus and which is Cupid?) as well as the "cord" [see **Object 1** for Pisces] that attaches tethers the two.

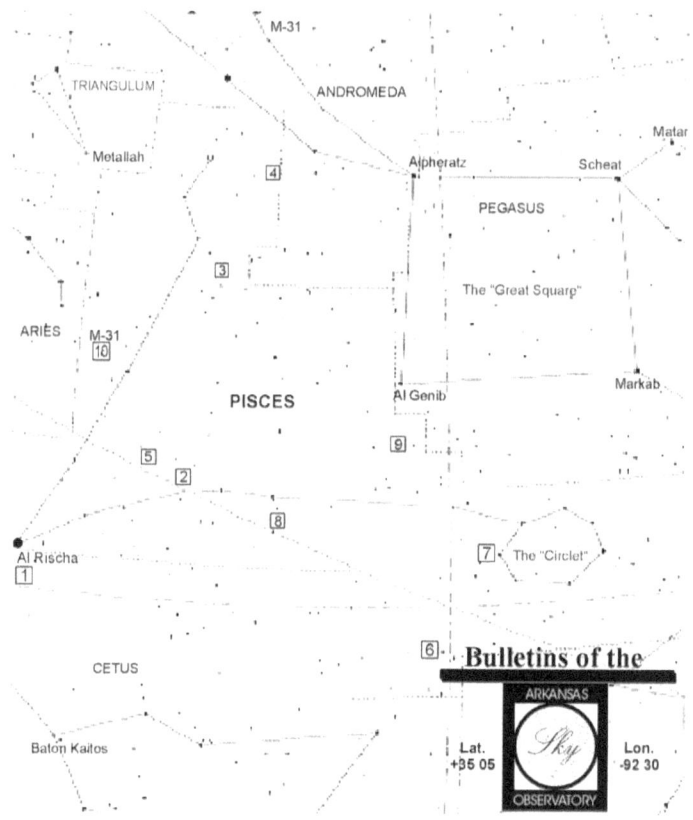

.....perhaps this is where the term "cold as a fish" originated?

As with all "GO TO" tour guide star charts, please use this accompanying chart above for a nice usable desk reference or good field chart as you proceed through the Pisces "GO TO" tour.

Also with every "GO TO" TOUR guide, each GO TO object in Pisces is discussed for your telescope regarding the type of conditions necessary for you to view it optimally for discern the very faintest details.........magnifications and aperture necessary for most objects, and much, much more. This is YOUR complete guide to get you on your way to exploring the best (and few!) objects in these two constellations. The following listing of "BEST" objects contains the finest or most interesting from my own observing experience and preference.

Use the star chart and the following guide as an excellent reference for your next star party itinerary, or a beginning for further study into the thousands of objects visible in this part of the sky. Truly these extensive Constellation Study Guides will most definitely put your PC sky program to work for you in the most efficient and enjoyable way possible! As a matter of fact, MANY AutoStar and other sky program users are now programming their own "Tours" based on these guides, using each constellation as a separate GO TO Tour for the AutoStar and other sky library that can be added in or deleted through the main edit screen on your PC or MAC computer.

We hope you enjoy these comprehensive GUIDES to touring the constellations via your computer-driven telescope. Each new installment is complete with diagrams, charts and illustrations that you will find nowhere else. Please let us hear YOUR feedback and your observations of each and every constellation after YOU have toured its vast reaches of our skies!

YOUR PISCES CONCISE DIRECTORY OF INTERESTING OBJECTS –

I have chosen the finest (or most interesting) **12 objects** in this PISCES "GO TO" tour; as with all guides, all objects listed below will be visible in most telescopes (some naked eye), from the small quality APO refractors, to the large reflectors of advanced amateurs. Of course, larger apertures may "show" an object a bit brighter and "better," but frequently a wide field and low power view is more desirable than aperture for FINDING the objects initially. Also true is that wide field, low power telescopes give many times a more contrasting view of the object against the background sky and are ideal for very large and extended sky objects.

Indeed, I strongly encourage you first FIND the target object, or its approximate location through your GO TO function with your lowest power and then - once IDENTIFIED positively - move up slowly in steps with magnification if necessary. Remember, not all objects "like" magnification. Sometimes better "field of view" (such as the wonderful wide fields provided by small refractors) is desired over light gathering (like the larger 8- to 10- inch and magnification.

The rule for determining "optimum magnification" is that: 1) too low power results in sky background glow detracting or diminishing the contrast against the deep sky object; 2) too high magnification darkens BOTH the sky background AND the object; 3) medium magnification can be achieved at which you have MAXIMUM contrast between the object and its darkened background sky. I have found through three decades of direct observing that about 15x per inch aperture, for deep sky observing is PERFECT for most objects. That being said, always remember that DOUBLE or multiple stars require whatever power you can crank out....the seeing conditions are the limiting factor here.

For my complete and comprehensive discussion regarding seeing conditions and sky transparency, see my discussion of this topic at the Arkansas Sky Observatories website: www.arksky.org .

With all deep sky objects, avoid attempting to observe when the moon is in the sky, even a very thin crescent, as its brightness in the sky will overshadow the very dim contrast afforded by even the brightest deep sky object; if you see the object at all against moonlight, you will NOT see the subtle outlying areas or the full detail of what is presented.

Pisces is dominant in late summer skies; riding midway on the Celestial Equator, it is suited for long-period observing for all telescope users both north and south of the equator. The "Circlet" in western Pisces rises in the east about dark (9 p.m. local time) on about July 25 and the center of the constellation "culminates " (passes over the meridian at midnight) around October 3, remaining in the sky throughout that night. All deep sky objects and difficult double stars are ALWAYS best observed when they are located nearly overhead (or as high in the sky as possible), thus requiring the observer to look through the thinnest portion of the Earth's "lens" of atmosphere and haze.

With all of the constellation lists, I recommend a good star atlas and/or chart which will list all the finest objects, constellation-by-constellation. One very handy reference guide is the *PETERSON FIELD GUIDE TO THE STARS AND PLANETS*, which features complete lists with declinations, right ascensions, magnitudes, and all pertinent information for you to expand your observing horizons beyond this brief GUIDE.

The exact process for bringing up a celestial object for your telescope to GO TO will vary with the sky program, telescope keypad, or App that you are using, but all fundamentally work the same logic to acquire your object as well as build your User Object Library. Always consult your HELP or User Guides of your specific program for the exact steps in using tabs and keystrokes to pull up and GO TO your objects!

The constellation tour Star Chart above (click on and save to a file on your PC; then open it and re-size to fit the page and print for a very handy at-the-scope star chart) will get you started on your journey for this constellation.

Following is the concise object list for your "GO TO" tour of PISCES; you may wish to find the majority of the objects from the AutoStar or your sky program Library (for example, you can easily go to the galaxy Messier 74 if you pull up "Object/Deep Sky/Messier/..then type in '74'...." and then press "Enter", followed by "GO TO" to access this "only" lonely galaxy in Pisces. On the other hand, if you want to experiment and become a "better computer user", try entering the exact R.A. and DEC coordinates of that object as described above after holding down the MODE key. You will find the accuracy of entered GO TO's to be pretty much comparable to your built-in library, but the capability of acquiring unlisted objects is fantastic!

OBJECT 1:   brighter star/nice double star - *Al Rischa* - Alpha Piscium - R.A. 01h 50' / DEC +02 31 - Mags: 4 & 5

OBJECT 2: good double star - Zeta Piscium - R.A. 01h 11' / DEC + 07 19 - Mags. 5.5 & 6.4 - good in all scopes!

OBJECT 3: great double - Psi Piscium - R.A. 01h 03' / DEC + 21 12 - Mags: 5.5 & 5.8 - good low power double!

OBJECT 4:   good double for ETX 90 - 65 Pisc - R.A. 00h 47' / DEC + 27 26 - Mags. 6.3 & 6.3 - 4.4" arc split!

OBJECT 5:   tough double for '90 - ADS 1254 - R.A. 01h 33' / DEC + 07 23 - Mag: 7.6 & 7.6 - A tough 1.6" arc!

OBJECT 6:   your challenge double - 27 Pisc. - R.A. 23h 56' / DEC (-) 03 50 - Mags. 5.0 & 10.8 - tough for 6-inch

OBJECT 7: nice variable for all scopes/binoculars - TX Pisc - R.A. 23h 44' / DEC +03 13 - Range 6.8 - 8 - nice!

OBJECT 8: "*Van Maanen's Star*" - R.A. 00h 47' / DEC + 05 09 - Mag. 12.4 (!) - An unusual white dwarf. 6-inch +

OBJECT 9:   doubled star test for LX 90 - ADS 287 - R.A. 00h 18' / DEC + 10 42 - Mags: 6.9 & 7.8 - only 0.6" split!

OBJECTs 10: nice galaxy - Messier 74 (ngc628) - R.A. 01h 34' / DEC + 15 32 - Mag. 10.1 - very large spiral

OBJECTs 11 through 17:   faint galaxies but brighter than visual 12.6

BONUS - NGC GALAXIES PISCES

(other than those listed above)
NOTE that all of these will be part of your PC program or GO TO telescope built in libraries, so they can be acquired easily from the Objects file!

Listing of "ngc" galaxies in Pisces in order of RIGHT ASCENSION (and NGC # order)
NOTE: These NGC galaxies may NOT be detailed in the following "Visual Guide"
as are those Objects 1-10 listed above...use the abbreviated descriptions as they follow the order:
NGC# / R.A. / DEC / MAGNITUDE / SIZE (in minutes arc -'-) / GALAXY TYPE , description

ngc0470 / 01 17 / +03 09 / 12.4 / 1.7 x 1.1 - spiral, very faint an tilted; can be seen in 6-inch
ngc0488 / 01 19 / +05 00 / 11.0 / 4.2 x 3.3 - pretty and large spiral, fairly bright....very round, concentrated
ngc0514 / 01 23 / +12 39 / 12.3 / 2.7 x 2.7 - face-on spiral, very faint; ETX 125 and up; fairly small
ngc0520 / 01 22 / +03 32 / 12.4 / 3.0 x 0.7 - very narrow and faint, LX-90 only; irregular/peculiar galaxy
n**gc0524** / 01 22 / +09 16 / 11.6 / 1.8 x 1.7 - tiny elliptical; starlike, can be seen in 4-inch but barely
ngc0718 / 01 51 / +03 57 / 12.5 / 1.2 x 1.0 - very small, faint spiral, object for large scopes only
ngc7619 / 23 18 / +07 55 / 12.4 / 1.0 x 1.9 - tiny elliptical galaxy, starlike in 6-inch+

....SO LET'S SWIM WITH THE FISHES IN THE CELESTIAL SEA and "sea" what comes up!! (refer to the Pisces Star Chart for all the objects described in detail in this *Constellations* guide.

A VISUAL GUIDE TO OUR DEEP SKY OBJECTS IN PISCES

[NOTE: since so many double stars are recommended in this tour, please refer to the LAST installment "GO TO" Lacerta for details on proper determinations of position angles for double stars. Since most of these included with the Pisces guide are relatively easy and spectacular doubles, most are referred to casually by "direction" rather than "position angle." Nonetheless, practice finding these via Position Angle (P.A.) to become a better student of astronomy]

Object 1 - Our "Starting" Brighter Star –
"AL RISCHA" (alpha Piscium) This is not only our starting "alpha" star for Pisces but a very nice double star right now. Its name comes from the Arabic Al Risha, which denotes "the cord", or fishing line that is connecting the two "fish" Venus and Cupid in their clever disguises! Moving in a very slow orbit of 720 years, the two stars are about as close now as it will be until the year 2085 when the two stars will be only 0.7" arc apart from our viewing post here on Earth. Presently the separation is just over 1.0" arc and MIGHT be split in a 4-inch scope under extremely steady skies and high power (about 200x). The primary star is magnitude 4.3 and is companion is slightly fainter at magnitude 5.2 almost exactly SW from the brighter star (Position Angle 235 degrees). Look for a nice color contrast on these two stars, yet no two observers seem to see the SAME colors....some see "green and white," while others may record them as "green and blue." This is an easy star to find, but the splitting of it might take a 6-inch and medium-high powers. Don't miss this nice double star!

Object 2 - Zeta Piscium - A Wonderful, Colorful

Double Star for the small telescope and Above - LOW POWER! This is a wonderful double star for low power and wide field observing, with an incredible color contrast of a light purple (fainter star) and pure white. At magnitudes 4.2 and 5.3 this is an easy and beautiful double for the 3-inch all the way up to the 8-inch....separation is a full 24" arc! Thus, low power and wide field afford a beautiful view of this relatively bright star immersed in a nice field of stars. However, to view the nice color contrast, use medium powers, or about 15x per inch aperture. Anything high or lower will lose the beauty of this star!

Object 3 - A Great Double Star - Psi Piscium –
Great Low Power Double No one gets left out of this constellation guide! Here is another gorgeous double star, both stars yellow in color and pretty bright and equal in magnitudes (5.5 & 5.8). The separation of these two stars is about 30" arc, or slightly smaller than Jupiter might appear in the same eyepiece. Look for the "fainter" star nearly due south of the brighter primary one. Low powers are ideal for this pretty star.

Object 4 - Another Double - 65 Piscium - a Really Nice Star Pair, and a TEST for a 60mm scope! Although this star is a MUST for all telescopes, here is a rather "rare" double star test for 60mm. This is a gorgeous star - both stars exactly magnitude 6.3(!) so they are visible in very low powers and even in most finderscopes. Two twin stars are separated by 4.4" arc, easily resolvable in a good 3-inch and larger scopes (this is a pretty sight....so you guys and girls with larger telescopes do NOT ignore this one!). However, it is not so easy with the wide field refractors, requiring some magnification (I suggest the most you can muster). Look for the stars oriented in a near east-west (Position Angle 269 degrees). In a 4-inch and larger scope I recommend about 20x per inch for the best views.

Object 5 - A Super Test Object for a quality 3-inch Double Star ADS 1254 This will be a challenge - but a fun one! - for the small scope; at separation 1.6" arc in position angle 50 degrees (almost exactly NE -SW orientation) here is a pair of EXACT equal brightness (magnitude 7.6) stars that are easy to find in all scopes....however, anything less than 3" aperture will not resolve this pair and a 3-inch should split this cleanly. Use pretty high magnification (at least 125x) for the best views in all scopes. The star will be quite faint in the 8 x 25 finders of the small scopes and fairly so in the larger 8 x 50 ...nonetheless, in the telescope at low "finding power" this star will be among the brightest in a wide field view. This is a VERY nice double (yellow, sun-like stars) in ALL scopes 3" and larger. A good double for larger telescopes.

Object 6 - 27 Pisc - WHAT??!! Another double star? Yep, and it get really tough here..can you handle it? Okay 3-inch scope owners....you did a good job on ADS 1254 above. Feeling pretty good? How about you 6-inch owners? Piece of cake? Well....let's try this one. At a closer 1.5" arc separation you would think that 27 Piscium would be just about as easy as ADS 1254, but hold on. The primary star here is a bright magnitude 5.1 (the brightest object in your low power field) but the secondary - at Position Angle 276 degrees, nearly

DUE WEST - is only magnitude 10.8, make it a very tough star for a 4-inch...but "still" attainable. Give it a shot. You have the resolving power (the 3-inch is rated at a separation "power" of 1.3" arc) and the light grasp (you should be able to see a star as faint as magnitude 11.7) so TRY it! Even a 5-inch might have some difficulty on this star with the proximity of the brighter star masking out the faint companion. Use the finder chart (I have drawn to match the "mirrored" orientation all compound scopes) to actually locate this star.

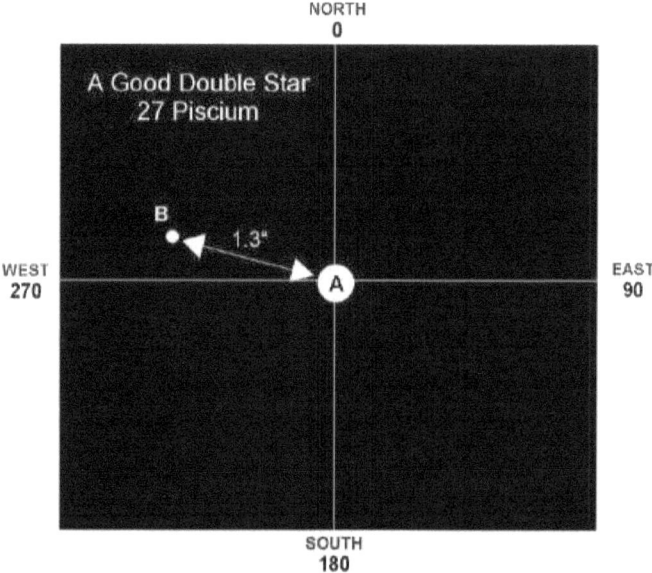

Object 7 - Variable Star - 19 Piscium - Also TS PSC for Variable Star Observers You can find this star by downloading the fine comparison chart from the American Association of Variable Stars (AAVSO) web site; the one here is the "b" low power chart which will be ample for both finding the star in your finderscope (or binoculars) AND measuring its brightness relative to nearby stars

provided on the chart. This is a good binocular variable, but still can be monitored in a wide field low power eyepiece. To obtain the chart, click on and save to file the link
https://www.aavso.org/apps/vsp/ .

Note for these charts, simply type in the NAME of the variable at top to generate your choice of chart – resize to page and print out for use at the telescope. TS PSC varies in magnitude from 5.3 to 6.2 which is NOT much, but this is a very interesting variable from a scientific standpoint. First, it has no set period and changes can be seen visual within a short span of fractions of a day; Second, this star varies differently in different colors of light! This can be seen even visually using both a blue and a red filter on the eyepiece. In addition, there will be very long periods in which the star will remain unchanged, only to begin suddenly fluctuating in minor amounts. Because of the unpredictable nature and infrequent changes of this star, I recommend giving it a quick look at every opportunity and reporting your magnitude estimates to the AAVSO at www.aavso.org about once a month.

LESSON NOTE: this is a **very red star** and there is a special technique necessary for estimating the magnitude of a RED STAR. All observers, when looking at a very red object for any length of time, will notice that the object will "appear" to get brighter the more it is observed. This is called the "*Perkinje Effect*" and is totally due to the eye's ability to accumulate and store wavelengths of light in the red end of the spectrum. Hence the more you look at 19 Piscium or even Mars for that matter, the brighter it will appear over time! To offset this effect when estimating magnitudes, the "trick" here is to DEFOCUS your red star from a point source (the Airy Disk) to a larger "extended" disk of light and compare the out-of-focus "donut" (small, mind you, not too large) to the out-of-focus images of nearby comparison stars!

To get an idea of JUST HOW RED this star really is...compare it to the blue stars nearby 21 and 25 Piscium!

Object 8 - A White Dwarf and Very Unusual Star - "Van Maanen's Star" - medium to large scopes only - Faint! The best way of locating this star is by looking about 2 degrees south of Delta Piscium and using the reversed star chart (for use with catadioptic telescopes) that I have specially prepared from the Smithsonian Astrophysical Observatory Star Catalog Charts. The limiting magnitude of this chart is about 14 and Delta Piscium is just outside the LOWER part of the view here....the circle is one degree in total diameter,north at top and EAST is to the right, just as it would be in your scope. The circle itself is "about" the field of view (slightly wider) in a 32mm eyepiece, and the "bright" star see to the lower right of Van Maanen's Star is magnitude 6.7, barely visible in the finder of our telescopes.

Now, about "Van Maanen's Star in particular.

This star obviously was discovered by a gentleman named **Mr. Van Maanen**, in 1917, to be specific.

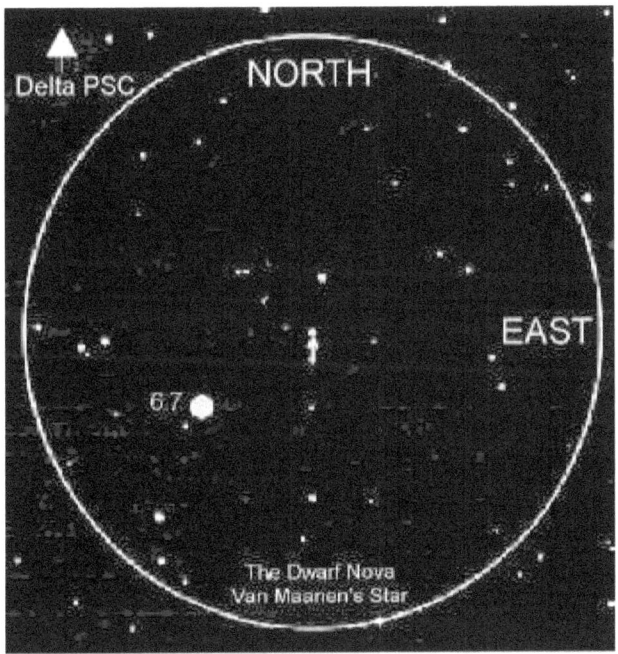

He noted the star particularly because of its large annual PROPER MOTION, or its rapid movement across the sky relative to more distant fixed stars; thus, he deduced that this must be very close star....so close that it actually progresses SE each year by almost 3" arc! Indeed, it has now been found that it is rivaled only by the companion to the bright star *Sirius* as the closest white dwarf to Earth.

The remarkable fact of this star is its diameter - IT IS ACTUALLY SMALLER than the Earth, at only **7,680 miles** across! HOWEVER - it is so dense that ONE CUBIC INCH of this star would weigh an incredible 20 TONS!

The star visually is **magnitude 12.4**, thereby "possibly" barely visible in a 6-inch, but certainly in 8-inch and larger scopes. The brighter star in the chart above is equally interesting as it TOO appears to be moving in the SAME DIRECTION as the faint star, but only a bit slower and about twice the distance from us. This is a must-see if your telescope can handle a 12th magnitude white dwarf that weighs **20 tons for each inch** of its mass!

Object 9 - ADS 287 - A Very Close Double Star for the 8-inch and larger scopes. Also known as Barnard 1093 this is a nice - but very close - double star, with a clear yellow-orange cast to both stars. At separation only 0.6", it will be a challenge to cleanly split these two 6.9 and 7.8 magnitude stars with the big 8" scope. Look for the fainter star in position angle 100 degrees, or just south of due east from the brighter star. About 300x will be required to separate the two components and over 400x will be better should the atmosphere be "steady" enough to handle that kind of magnification.

Object 10 - Actually a Deep Sky Object!! - Messier 74, a Fine Spiral Galaxy! In this huge constellation this is the ONLY deep sky object of merit, and certainly the only one suitable for modest amateur telescopes. At Magnitude 10.1, this galaxy appears actually a bit brighter and is certainly visible in small telescopes as a very large circular glow, 11' arc across!

The photograph above, taken through the famous Palomar 200" reflector, shows the incredibly symmetrical and circular shape of this face-on galaxy. Notice how tightly the very "rough" arms are wound into the broad "hub" of this galaxy. Hence, the spiral arms are visible in any moderate to large amateur telescopes and visible as very non-uniform "clumped" appendages due to the beautiful and dense star groups which can plainly be seen in this photograph. Although this galaxy appears very large in our telescopes, it is an incredible 30 million light years away!

Regarding viewing conditions, always use LOW POWER and a wide field of view to best observer this galaxy....the light of it is so spread out over such a large area that very wide fields (and smaller telescopes!) actually reveal much more detail that high powered views.

WANDERING ABOUT....YOUR NEW "USER OBJECT" IN PISCES

We simply must make **Van Maanen's Dwarf Nova** our GO TO USER OBJECT for Pisces (we do because I understand there is a law and they do enforce it where Mr. Maanen is from...).

This is likely the ONLY dwarf novae that you will actually be able to not only view for yourself, but also share with others at star parties and special events. Because of its proximity to the bright star Sirius, that white dwarf companion is a much more difficult object (seemy discussion: GUIDES - *Observing the Companion to the Bright Star Sirius*). Thus, this white dwarf star is an ideal "show and tell" object for 4-inch and larger telescopes at ANY star party.

I highly recommend reading up on this fascinating breed of stars it **Robert Jastrow's** "*Red Giants and White Dwarfs"* so that you will be well versed in the unusual nature of these incredible dense stars and their evolutionary tracts.

For AutoStar, go to: "Select/Object [enter]...." scroll down to "User Object" [enter]. Now enter the coordinates given above for "Van Maanen's Star", using the number keys on AutoStar. After entering the coordinates and pressing "Enter" yet again, scroll down one and you can list the magnitude of the object as "12" [Enter].

Remember that your keypad, App or sky program will vary from the techniques for Autostar....but all are logically the same. Please read your User Guide or HELP to learn the process to produce your own User Object library for your computerized scope!

So with the addition of a white dwarf star, you have a growing User Object library which includes a black hole, dark nebulae, curious multiple stars, variable stars and even more weird deep sky objects. All of these are wonderful for conversation-starters and crowd-stoppers at the big astronomical events!

* * *

If you are observing Pisces, then likely you are lured to the interesting constellation of ANDROMEDA, just to the north. Because of the wonderful Messier 31 galaxy and a few other very special objects, requests for a GO TO guide for this constellation were outranked ONLY by those wanting loads of exciting explorations of Orion and Taurus! So in a couple of weeks we will visit the wonder world of the Andromeda galaxy and other worlds beyond.

See Volume One of Sherrod's The Constellations for your guide for this wonderful maiden of the sky.

* * *

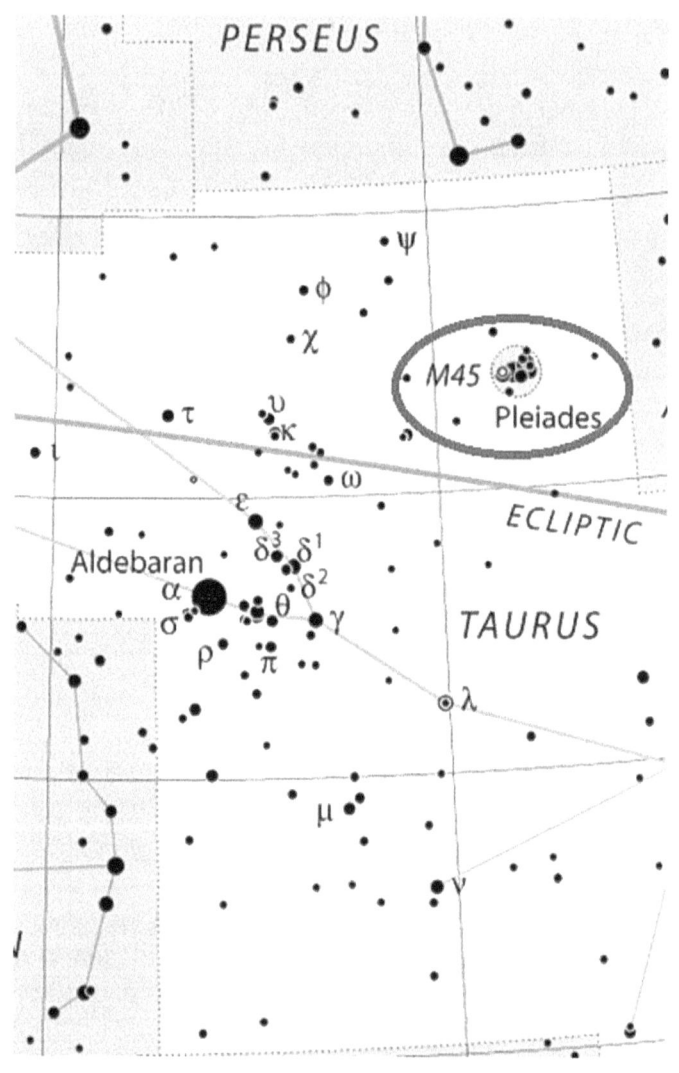

Location of the Pleiades star cluster
Chart courtesy
IAU / Sky & Telescope

Chapter 29

OBSERVING THE STAR OF THE PLEIADES
See Also
Constellations/*Taurus* Observing Guides
for detailed overview

INTRODUCTION

At only 410 light years away, the *PLEIADES* or "Seven Sisters" is the second-closest star cluster to Earth, its neighbor the *HYADES* (both in the constellation of **Taurus**) being the closest. Known also as **Messier 45** (M-45), the Pleiades is a very memorable naked-eye sight and often confused as "the little dipper" by those new to the nighttime skies.

The Pleiades (pronounced "PLEA-a-DEES" is NOT a constellation and never has been, even to the ancient sky gazers who started all this constellation myth and lore. But it has been an important "asterism" of the night sky since before history began….a curiosity of a tiny grouping of stars the simply beckoned legend and lore.

Please see our Constellations guide to Taurus for a full discussion of the mythology associated with this wonderful star cluster.

Some Personal Thoughts

From a personal standpoint, the Pleiades is one of the primary reasons that I fell in love with the sky and pursued astronomy for at least 60 years.

As a child of six, I was already fascinated with science and particularly the dark nights that we still had even in the confines of city boundaries. My mind wandered as it wondered of the distances to the stars in space....why some were so dim, yet so many were sparkling bright.

My maternal grandfather, Garland Conatser was a scholar and teacher who had studied for his education degree the science of Astronomy in college. We frequently sat on a picnic table outside his small farm at Ozark, Arkansas, watching meteors and looking at the stars.

My first lesson six decades ago in constellations was on one such night.

"Hey! There's the little dipper." I pointed out to him, this star pattern directly overhead in the sky.

As grandparents so well do, he corrected me without insulting me in doing so.

"Oh....those are the stars of The Seven Sisters....The Pleiades," he explained in an excited tone, knowing that he would lead me want an explanation. And then – he pointed out "The Little Dipper", the real one...and I have been disappointed in that one ever since.

But he took the time to explain – as you should do to your children and neighbors who hang around while you are observing – that those seven stars have been around longer than mankind. He told me of the legends of the Chinese, the Anglos....the Native Americans. And all those legends of seven industrious women were oddly alike, yet culturally different.

"HOW…." I thought even as a young child, "did all those lands separated across our world and no communication between them in ancient times, have the same story about a group of stars in the sky?"

And I still wonder that today.

Nature of the Pleiades

This tiny asterism in the sky does, indeed, look exactly like a "little dipper" complete with bowl and handle….far more than Ursa Minor which is the official "Little Dipper" of the sky.

Not a constellation, this is an actual STAR CLUSTER, with hundreds of stars all held in place and moving in tandem in our Milky Way galaxy as one entity. The stars of the Pleiades are literally bound together by gravity, just as all other star clusters throughout the heavens.

The only difference in this one is that it is so close – only the *Hyades* in Taurus is closer.

Recent studies have suggested that there are as few as 250 stars and as many as up to 500 stars that are actually part of this remarkable cluster. From our vantage point, the cluster is moving as a large group about **25 miles per second** to our southeast.

The name "*Pleiades*" has several translations but the most common two are derived from ancient

Greek text, one which interprets the meaning as "*..to sail away,*" while the other translates "*...full*", likely a reference to the many stars crammed in such a small area of sky!

One common name for the Pleiades that has been around for a long time is "*the Seven Sisters*" implying that perhaps there is one star - "*the lost Pleiad*" as it has become known - that has faded considerably since early times, since only SIX are now conspicuously equal in brightness (see accompanying chart).

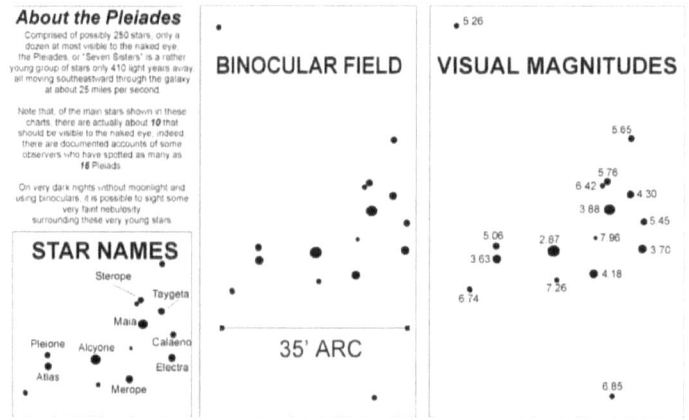

OBSERVING THE PLEIADES

There is no trouble finding the Pleiades; rather, the problem is getting all the stars in one field of telescopic view! Perhaps binoculars show the cluster best, and with 7 x 50's or a wide field, low power telescope (the 3 and 4-inch scopes are ideal for this) it is possible to see some of the famous nebulosity surround some of the stars (especially Merope) on a very dark night in a remote location. Lowest power is always best for this, and allows all stars to be viewed simultaneously.

A long focal length telescope, even with lowest power, cannot get all of the bright stars in the same field of view, even using a 40mm wide field Plossl eyepiece; the view in wide field telescope finders, however, is excellent! In a wide field scope, with the same eyepiece, the view is much better with such a wide field that the true character of the cluster can be appreciated.

Some larger telescopes with wide field eyepieces can get the entire cluster in a single field (in an 8" f/10 Schmidt!) using the big 2" 53mm wide angle Erfle or similar.

There is a **LIMITING MAGNITUDE test chart** in the chapter of *The Constellations*, Volume Two that utilizes the star field surround the bright star "*Electra*," in which you can determine the absolute faintest star perceptible with your telescope under very dark conditions. I urge you to attempt to try this test so that you will be better familiar with the capability of your telescope.

Concerning naked eye observation of the Pleiades, the chart here clearly indicates that LEAST 10 of the Pleiades stars should be visible to the naked eye; indeed, at a very dark observing location, the dark-adapted eye should be capable of viewing stars down to magnitude 6.1; some observers in ideal desert high altitude sites record limiting magnitudes of the eye at down less than 7.0! Nonetheless, because of the closeness of each bright star to the next, the actually "glare" from this group prevents us from viewing some of the fainter naked eye stars. **William Dawes** (the early astronomer famous for "Dawes' Limit, the resolution factor of telescopes as a function of their apertures) was said to have been able to see 13 of the Pleiads with his naked eye.

A 3-5 inch scope should reveal between 80 and 100 of the cluster stars easily, with more and more stars appearing as larger telescopes are used.

INTERESTING ASPECTS OF THE PLEIADES

There are several things that observers should be aware of concerning the Pleiades, and some "factoid" information that makes sharing at a star party interesting.

1) Because the Pleiades cluster is located within the ECLIPTIC - the narrow band of the Zodiac through which the planets, moon and sun appear to move relative to the Earth - the chances are very good that one of those bodies will pass THROUGH the star field occasionally. Indeed, the MOON is a frequent visitor to the Pleiades when the earth-moon-sun alignment is right. During such passages, the lunar disk OCCULTS, or covers up many of the Pleiades stars in sequence as the moon moves slowly toward the east during the course of the night. Upcoming Pleiades occultations are forecast in Sky and Telescope and are often updated for observers long beforehand on the Internet.

2) For your information, the Pleiades is one of the youngest - *20 million years old* - of all star clusters

and all stars of the cluster are from the same formation; it is the bright gas clouds that can be glimpsed and photographed that is left over from the early star formation.

3) The entire span of the Pleiades stars is almost one (1) degree, or the extent of two full moons; hence, you can use the cluster to gauge your field of view in any eyepiece.

The Pleiades from Arkansas Sky Observatories TMB 130SS APO wide field refractor

4) The Pleiades is a wonderful target for beginning astrophotographers, as it "likes to have its image recorded" on PIGGYBACK photographs, taken with a camera and simple lens of any focal length which rides atop the telescope (hence, "piggyback"). Of course, the longer the focal length of the camera lens, the greater the image scale. With longer (i.e, 200mm) lenses, the nebulosity

surrounding some stars can be easily recorded on modern sky camera. Note the nebulosity around "*Merope*" and the other five primary stars in the accompanying photograph. This photo was recorded from a small telescope; there was no guiding; the telescope tracked for 120 seconds while I drank my cup of coffee!

*   *   *

One of the most interesting aspect of the Pleiades - perhaps the most interesting of all sky objects - is the cluster's history and lore and I encourage observers (on a dark and stormy night) to read Robert Burnham's account of this history in his "Celestial Handbook," Vol. 3. Throughout history this little cluster of stars glimmering against the dark of autumn and winter skies has fascinated and provoked observers of all nations, of all cultures and all beliefs....

"Many a night I saw the Pleiads, rising thro the mellow shade....

Glitter like a swarm of fireflies tangled in a Silver braid...."

                                             Alfred Tennyson

Chapter 30

SAGITTA and VULPECULA
*The Archer's "Errant Arrow"
and the "Cunning Sky Fox"*

In our next "double" constellation study, we will re-visit TWO constellations that were mentioned previously in my guide to the wonderful and large constellation of *Cygnus* (*The Constellations*, Vol One) Unlike the also small constellations of Delphinus and Equuleus in which deep sky objects were either at a premium or sparse at best, these two likewise SMALL constellations in the high northern sky provide an abundance of wonderful and challenging deep sky objects as well as an array of good variable and double star systems!

So it's time to dust off the old computerized telescope and put it to work....your challenge is more rewarding on these constellations than their sizes might indicate!

Sagitta (pronounced: "SAJ-eh-ta") is one of the few constellations that actually resembles what it is supposed to be. With constellations such as Sagitta and Vulpecula I envision early Greek stargazers and constellation-namers looking up at the dark night sky and realizing that they actually "....had stars LEFT OVER!" after assigning all the Greek gods and mythological heroes. What do you do with these stars, particularly those that form such an obvious asterism as SAGITTA? Well, you find some way to "associate" those stars with an already-immortalized Greek hero that you HAVE named to the celestial firmament.

Sagitta is an "*arrow*" and that is indeed, what it looks like! The figure below shows the errant arrow, supposedly launched by our celestial archer SAGITTARIUS toward the threatening SCORPIUS (see my TOUR in GUIDES/Constellations/Scorpius), but as you can plainly see when standing outside on a dark night, he clearly missed his mark!

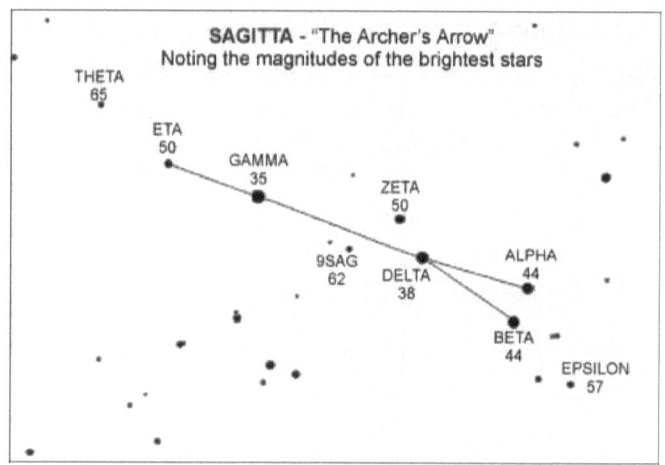

Note from the chart above that the stars ALPHA, BETA, DELTA and ZETA clearly define the arrow's feathered tail, while the shaft of the arrow is clearly delineated by the stars GAMMA and ETA. The magnitudes of the conspicuous "**asterism**" (star pattern) of Sagitta are give aside each star with the decimal point omitted to keep you from confusing that "dot" with a faint star (45 = magnitude 4.5). To be the smallest total area for any constellation, little Sagitta contains a wealth of fine double and multiple stars and some remarkable variables in addition to many fine deep sky objects!

To its north and east is the larger (but still small)

constellation of VULPECULA (pronounced: vul-PECK-u-la) which, for whatever unknown reason, represents the cunning and ever-so-elusive Fox who prowls throughout the land and sky throughout the dark of night. I YOU can make a "fox" out of the only THREE (3) fairly dim stars seen with the naked eye of this constellation, you really have an imagination equally ONLY by those early Greek stargazers.

Clay Sherrod's
CONSTELLATION GUIDES:
"Go To" SAGITTA & VULPECULA"
An Errant Arrow and the Crafty Fox

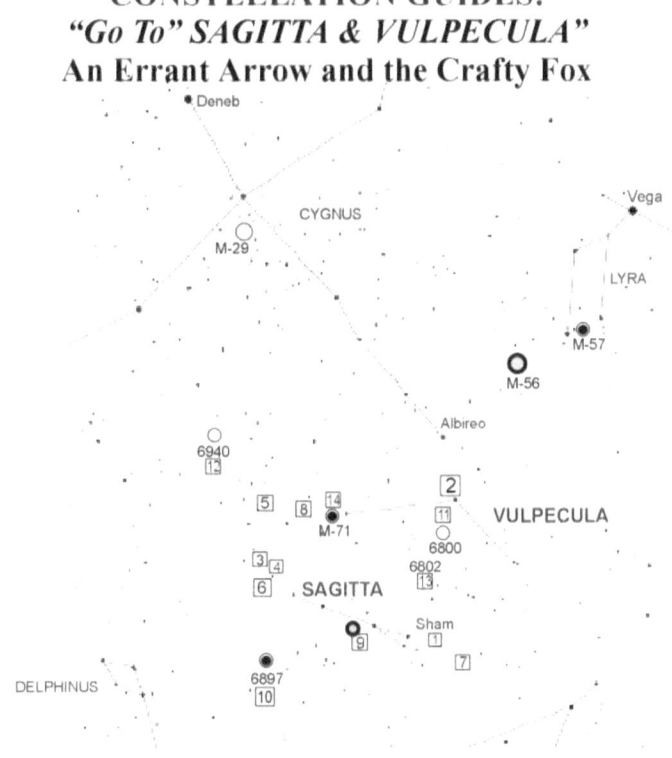

Nonetheless, Vulpecula is PACKED with objects, and only a smattering of them are presented here for lack of room. It is in this constellation that the famous "Dumbbell Nebula" (Messier 27) is found; this objects was discussed in the Cygnus TOUR noted above, but we will revisit this in length here.

Each GO TO object is discussed for your telescope regarding the type of conditions necessary for you to view it optimally for discern the very faintest details....double star challenges for each size telescope .....magnifications and aperture necessary for most objects, and much, much more.

This is YOUR complete guide to get you on your way to exploring this large and interesting constellation. ONE THING that I might add about the region in and surrounding Sagitta and Vulpecula: Although some interesting and "test" double stars as well as VERY exciting variable stars are always included in the "GO TO" tours, there is a particularly wide range of very interesting double and multiple stars too numerous to list or mention here.

Consult a good handbook, such as the "*Burnham's Celestial Handbook*," Vol. 3 for a very comprehensive list of locations, magnitudes and angular separations of these wonderful stars. There are many stars for EVERY telescope size and type. IN ADDITION, consult the wonderful wealth of information at www.aavso.org for complete details concerning all the fabulous variable stars and novae known for both Sagitta and Vulpecula; not only can you get great free observational information about these stars, but there are thousands of wonderful star charts FREE for downloading off that site; just go to "Standard Charts" then click on the appropriate RIGHT ASCENSION for each star that you would like to download comparison charts for, and they are yours for the effort!

Use this attached star chart and the following Guide as an excellent reference for your next star party itinerary, or a beginning for further study into the thousands of objects visible in this part of the sky. Truly these extensive Constellation Study Guides will most definitely put your sky program to work for you in the most efficient and enjoyable way possible! As a matter of fact, MANY sky program users are now programming their own "Tours" based on these guides, using each constellation as a separate GO TO Tour for the AutoStar or PC sky program library that can be added in or deleted through the main edit screen on your PC or MAC computer.

We hope you enjoy these comprehensive GUIDES to touring the constellations via your AutoStar or sky program and its computer-driven telescope. Each new installment is complete with diagrams, charts and illustrations that you will find nowhere else. Please let us hear YOUR feedback and your observations of each and every constellation after YOU have toured its vast reaches of our skies!

YOUR SAGITTA and VULPECULA CONCISE DIRECTORY OF INTERESTING OBJECTS –

The tiny sizes of these two constellations are deceiving when considering the great numbers of very fine objects within them. I have provided a comprehensive LISTING (similar to those for the many ngc galaxies visible in at least some of our for the constellations of Coma Berenices and Virgo) This listing is found at the end of our "concise list" of my selected top objects and gives an abbreviated description, position, magnitude and size of each of the brighter NGC (or other) object found in the two constellations.

For your specific "GO TO" tour I have chosen the finest **14 objects** in Sagitta and Vulpecula; as with all GUIDES, all objects listed below will be visible in all telescopes (some naked eye) from 3-inch to 6-inch; of course larger apertures may "show" an object a bit closer and "better," but frequently a wide field and low power view is more desirable than aperture. You are encouraged to venture to your "dark sky site" for the BEST views of the many very faint **NGC** (New General Catalog) objects listed here. Even binoculars will show dozens of deep sky objects that cannot be fully appreciated in large telescopes with limiting fields of view! Remember that these two constellations - although not within the Sagittarius arm of the dense Milky Way cloud - are very close to the rich star clouds of our summer months and have some outstanding very low power star fields that are seen very well in the smallest telescopes!

The convenient sky placement of these two constellations lends itself well to very good and long-period observing for observers north of the equator. When rising about dark in the east (mid-spring) it will remain in the sky throughout the night, transiting the meridian at about midnight during that season. All deep sky objects and difficult double stars are ALWAYS best observed when they are located nearly overhead (or as high in the sky as possible), thus requiring the observer to look through the thinnest portion of the Earth's "lens" of atmosphere and haze.

As with all of the "GO TO" TOUR constellation lists, I recommend a good star atlas and/or chart which will list all the finest objects, constellation-by-constellation. One very handy reference guide is the *PETERSON FIELD GUIDE TO THE STARS AND PLANETS*, which features complete lists with declinations, right ascensions, magnitudes, and all pertinent information for you to expand your observing horizons beyond this brief guide. For the computer or media device, my favorite quick reference and very comprehensive descriptions of objects are the many Smart Phone APPS that are available such as Voyager, etc.. Note that this very affordable software package ALSO allows very quick and accurate construction of "Go To" Autostar tours for these constellations as well as allows a very nice laptop control of any telescope via Bluetooth or other wireless connection!

FOR THE AUTOSTAR: Note that your AutoStar or sky program may NOT have every object listed on every constellation GO TO tour....this is intentional. You can access some of the most interesting objects of the sky directly from their coordinates.

The exact process for bringing up a celestial object for your telescope to GO TO will vary with the sky program, telescope keypad, or App that you are using, but all fundamentally work the same logic to acquire your object as well as build your User Object Library. Always consult your HELP or User

Guides of your specific program for the exact steps in using tabs and keystrokes to pull up and GO TO your objects!

The constellation tour Star Chart above will get you started on your journey for this constellation.

Following is the concise object list for your "GO TO" tours of Sagitta and Vulpecula; you may wish to find the majority of the objects from the Sky Library (for example, you can easily go to the "Dumbbell Nebula" if you pull up "Object/Deep Sky/Messier Object/..type in '27'...." and then press "Enter", followed by "GO TO" to access my favorite beautiful galactic cluster. On the other hand, if you want to experiment and become a "better computer user" try entering the exact R.A. and DEC coordinates of that object as described above after holding down the MODE key. You will find the accuracy of entered GO TO's to be somewhat less than those stored in AutoStar, (other programs operate similarly) but the capability of acquiring unlisted objects is fantastic!

With more popular named objects, you can also find the Dumbbell and similar objects by going to: OBJECT / DEEP SKY / NAMED....and scrolling until you simply find "Dumbbell Nebula" and press enter, then GO TO.

OBJECT 1: your not-so-bright index star - *SHAM* (alpha Sagittae) - R.A. 19h 38' / DEC + 17 54 - Magnitude: 4.4

OBJECT 2: your index star in Vulpecula - (beta Vulpeculae) - R.A. 19h 27' / DEC + 24 34 - Magnitude: 4.6

OBJECT 3: (Sagitta) super variable - V Sagittae - R.A. 20h 18' / DEC + 20 57 - Magnitude: 9.5 to 12, VERY erratic - FUN!

OBJECT 4: (Sagitta) another great variable - FG Sagitta - R.A. 10h 10' / DEC + 20 11 - Magnitude: 9?? - one-of-a-kind!!

OBJECT 5: (Vulpecula) good variable star - W Vulpeculae - R.A. 20h 08' / DEC + 26 08 - Magnitude: 8.2 to 10.2, semi-regular

OBJECT 6: (Sagitta) great triple star! - Theta Sagittae - R.A.20h 08' ' / DEC + 20 46 - Mags: 6.2, 8.5, 7 very wide, EASY!

OBJECT 7: (Vulpecula) great double star! - 2 Vulpeculae - R.A. 19h 16' / DEC + 22 56 - Magnitudes 5.4 & 9.5, great 3-INCH test!

OBJECT 8: (Vulpecula) tough double! - 16 Vulpeculae - R.A. 19h 59' / DEC + 24 48 - Mags: 5.9 & 6.2 - REALLY close in 8-INCH; TOUGH!

OBJECT 9: (Sagitta) globular? galactic cluster? - Messier 71 (ngc6838) - R.A. 19h 52' / DEC + 18 39 - Mag: 9 – NICE

OBJECT 10: (Sagitta) planetary nebular - ngc6897 - R.A. 20h 08' / DEC + 16 46 - Mag. 11 - tough but do-able in the 6-inch+

OBJECT 11: (Vupecula) very open cluster - ngc6800 - R.A. 19h 25' / DEC + 25 05 - Star Mags. 10th, nice, about 25 stars, open

OBJECT 12: (Vulpecula) galactic cluster - ngc6940 - R.A. 20h 33' / DEC + 28 08 - Mag. 8.2, about 100 stars, each mag. 9.0!

OBJECT 13: (Vulpecula) galactic cluster- ngc6802 - R.A. 19h 29' / DEC + 20 10 - Mag.11, bar shaped, 60 stars...nice in 6-inch+

OBJECT 14: (Vulpecula) *THE DUMBBELL NEBULA* - Messier 27 (ngc6853) - R.A. 19h 57' / DEC + 22 35- Mag. 7.6, the BEST!!

OBJECTS 15 THROUGH 41 - NGC OBJECTS IN SAGITTA AND VULPECULA (other than those listed above)listing of "ngc" objects in order of RIGHT ASCENSION (and NGC # order) visible in modest scopes:

NOTE: These NGC / IC objects are NOT detailed in the following "Visual Guide" , only listed for you to explore. Objects 15-41 are listed above...use the abbreviated descriptions as they follow the order:

NGC# / R.A. / DEC / MAGNITUDE / SIZE (in minutes arc -'-) / DESCRIPTIONS (abbreviated)

IN SAGITTA:

ngc6838 / 19 52 / +18 39 / 9.0 / 5.5' – globular cluster, M-71 (see below)
ngc6879 / 20 08 / +16 46 / 11.0 / 5" - very tiny planetary nebula, looks like star - central star mag. 15
ngc6886 / 20 11 / +19 50 / 11.1 / 9' x 6' - large, but very dim. 6-inch scop only (16th mag.centr.star)
IC4997 / 20 18 / +16 35 / 11.0 / 2" arc - starlike planetary, TINY. Central star is 14th mag (LX 90?)
H 20 / 19 51 / +18 13 / 9.6 / 10' diameter - large, about 20 11th mag. stars, very faint

IN VULPECULA:

ngc6800 / 19 25 / +25 05 / 9.7 / 15' - large (half moon size!) cluster of about 25 stars, all mag. 10
ngc6802 / 19 28 / +20 10 / 11.0 / 5' diameter - open cluster, small, about 60 13th mag. Stars
ngc6823 / 19 41 / +23 12 / 10.0 / 5' - small galactic cluster with about 30 stars to mag. 11, nice

ngc6830 / 19 49 / +22 58 / 9.4 / 8' diam. - nice galactic cluster, 20 11th mag. stars, good object for small telescope
ngc6834 / 19 50 / +29 17 / 10.1 / 4' diam -- small cluster, 50 stars at mag. 11 to 15 – 8-inch object
ngc6885 / 20 10 / +26 20 / 9.3 / 20' diam. - very large cluster, 30 stars + 6th mag one, good in low power.
ngc6940 / 20 33 / +28 08 / 8.2 / 20' diam. - NICE large cluster, 100 stars, mag 9 - great low power object!
ngc6842 / 19 53 / +29 09 / 12.5 / 50" diam. - very large planetary nebula, 8-inch or larger only object – FAINT
ngc6853 / 19 57 / +22 35 / 8.0 / 8' x 5' - Dumbbell Nebula, M-27, see below!
IC1305 / 19 37 / +20 06 / 11.0 / 2' x 1' - nebula - very difficult and faint; good photographic object
IC4954 / 20 03 / +29 06 / 11.6 / 1' x 1' - nebula - small and VERY faint - maybe in a 6-inch, low power.

....AND NOW ON WITH THE SHOW!! (refer to the Star Chart for all the objects described in detail on the "Guide")

A VISUAL GUIDE TO OUR DEEP SKY OBJECTS IN SAGITTA and VULPECULA

Object 1 - Our Starting Star in Sagitta - "Sham" (alpha Sagittae)
At magnitude 4.37, there is nothing particularly spectacular about this star; it is an "F" spectral type star, similar to our own sun and appears to be a single star as well. In the chart above you can see that Sham denotes the northernmost "feather" stars

of the Sagitta arrow. You can keep looking.....but there really is not a whole lot more to see! (but don't you think THAT is exactly what astronomers on the planet circling Sham are telling folks looking our way at the Sun right now??).

Object 2 - Alpha Vulpeculae - Our Target Star for Vupecula
It's NOT a double star...it's not a variable star....it's not a bright star. But at magnitude 4.6 it IS the brightest object in the cunning fox Vupecula. Twice as close to us as our previous star, alpha Vul is a more evolved star than our sun, being a "G" spectral class. Keep looking.....nothing changes. However, you DO now know where Vulpecula is, don't you? I'll bet you could not have pointed it out before this "GO TO" TOUR!

Object 3 - A Very Interesting and Rewarding Variable Star - V Sagittae
This is an incredible and greatly erratic star as the light curve shown below can plainly demonstrate.

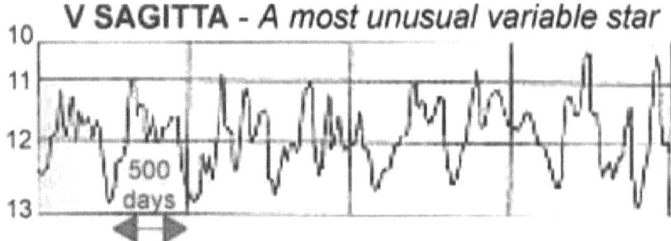

Known as one of the "cataclysmic variables" (see: GUIDES/Observational/Variable Stars on the ASO website), V Sagittae is one of the most unusual of all irregular variable stars. Here are the facts: 1) the star is likely, like the famous SS Cygni, a double star which shares volatile gases with its companion, thereby resulting in "some" of the unpredictable light changes; 2) it can suddenly (within a day!) change brightness, up or down, by as much as THREE (3) magnitudes; and 3) it can flicker in magnitude within hours by as much as one magnitude. V Sag has no known "period" that is predictable, although some pattern might be imagined looking at the light curve above, with some rapid fluctuation occurring about every 96 days. It is an IDEAL object for 4-inch and larger scopes, as it fluctuates between magnitude 9.1 and 12 without notice. The irregular variable stars deserve MUCH attention on a NIGHTLY basis, and they are FUN to observe! You can actually see the Universe at work with these stars in only a short period of time.

This star is located about 9,000 light years away from Earth.

To locate the star, use the "a" finder chart from the AAVSO at
https://www.aavso.org/apps/vsp/ . Note for these charts, simply type in the NAME of the variable at top to generate your choice of chart,
and once found (particularly when dimmer) use the narrow field "g" chart for comparison by downloading to a file, resizing to print and print out a field star chart for this star! This is one variable star worth keeping up with!

Object 4 - A Totally Unique Variable(?) Star - A
This may well be the most unique of all variable stars that you will observe; it does not change much, but just watching it routinely may pay off in the

long run. You see, this star "started out" back in around 1870 as a very dim, perhaps 14th magnitude star of little interest. For many years it remained at 14th magnitude....then, around 1900, the star began to INCREASE its brightness very predictably (see the light curve below).

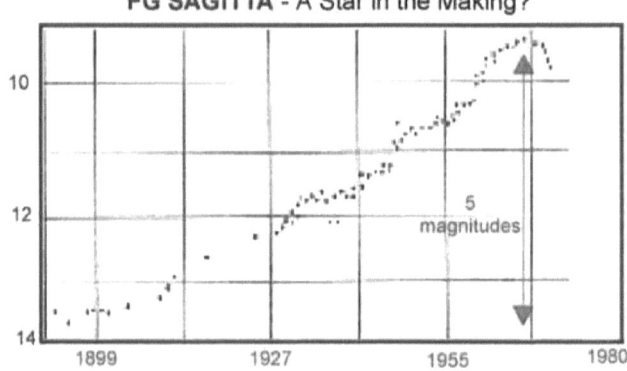

So far, it appears spectroscopically that this is a supergiant star, and it most definitely has been confirmed to have a large disk of nebulosity around the star, much like a planetary nebula. From the velocity of the disk's expansion and the distance of 8150 light years for the star, the shell - and the image of the star that produced it - is only about 3000 years old. Whether or not this is a star in formation.....or an extremely slow nova....there is nothing else out there quite like FG Sagittae.

This is one variable star that the AAVSO does not have a confirmed comparison/finder chart for; this SAO adapted chart shows the position of FG.

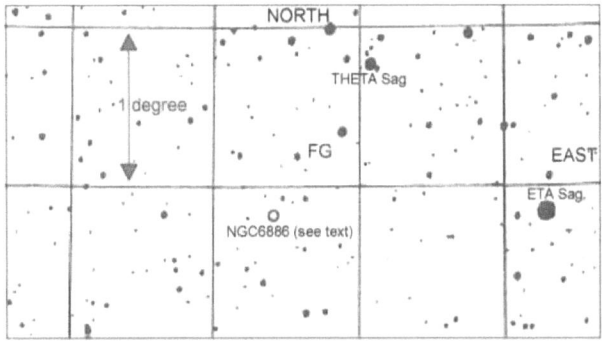

The most unusual "nova"? / variable star FG Sagittae

Note how it is located relative to Theta (up and to the right) and Eta (right and down a bit); this will greatly help you confirm this star since it forms a neat little triangle with these two brighter stars. REMEMBER....north is at the TOP and EAST to the RIGHT in catadioptic scopes, so the chart is oriented correctly "up and down," but is a mirror image left-to-right.

Object 5 - W Vulpeculae - A nice Semi-Regular Variable Star

This star is well within reach of all telescopes and is an ideal object for study with small telescopes, with its magnitude range of 8.3 to 10.2. Although it has a "period" from brightest to brightest again of some 234 days, it has major fluctuations in between and many times the period changes by many days, weeks or even months. This is a VERY "old" M-type Red Giant star and at brightest this color is clearly evident a medium power, particularly when comparing with many others stars in the same field of view in a wide field eyepiece. Since there is curiously no AAVSO chart for this important star, use the chart found at:

https://www.aavso.org/apps/vsp/
Note for these charts, simply type in the NAME of the variable at top to generate your choice of chart – for "R Vulpeculae" for comparison as this is very close to the position of this target star.

Object 6 - Theta Sagittae - A Very Nice "Triple" Star

Although listed in many sources as a "double star", Theta Sagittae has a third OPTICAL companion that makes this a really nice and easy sight for our telescopes. The true "physical" double star pairs Theta (magnitude 6.5, at the "end" of the "arrow" in the figure at the beginning of this "GO TO" tour) and an 8.5 magnitude star about 12" arc distant, and easy target both in separation AND in magnitude for a 3-inch and above telescopes! Use about 15x per inch for best views in all scopes. Although you will not have to "hunt" for it, look for the 8th magnitude companion star NW of Theta. The third, and more distant (optically) star is a 7th magnitude star that will be easy to spot 1.2' (minutes arc) SW of Theta; all but the last star are "F" type solar stars.....the 7th magnitude star is a very "late" spectrally RED star (it can be seen as such). For best views of the yellow-red color contrast between this and Theta, use about 20x per inch, or medium magnification.

Object 7 - Another Very Nice Double Star - "2 Vulpeculae" Great test a 3-inch scope

This is a good test double star, with a separation of ONLY 2.0" (the theoretical resolution limit - "Dawes' Limit" - for a 3-inch is 1.3" arc). The star 2 Vulpeculae is a relatively bright magnitude 5.4,

while its optical companion (not a true double) is a relative faint 9.5, therefore making it a
challenge NOT ONLY for a 3-inch, but perhaps a 5-inch scope as well. Can YOU see the secondary star? The faint companion is due SE of the brighter star; try masking out the brighter star with a crosshair element eyepiece and you should see it just fine. The primary star of this is ALSO a variable star: ES Vulpeculae with an extremely rapid period of ONLY 0.609 days, changing from 5.4 to 5.5 rapidly. These changes are NOT visible to the naked eye, only through sensitive photoelectric and CCD electronic sensors. This is a very nice target for an 8-10 inch telescope.

Object 8 - Another Double - "16 Vulpeculae", but this one challenges even an 8-inch! Not for the impatient!
Okay, all you "big dogs" with an 8-inch....try this one. Here is 16 Vulpeculae, a VERY difficult but definitely resolvable, double star. Its current separation is ONLY 0.8" arc, right at the limit of the 8" scope if you are very well collimated. The primary star is magnitude 5.5 and the secondary is very close to that, magnitude 6.2, and located just south of due east from 16 Vul. You will need the steadiest of nights and a rock-solid mount since you will have to "jack up" the magnification to around 400x to split this difficult object! By the way.....I know someone who has done this with 5-inch at 405x! Smaller scopes: don't even bother with this one.

Objects 9 - A Globular? A Galactic? A Globular? Messier 71 - An Unusual Star Cluster!

Here we have Messier 71. Look for this object halfway from Delta Sagittae to Gamma and only about 10 degrees due NORTH of the bright star Altair, in Aquila the Eagle. At present I cannot tell you that this is a globular (see photo from the Mt. Wilson 120" reflector below) OR perhaps a VERY compacted galactic cluster that "appears" to be a distant globular.

Messier 71 – Considered to be a globular star cluster, but possible a very distant and dense open star cluster
*   *   *

With a 3-inch it appears as a fairly large and faint hazy patch of light, clearly visible on a dark night. At magnitude 8.9 and a fairly large 6' arc, this should be an easy object in that scope and also should be visible in smaller scopes. However, using a 6-inch and larger, many stars of magnitudes 11 to 15 can clearly be seen in this interesting object. Look for another very interesting cluster - Herschel

20 - only about 1.5 degree SSW of M-71, clearly in the same field of view in a low power wide field eyepiece. H-20 is ALSO visible (see the NGC listing above) in a 4-inch at low magnifications....it is about 10' arc across so it is very spread out! Although it is NOT known if this is a globular or galactic cluster, its distance of 18,000 light years suggest to me that this is indeed one of the many Milky Way companion globular clusters at the same distance and location.

Object 10 - A Very Faint and TINY Planetary Nebula - ngc 6897 in Sagitta

Here is one strictly for the larger telescope, ncg6897. This planetary nebula still exhibits its central star at magnitude 15.1. The nebula itself, at a brightness of only 11.2 (visual) is a very difficult object even in a 6-inch and 8-inch, as it extends only a 5" arc diameter! Tiny! Thus, what you can expect to "see" here is something smaller than the planet Uranus looks at the same power. Look with the larger scopes for a very faint "out-of-focus-appearing" star with not much size to it; it will be clearly distinguishable from the many other stars seen in the field at an optimum magnification of about 180x to 225x since it will appear to have a small "glow" around it....other than that, there is not much detail visible. Remember, however, you are looking at an incredibly distant object - over 2,400 parsecs - that seems to be somehow "paired" with another planetary nebula only 3 degrees to its north, ngc6886 (see above listing) which also is visible in the scopes 6-8 inch range; both are located at the same distance from Earth.

Object 11 - A Very Large and Sparse Galactic Open

Cluster - ngc6800 - wide field small APO is ideal scope

Here is an object that is magnitude 8.5 overall, but is spread out over such a large area (15' arc - about half the moon's diameter), that it actually is a very difficult object, particularly in large telescopes with narrow fields. The ideal scope for this might be the small, compact APO refractor, with its wide field of view and a magnification of around 50x. Only here can you get the "cluster effect;" in scopes with a more restricted field, the object appears to merely be many stars of about 10th magnitude spread out across the field of view. Look for this nice low-power cluster only 30' arc NW of alpha Vulpecula (see above), and note that there are about 25-30 stars ALL of around 10th magnitude that should be visible in all scopes.

NGC 6940 – very large and dense open star cluster

* * *

Object 12 - A Very Nice Galactic Star Cluster in

Vulpecula - ngc6940 - Why is this NOT a Messier object?

This is a nice and usually overlooked object for small amateur telescopes, a large galactic star cluster containing at least 100 stars! All stars are in the 9th magnitude range so this is an EXCELLENT object for a small telescope as well as the larger telescopes using a very wide field of view since the cluster is spread out over a 20' arc area. Overall this is a very bright cluster, with a combined visual magnitude of 8.2; the stars and the true cluster effect should be witnessed in all telescopes, but like the previous object, low power and wide fields are desired for the best view....if you do NOT have dark skies, this object cannot be fully appreciated!

Object 13 - An Interesting Cluster with an Interesting Shape - ngc6802

Yet another galactic cluster in Vupecula, but this one is quite interesting, particularly in the 6-8 inch telescopes....this one is shaped distinctly like a "bar," or an elongated rectangle. It is not at all as large as ngc6940 above, show an overall expanse of only 3.5' arc. Thus it requires some magnification to make out this shape as well as the very faint stars of magnitude 13 through 18. Although a few of the "brighter" stars will be glimpsed in a five-inch, the rest will appear as a very faint 11th magnitude glow at medium-high (i.e., 160x to 200x) powers; the LX 90's greater light gathering and faster focal ratio will show more stars and perhaps be better suited to demonstrate the "bar" shape; I can barely detect this cluster as "....an oval glow" in 3-inch scopes under very dark skies.

Object 14 - *THE DUMBBELL NEBULA* –
Messier 27

Yes, we discussed this object in our "GO TO" TOUR of Cygnus in Volume One, since most people "think" this beautiful object is in that constellation, just off the "bottom" (southeast) of the Northern Cross star, Albireo. So for a full discussion of this BEST OF PLANETARY NEBULAE, I refer you to that constellation's guide. However, I was able to revisit this wonderful object on a very dark night at the Observatory and used a variety of telescopes on it.

Messier 27 – The Dumbbell Nebula
Taken by the author with a Celestron 14 – 120s
*   *   *

I was able to get an exceptional look at this object, which draws you back and back like a magnet in late May on a moonless night from the Mountain. With an 8-inch telescope the magnitude 13.4 central star was clearly visible as was another slightly brighter star embedded (actually in the foreground) in one of the gaseous extensions. This star was clearly visible in the smaller telescopes in moments of very good seeing, but the central star was never seen. There was detail within the gas cloud itself seen in both scopes at 100x (20mm eyepiece), making a gorgeous impression. I found that with all telescopes about 100x was ideal for this very large nebula. In the 24" at Newtonian focus this objects just blows you away, with countless striations of filamentary gases visible coming from the central star. M-27 literally fills the field of view of a 35mm Televue Panoptic eyepiece at f/4! Much of the visible detail, and the stars discussed can be seen in the photograph snapped quickly one night with a Celestron 14 telescope and CCD camera.

In my opinion, Messier 27 is by far the most outstanding object of the summer skies (and there are plenty of contenders to choose among!)

WANDERING ABOUT....YOUR NEW "USER OBJECT" IN VULPECULA

One of the skies BEST variable stars is found in the modest constellation of Vulpecula.
R VULPECULAE is located at: **R.A. 21h 02m / DEC** and is included in your variable star database library on Autostar and other quality star programs. Nonetheless this is a wonderful star for casual variable star observing that allows you to make periodic brightness estimates while cruising around the summertime skies! So for this installment of your "GO TO" TOURS we are going to add this (using the coordinates above) to the USER

OBJECT library which we are compiling from these tours.

For the AutoStar, go to: "Select/Object [enter]...." scroll down to "User Object" [enter]. Now enter the coordinates given above for "R VUL", using the number keys on AutoStar. After entering the coordinates and pressing "Enter" yet again, scroll down one and you can list the magnitude of the object as "7.5"[Enter].

Remember that your keypad, App or sky program will vary from the techniques for Autostar....but all are logically the same. Please read your User Guide or HELP to learn the process to produce your own User Object library for your computerized scope!

Now a bit about the object. For more detailed descriptions of it, consult the AAVSO at www.aavso.org for all their observing literature. To LOCATE among the field of stars and to COMPARE its changing brightness when "brightest", use: https://www.aavso.org/apps/vsp/ . Note for these charts, simply type in the NAME of the variable at top to generate your choice of chart – from which is the "a" low power and wide field chart to actually locate the actual star; once found and particularly when at its dimmest, use chart "g" for fainter comparison stars. Both charts should be saved to file, opened and resized (and saved again at that size if you wish) and then printed for use at the telescope.

R Vulpecula is a spectacular "short" long period variable star of only 136.82 days, ranging from a relatively bright 7.4 magnitude to 13.4, still easily visible in a 6-inch on a very dark night; therefore the entire cycle of this wonderful star can be observed in about four months time! Even with a good 3-inch telescope most of the cycle can be monitored, with the approaches to and brightening after minimum the only parts lost (about 14% of its cycle). One reason I have selected this star for your study is its COLOR: RED! This is an M-class very late star on the spectral evolutionary scale is demonstrates an incredibly scarlet color during most of its cycle and particularly during its brightest.

So with the addition of a this wonderful and rapid variable star, you are adding yet another among the curious multiple stars, variable stars and weird deep sky objects. All of these are wonderful for conversation-starters and crowd-stoppers at the big astronomical events!

While you are in this area of the sky, why not visit the incredibly LARGE and LONELY constellation of CAPRICORNUS, the "sea goat" (what??). Unless you may have been "born under the sign" of this faint and obscure constellation it is likely that you have not spent much time there. In the next installment.....you will see why. JUST KIDDING! We are going to concentrate mostly on DOUBLE and MULTIPLE stars in Capricorn as there are many fantastic such stars there! And one lowly Messier Object, the beautiful Messier 30, one of the finest globular clusters in the sky!

See Volume One for your complete guide to the large offering of Capricornus.

*   *   *

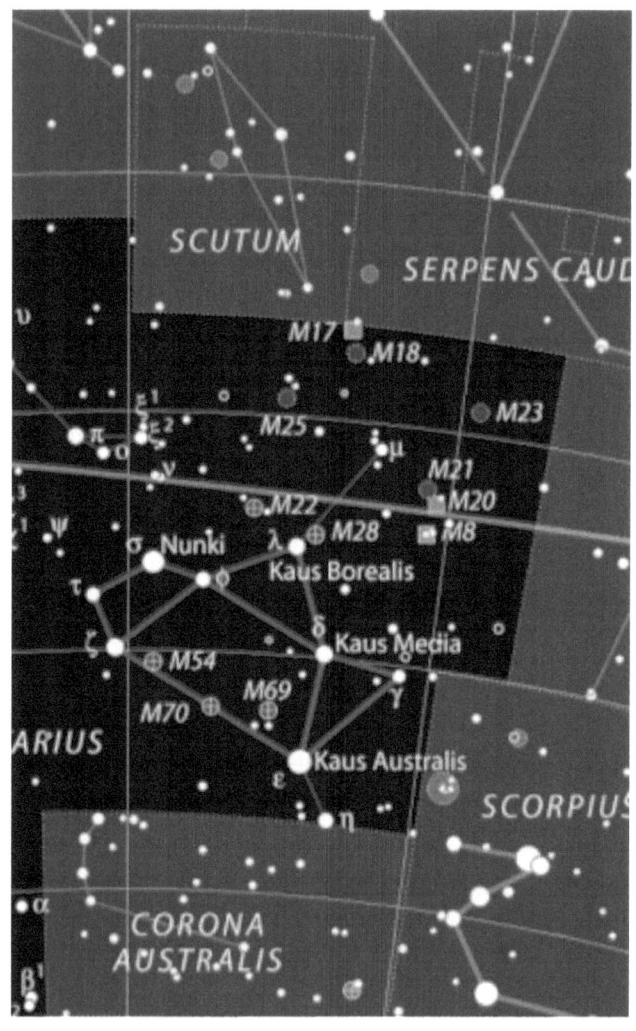

Sagittarius/Scutum Star Chart
Adapter courtesy from the
IAU / Sky and Telescope Charts

Chapter 31

SAGITTARIUS and SCUTUM
The Keen (...or perhaps near-sighted) Archer of the Sky

In will now "GO TO SAGITTARIUS" and explore those incredible sights while we are penetrating the very **core** of our Milky Way galaxy of over 200 billion suns. We are looking deep into the very nuclear center of the galaxy, through dense and rich star clouds, heavy concentrations of dark light-obscuring nebulae, and many wonders of deep space. As we approach this center, the density of the stars per parsec increases and the population figures become staggering. It is through this rich stellar field that we should take time to spend time, scanning slowly with our eyes, binocular, and our telescopes at the magnificent wonders that only this vantage point - Earth - has to offer.

It is worth your time and efforts to find yourself a "rock" somewhere in the secluded darkness of a moonless night and allow your mind to rest and wander as your eyes become more and more adapted to the subtle contrasts of a night only illuminated by the soft glow of the stars within our Milky Way; soon, multitudes of the most minute of images begin to emerge in your sight and in you mind as the brilliant Sagittarius Star Cloud takes shape and form. Soon....it is like watching the sunlight reflect from the sparkling grains of sand on the cleanest of beaches.

To many, the familiar brighter stars of Sagittarius outline what is commonly referred to as the "*teapot*," which can be quickly seen in my outlined diagram below which compares it to the traditional constellation pattern of the "celestial archer." The brighter stars (see the list below under "Object 1, Nunki") outline a perfect traditional teapot, complete with a pointed lid, spout and handle!

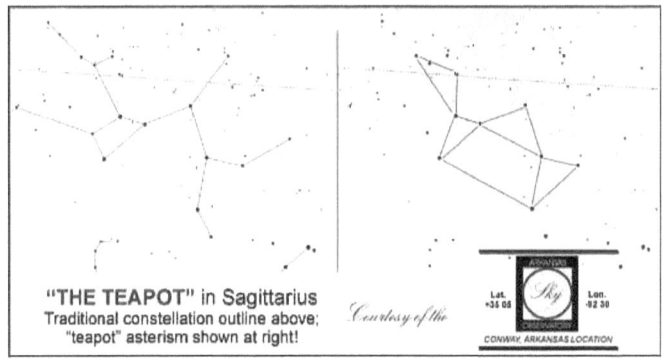

"THE TEAPOT" in Sagittarius
Traditional constellation outline above; "teapot" asterism shown at right!

This area of the sky is so rich with clusters of stars ("galactic" or "open star" clusters), bright diffuse nebulae from which stars are actually "born," dark dust that blocks light and energy form the galaxy's core, hundreds of globular clusters.....and infinitely more.....that it boggles the imagination even to begin our TOUR. With so many objects to observe, most of them beautiful beyond belief, how do you choose which to spend your time on? Answer: do not choose, observe them all....they will be there for a lifetime!

So....this "GO TO" tour of Sagittarius, the great (or perhaps not-so-great) celestial archer, will be somewhat different than previous tours....we will have be a very short introduction into the overall history and views of this magnificent star field, and then we will concentrate exclusively on DEEP SKY objects for this tour - no doubles, no variables....just interesting DEEP SKY objects.

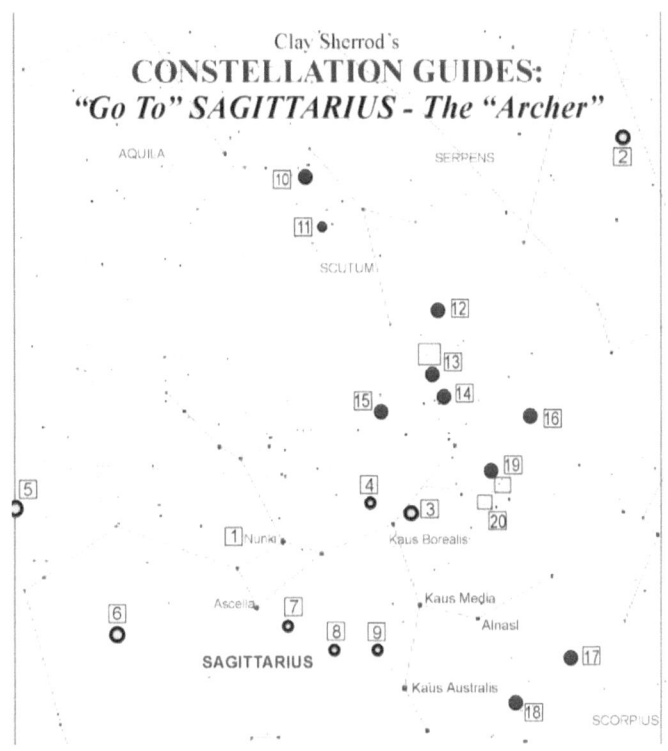

A finder chart for locating many of the GO TO objects in the constellation of Andromeda; if using a computer planetarium program, you are encouraged to plot the objects on your screen for higher resolution than this chart provides.

*   *   *

Sagittarius is PACKED deep sky objects and I have been able to whittle down my list to only the **20 finest** or most unusual objects. As you can see from the Arkansas Sky Observatory Constellation Chart above, there is a very high concentration of objects for us to view....but this is only the best objects that I have selected. For every ONE that I have listed, there are hundreds of NGC, Caldwell, named objects, double stars, tremendous variable stars, dark nebulae and on and on.... Your AutoStar directory has many such NGC objects listed....double and multiple stars as well!

Since we are essentially looking right into the core of the Milky Way Galaxy, you would expect the greatest number of all sky objects to be packed into the hub of our great system of hundreds of billions of stars....and indeed they are.

Note from the sky chart included here that the *ECLIPTIC* passes through the northern portions of Sagittarius; hence this is one of the twelve *ZODIACAL* constellations in that the planets, moon, sun and earth appear to move centered on the defining "middle point" that we call the "ecliptic."

This is NOT to be confused with the *CELESTIAL EQUATOR, which* passes far to the north of Sagittarius. That line is the reading "0" degrees on your properly adjusted declination setting circle. All angles NORTH of this equatorial line are positive ("+") and all angular measures (declinations) south of the celestial equator are negative ("-"); hence you will see references in this "GO TO" GUIDE to both "+" and "-" declinations for celestial objects.

Hence, even the HIGHEST OBJECTS in Sagittarius appear very low in the southern skies for northern hemisphere observers.....and ideally situated for our friends south of the Earth's equator.

Each GO TO object is discussed for your telescope regarding the type of conditions necessary for you to view it optimally for discern the very faintest details....double star challenges for each size telescope .....magnifications and aperture necessary for most objects, and much, much more. This is YOUR complete guide to get you on your way to exploring this large and interesting constellation.

Although there are MANY wonderful and colorful double and multiple stars that should be examined in your scopes, I have limited this "GO TO" tour to only deep sky objects. So, to find the names and locations of the best double stars, consult a good handbook, such as the "*Burnham's Celestial Handbook*," Vol. 3 for a very comprehensive list of locations, magnitudes and angular separations of these wonderful stars. There are many stars for EVERY telescope size and type.

Use this attached star chart and the following Guide as an excellent reference for your next star party itinerary, or a beginning for further study into the thousands of objects visible in this part of the sky. Truly these extensive *Constellation* study guides will most definitely put your sky program to work for you in the most efficient and enjoyable way possible! As a matter of fact, MANY AutoStar and PC sky program users are now programming their own "Tours" based on these guides, using each constellation as a separate GO TO Tour for the sky library that can be added in or deleted through the main edit screen on your PC or MAC computer.

Many observers have now equipped their telescopes to couple with smart phones on pads via wireless to completely control their observing at the touch of a tiny smart devices screen via sky Apps such as Sky Safari and others!

We hope you enjoy these comprehensive GUIDES to touring the constellations via your AutoStar or other sky program and its computer-driven telescope. Each new installment is complete with diagrams, charts and illustrations that you will find nowhere else. Please let us hear YOUR feedback and your observations of each and every constellation after YOU have toured its vast reaches of our skies!

Introduction to Sagittarius

In many cultural tales of old, the great archer Sagittarius is pursuing the huge *Scorpion* (Scorpius) who struck the fatal sting to the exposed foot of the great hero **Orion** and thus ended his seemingly never-ending ability at immortality. But in other tales, the archer clearly missed his target, as we see the arrow (the constellation "**Sagitta**" clearly gone astray, well above **Aquila** the eagle....in the opposite direction!

This is the ninth constellation of the Zodiac, following Scorpius and preceeding Capricornus. The mythological image of Sagittarius is that of a *Centaur*, apparently somewhat in conflict with our modern southern hemisphere constellation of **Centaurus**....who obviously is: also a Centaur!

It has long been speculated that there would be at LEAST twice as many Messier objects if *Charles Messier* had been observing and recording his "faint fuzzies" much further south in latitude than his homeland of France. Of the hundreds of naked eye and binocular objects throughout the Sagittarius Milky Way that are literally wonders to behold, he noted only 16 objects in this constellation, clearly as a result of atmospheric extinction to the south. "Atmospheric Extinction" is merely dimming of celestial objects as their light passes through the dense "lens" of air which increases in thickness as we level out near each horizon.

You will clearly see as you TOUR through this constellation that there are MANY MORE THAN 16 remarkable objects that you pass as you GO TO this guide's selected objects.....many objects that are seemingly ignored simply because they do not have that "Messier blessing...." Most are referred to in the **NGC** list (New General Catalog) and can be found in a good star atlas or reference list, and nearly all - once identified properly - are found by NGC number on your Autostar or sky program under "Select/Object/Deep Sky/NGC.....[scroll through list]." Many such objects can be found with the naked eye, or scanning with binoculars, very deep and low in Sagittarius.

YOUR SAGITTARIUS CONCISE DIRECTORY OF INTERESTING OBJECTS –

In Sagittarius are more objects visible in all scopes - from the wide field APO refractors, to the light-grabbing 8-inch and larger telescopes - than you could possibly see in a lifetime. Interestingly, these are NOT mundane objects, either as you might imagine could fill some otherwise "empty" pages of a constellation like....well, Aries, devoid of a wide range of interesting objects. It was very difficult to select (and limit) the number of objects for our "GO TO" TOUR. Only the best and in some cases, most challenging, objects are chosen for this brief tour.

I have chosen the finest 20 objects in this Sagittarius "GO TO" tour; as with all guides these objects listed below will be visible in all telescopes (some naked eye) from a 3-inch through an 8-inch; of course larger apertures may "show" an object a bit closer and "better," but frequently a wide field and low power view is more desirable than aperture. This is the case for MANY of these objects since we are looking directly into the very star-cloud-rich areas of our Milky Way galaxy. Indeed, I strongly encourage you to step away from the telescope often and scan the beautiful open skies and star fields with a good pair of 7 x 50 or 10 x 50 binoculars. You will be tempted to venture to your "dark sky site" for a full evening of laying back on a blanket and scanning the skies with a good star atlas, a red flashlight, and a dependable "observing buddy."

I have many times just stood in awe for over several hours and peered at the beauty that is the space toward Sagittarius....you can never see this sight too often. Once your eyes are fully dark adapted, you will even be able to see *The Great Sagittarius Star Cloud* (see the wide-field photo below) appearing as its namesake a very open and bright cloudy object that is huge in the open sky, sprinkled with thousands of tiny naked eye stars. Even the binoculars will show dozens of deep sky objects that cannot be fully appreciated in large telescopes with limiting fields of view!

The darker adapted your eyes become....the more you will be able to see; in fact the Milky Way through Sagittarius has been know to even cast a shadow on the ground from observers in very dark skies!

The Great Sagittarius Star Clouds and objects within. Chart courtesy Sky & Telescope

* * *

You will note that this list of objects for Sagittarius is somewhat unique from other "GO TO" tours in that the objects are all grouped by category: 1) your "starter star", as always starting with a prominent star in the area to get you off and running; in this case it is the star "*Nunki*", or sigma Sagittarii; 2) prominent globular clusters; there are some wonderful globulars mixed among the dense star-rich skies of Sagittarius; we will START with a familiar globular (Messier 14) which we have examined in Ophiuchus and proceed to check out all eight (8!) fine globulars in Sagittarius....Messier 14 will be your "marker" for comparison to the somewhat more difficult globulars in Sagittarius; and, 3) galactic star clusters and nebulae....so many clusters in this region have associated nebulosity with them that I have grouped this together!

For example, the fine star cluster Messier 18 is adjacent to the fabulous "Omega" nebula.....famous "*Trifid Nebula*" which can clearly be seen in our scopes is right next to the star cluster Messier 21!

Thus, for this "GO TO" tour guide for Sagittarius, we are breaking rank and NOT presenting the Messier objects in sequential numerical order as normally done.

As with all of the "GO TO" TOUR constellation lists, I recommend a good star atlas and/or chart which will list all the finest objects, constellation-by-constellation. One very handy reference guide is the *PETERSON FIELD GUIDE TO THE STARS AND PLANETS*, which features complete lists with declinations, right ascensions, magnitudes, and all pertinent information for you to expand your observing horizons beyond this brief guide.

FOR THE AUTOSTAR and similar scope controllers: Note that your handbox will NOT have every object listed on every constellation GO TO tour....this is intentional. You can access some of the most interesting objects of the sky directly from their coordinates. It is quite simple as you merely enter these coordinates as instructed on your HELP or User Guide for your particular program.

The constellation tour Star Chart will get you started on your journey for this constellation.

Following is the concise object list for your "GO TO" tour of Sagittarius; you may wish to find the majority of the objects from sky Library (for example, you can easily go to the "Lagoon Nebular (Messier 8) " if you pull up "Object/Deep Sky/Messier Object/..type in '08'...." and then press "Enter", followed by "GO TO" to slew to one of the most beautiful bright nebulae in the entire sky. On the other hand, if you want to experiment and become a "better computer user" try entering the exact R.A. and DEC coordinates of that object as described. The capability of acquiring unlisted objects is fantastic!

The exact process for bringing up a celestial object for your telescope to GO TO will vary with the sky program, telescope keypad, or App that you are using, but all fundamentally work the same logic to acquire your object as well as build your User Object Library. Always consult your HELP or User Guides of your specific program for the exact steps in using tabs and keystrokes to pull up and GO TO your objects!

OBJECT 1: bright star - NUNKI (sigma Sagittarii) - R.A. 18h 52m / DEC (-)26 22 - Magnitude 2.1
OBJECT 2: (reference starting object in Ophiuchus!) globular cluster - Messier 14 (ngc6402) - R.A. 17h 35m / DEC (-)03 13 - Magnitude 8.5
OBJECT 3: globular cluster - Messier 28

(ngc6626) - R.A. 18h 22m / DEC (-)24 54 - Magnitude 7.3 (nice!)
OBJECT 4: globular cluster - Messier 22 (ngc6656) - R.A. 18h 33m / DEC (-)23 58 - Magnitude 5.9 (super one!)
OBJECT 5: globular cluster - Messier 75 (ngc6864) - R.A. 29 03 / DEC (-)22 04 - Magnitude 8.0
OBJECT 6: globular cluster - Messier 55 (ngc6809) - R.A.19h 37m / DEC (-)31 03 - Magnitude 4.7! (but very low!)
OBJECT 7: globular cluster - Messier 54 (ngc6715) - R.A.18h 52m/ DEC (-)30 32 - Magnitude 7.1
OBJECT 8: globular cluster - Messier 70 (ngc6681) - R.A. 18h 40m / DEC (-)32 21- Mag: 9.6 - tough one! small!
OBJECT 9: globular cluster - Messier 69 (ngc6637) - R.A. 18h 28m / DEC (-)32 23 - Mag: 8.9 - not easy, very small!
OBJECT 10: (in Scutum - YOUR REFERENCE GALACTIC CLUSTER) the best open cluster - Messier 11 (ngc6705) - R.A. 18h 48m / DEC (-)06 20 - FANTASTIC!!
OBJECT 11: (in Scutum) reference open cluster - Messier 26 (ngc6694) - R.A. 18h 43m / DEC (-)09 27 - small, 20 stars
OBJECT 12: (in Serpens....to compare with Trifid nebula and associated cluster!) galactic cluster with nebula - Messier 16 (ngc6611) - R.A 18h 16m / DEC (-)13 48 - 6.4 mag. 60 stars!
OBJECTs 13: Messiers 17 / 18 (nebula and cluster "*Omega Neb.*", ngc6618) - R.A. 18h 18m / DEC (-)16 12 - beautiful!

OBJECT 14: galactic cluster - Messier 24 (ngc6603) - R.A. 18h 15m / DEC (-)18 27- bright! 45 stars!

OBJECT 15: another galactic cluster - Messier 25 (IC4725) - R.A. 18h 29m / DEC (-)19 17 - very similar to Messier 24

OBJECT 16: What... Another galactic cluster? - Messier 23 (ngc6494) - R.A. 17h 54m / DEC (-)19 01 - 120 stars!!!

OBJECT 17: (in Scorpius - for comparison, lest you forget....) Not satisfied yet?....galactic cluster - Messier 6 (ngc6405) - R.A. 17h 37m / DEC (-)32 10 - nice, very LOW!

OBJECT 18: (also in Scorpius, a repeat visit....) galactic cluster - Messier 7 (ngc6475) - R.A. 17h 51m / DEC (-)34 48 - bright but tough on horizon.

OBJECT 19: wow! *TRIFID* nebula (Messier 20 - ngc6514) and cluster M-21 (ngc6531) - R.A. 17h 59m / DEC (-)23 02

OBJECT 20: *The "Lagoon Nebula"* - a MUST SEE object! Best of the best - Messier 8 (ngc6523) - R.A. 18h 02m / DEC (-)24 20 - VERY large, bright, breathtaking!

A VISUAL GUIDE TO OUR DEEP SKY OBJECTS IN SAGITTARIUS (with help from Serpens and Scutum!)

Object 1 - Bright Star - "*Nunki"* (sigma Sagittarii) WHAT happened here....where's "Alpha??"
Every tour stars out with the brightest star in the constellation or area we are exploring.....normally that is the "Alpha star" from the early Bayer designation of Greek alphabet letters to designate the BRIGHTEST (alpha) to the dimmest (omega). But something has happened in Sagittarius. We are

starting our "GO TO" tour in this case with the star *NUNKI*, or "sigma Sagittarii." So why not Alpha (??) the star known as "*Alrami*"?

Here is a quick look at the actual visual magnitudes of the brightest stars of Sagittarius (those mostly comprising the "teapot") are:

alpha -	Mag. 4.11 - "Alrami"
beta (1) -	Mag. 4.24 - "Arkab" ;
beta (2) -	Mag. 4.51 - "Arkab Posterior"
gamma -	Mag. 3.07 - "Nash" / "Nushaba"
delta -	Mag. 2.84 - "Kaus Medius"
epsilon -	Mag. 1.82 - "Kaus Australis"
zeta -	Mag. 2.71 - "Ascella"
lambda -	Mag. 2.94 - "Kaus Borealis"
sigma -	Mag. 2.14 - "Nunki"

So you can clearly see that SIX of Sagitarrius' brightest stars are DIMMER than Nunki, which actually outshines all the named stars of the constellation except "Kaus Australis!"

Nunki's dominance in early mythology and cultural lore makes it even more puzzling as to why **Bayer** did not designate this brighter star higher (at least "beta") in the Greek alphabet where the brighter the star the "earlier" its designation in the alphabet. From the **Babylonian sancripts**, Nunki was the star that signaled the rising of the waters: the "*water constellations*" of Capricornus ("sea goat"), Aquarius ("water bearer"), Pisces ("the fish"), Cetus ("the whale") and Pisces Austrinus ("the southern fish"). Referring to the "teapot" shape of the constellation, the ancient Chinese referred to Nunki as the "point star", where the handle connects to the "bowl" of the ladle that also can be imagined out of this asterism.

Object 2 - Globular Cluster in Ophiuchus (something to whet your appetite - Messier 14
To me, one of the most uninteresting of all globular clusters has been the relatively bright, but un-resolvable globular cluster Messier 14 in Ophiuchus; hence, for the element of torture, we will start out globular cluster parade for this "GO TO" TOUR with that object so that you might compare with those seen following. For a full description of Messier 14 and its locale, please refer to my Ophiuchus guide here in this volume.

Object 3 - Messier 28 - A Fine Globular Cluster in Sagittarius
This is a very bright an highly concentrated globular located in the same wide field view as bright lambda Sagittarii, only 3/4 of a degree NW. It is extremely compact and will appear very much like M-14, above, except at magnitude 6.8 will appear brighter. It is VERY small compared to our next object, M-22, as it extends only 4.7' arc! Do not expect to see any stars resolved, even with the 8" scope; however, this is a bright and fine object - a bright "snowball" - in all telescopes; it makes a particularly outstanding sight in very low powers amidst the bright stars, particularly with the wide field APO refractor scopes.

Object 4 - Now you are spoiled - Messier 22
Only 3 degrees (same binocular field!) from the rather mundane Messier 28 is the bright and large globular cluster Messier 22, which is twice as close to us as the fainter globular. At over 21' arc and

magnitude 5.9, this is truly one of the BEST globular clusters for viewing! It is huge in low power and wide fields, and is a memorable sight on a very dark and moonless night. It is equal in every respect to the "granddaddy" of globulars, Messier 13 in Hercules.

Some star hints can be appreciated even with the small refractors will clearly resolve stars around the perimeter of this large cluster. The 6-inch shows stars all the way to the core, while the 8-inch and larger can actually reveal the colors of the brightest red giant stars toward the middle of this remarkable cluster. For the record, most of the stars we see in this cluster of about 100,000 stars are magnitude 11, clearly definable in the 3-inch. It is one of the most "resolvable" of all globulars, perhaps even easier than Messier 13, except for its low altitude on the southern horizon.

Object 5 - Another Globular Cluster - Messier 75 – "tiny and tough!"
By astronomical standards this may be the most tightly packed - or seemingly so - of all globular clusters. Only Messier 80 in Scorpius (*Constellations* Vol. Two - Scorpius) comes close to this degree of star concentration. However, it may be an illusion because of the TREMENDOUS distance of this object - almost 100,000 light years away! Messier 75, just on the Sagittarius/ Capricornus border is beyond resolution in our telescopes and requires greater than an 18" to begin to see stars distinctly at magnitude 17. Nonetheless, this very distant cluster bears observing as one our most distant objects within the Milky Way galaxy; be careful with smaller telescopes to not confuse this 8th magnitude cluster with a faint star with haze.

Object 6 - A Good, Large and Very Open Globular Cluster - Messier 55
Messier 55 is an unusually large and open globular; although its total magnitude is a bright 4.4, its brightness is spread out over a large 10' arc area, making it more difficult than it would appear. Nonetheless, look for some resolution with a 3-inch and very good resolution toward the center with the 6 and 8-inch scopes. Even the small refractors should show some "mottling" of stars around the edges. The sharp-eyed observer will clearly notice that M-55 is elongated in a north-south direction. Although most visible stars are about 13th magnitude, the "glow" that you will see in small

telescopes is a result of thousands of stars to magnitude 17 and beyond in this very large object.

Object 7 - Easy-to-Find Globular Cluster Messier 54

Here is an easy object and a very spectacular sight in the wide field views of the small APO refractor, as well as with the good wide field oculars of the larger telescopes. It is only 1.5 degree west and slightly south of zeta Sagittarii (see chart). This cluster is a bit different than most and certainly different than M-55 above in that it is a fairly uniform brightness across, but suddenly increasing dramatically in brightness toward the center, almost like a large nucleus. It is a VERY small globular, only 2.1' across, and very difficult to resolve; look for only peripheral stars to be seen using averted vision with the 8-inch; smaller scopes show only a diffuse glow with a clear bright "core."

Object 8 - Another Globular....and this one is tough! Messier 70

This object and Messier 69 (below) make an interesting wide-field pair in the wide field refractor telescopes. Both globulars are very small objects. Messier 70 is only 2.5' across and a tough magnitude 9.1, making it almost star-like in low powers and wide fields. Medium magnification is best on both this and M-69, which is also only 2.9' across and magnitude 8.9. These clusters, like M-75 are very distant - some 69,000 light years - hence their faintness and lack of detail. Do not expect any detail in scopes under 18-22" diameter.

Object 9 - Messier 69 - Twin to M-70, above

Look for this globular, appearing almost identical to M-70, about two degrees west. Only about 4' arc north and a bit west of the cluster is a very reddish-appearing 9th magnitude star that makes an interesting contrast to the diffuse appearance of this roundish globular. For details of M-69, see M-70 above.

Object 10 - No More Globulars!
The 'Gold Standard' for Open Clusters –
Messier 11 (in Scutum)
If you can find a finer and more breath-taking open star cluster than Messier 11 (the "*wild duck cluster*"), then you are observing from a different planet. Messier 11 in Scutum sets the standards for all galactic star clusters, so I wanted you to remember this one first before moving on to the nice open clusters (but none to compare) in Sagittarius. For a full discussion of Messier 11, see *The Constellations*, Vol One, *Aquila* in this series.

Object 11 - Another Open Cluster in Nearby Scutum - Messier 26
Not particularly exciting, I wanted you to first glimpse Messier 26 AFTER seeing the magnificent M-11 and BEFORE going onto the clusters in Sagittarius to make you more appreciative of what you are about to see in the beautifully-star-packed wide field views ahead. A full discussion of Messier 26 is also found in the ASO Constellation Guide of Aquila here on this website under GUIDES.

Object 12 - Messier 16 - Galactic cluster in Serpens
This is included since it is part of the magnificent "arm" of stars of this region of the Milky Way which includes the marvelous Messier 24, the

Lagoon Nebula, the Trifid Nebula, Messier 23, and so many more packed areas of stars. In Serpens, this object is very tightly compacted, yet clearly resolvable in all of our telescopes. It appears as a very star-poor globular with stars resolvable even in a low power APO, most of which are within the same magnitude range. This makes a very fine low power object for larger telescopes using wide field eyepieces as it is in a great star field of the Milky Way; look for the incredible large cluster Messier 24 due east of this tiny group of stars, and the famous "Lagoon nebula" (M-8) almost due south. This object is just inside the right field of the following wide-field photograph.

Objects 13 - Galactic Star Cluster Messier 18 and "*The Omega Nebula*", Messier 17

The star cluster known as Messier 18 may be somewhat of a disappointment. At magnitude 7.5, it contains about 15 stars spread over an area of 12' arc. In a field wider than one degree, you can get BOTH Messiers 17 and 18 in the same field of view, with the cluster being about 1 degree south of the beautiful Omega Nebula, Messier 17. (SEE PHOTO BELOW) These objects are actually about 1000 light years apart in space, with the cluster being in the foreground from our vantage point. Messier 17, also known as the "Swan Nebula" and the "Horseshoe Nebula" is about 7' arc long and some 2' arc across with a visual magnitude of only 8.9.....misleading. This nebula can clearly be seen on a very dark night in even binoculars and is an outstanding object. If the sky is dark enough, be sure to look for the distinctive "hook" shape, conforming to the "swan's head." Like so many bright nebulae, M-17 appears to be associated with about 20 nearby stars, perhaps not unlike the association of the famous Orion nebula to the stars within its vicinity. Use very low powers to observe Messier 18 (the cluster) and about 75x in all scopes for the Omega nebula to increase the contrast of this very faint cloud of gas.

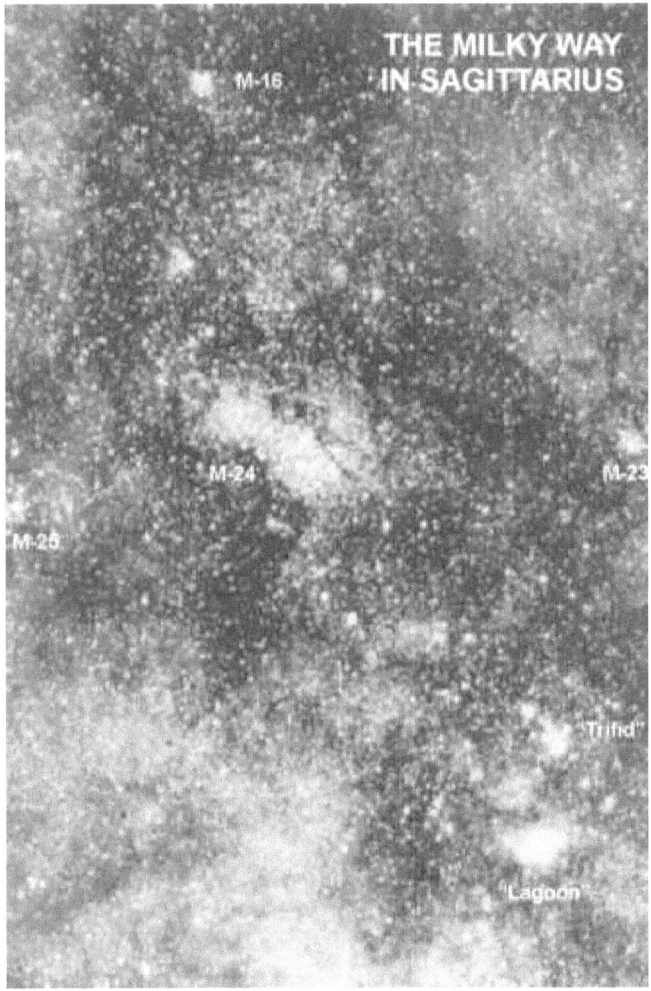

Object 14 - Huge Open Cluster - Messier 24 (see wide field photograph preceding)

Also known as the "*small Sagittarius star cloud*" this is an outstanding wide field object, and is clearly visible to the unaided eye on a moonless night. Although not accurately described as such by Charles Messier, this has been assigned as his 24th object....what he recorded was a tiny 4' arc cluster at this position, and surely this galactic cluster has not grown so greatly in 250 years! M-24 has a diameter equal to three full moons - 1.5 degrees! It is resolvable in small telescopes as a magnificent cloud of stars and larger telescopes do not do it justice. It is a true wide field object. Scanning through this cloud with very low powers reveals hundreds of thousands of seemingly tiny stars, all of nearly equal magnitude. Compare the prime location of this object relative to the other fine objects in the Sagittarius Milky Way.

Object 15 - Messier 25 - Galactic Cluster (see photo for location!)

Located immediately east of M-24, this may seem quite a letdown after the magnificent array of stars within Messier 24. However this is a very rich and deserving cluster of stars, easily resolvable in the ETX 60 and 70 and a beautiful sight in very low power views with the 4-inch and larger instruments. Note in low powers on a very dark night the "strings" of stars that are visible from the center of this cluster, notably in a NW direction (to upper left in your catadioptic eyepiece orientation), SW (lower left), and in a near eastward path (to the right). Since all 50 stars of this cluster are 12th magnitude and brighter, all should be seen on a very dark night with the 4-inch, and should be very easy in the 6-inch. There are perhaps 12 stars magnitude 9 and brighter and that gives a remarkable 3-D appearance to this 2000 light-year-distant cluster. One of the brighter stars is U Sagittarii, a notable Cepheid variable star (see my guide of Draco/ Cepheus in Volume One). This star, varying between magnitude 6.3 and 7.1 in a period of only less than seven (7) days, is an ideal star for beginner variable star observers. The finder chart I have adapted below has been REVERSED for orientation to match the field of view of catadioptic telescopes with NORTH at TOP, EAST at RIGHT.

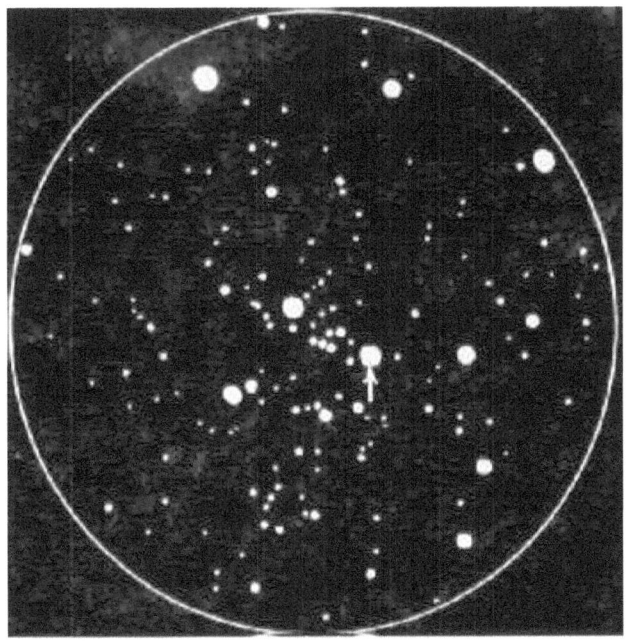

A FINDER CHART FOR THE VARIABLE U Sagittarii
NORTH AT TOP, EAST AT RIGHT

This finder chart is about one-half degree (about the size of the moon's disk) across, and very nearly

matches the field of view of a 6-inch with a 26mm Plossl eyepiece. For a more detailed observing chart complete with magnitudes of comparison stars, download, save to file, resize and print the American Association of Variable Star Observers chart for R and 350 Sagittarii (which will include U Sagittarii in the wide field of this "a" finder field chart) at:
https://www.aavso.org/apps/vsp/ . Note for these charts, simply type in the NAME of the variable at top to generate your choice of chart.

Object 16 - Another Galactic Cluster - Messier 23 – A MEMORABLE SIGHT! Do Not Miss This One!
This is a beautiful star cluster, very rich and located immediately west of M-24 about four degrees. Wide field eyepieces are a must for this grand and large cluster which contains no less than 120 stars and covers one-half degree of sky! If you observe with the larger 6 and 8-inch scopes, be sure to use your lowest magnifications and widest fields possible. This is an ideal object for the wide field a small APO refractor, with stars scattered like jewels strewn across black velvet of the sky. Star magnitudes range from about 8.5 to 12th, so nearly all stars are visible in at least a 3-inch. There is one brighter star of magnitude 8.1 to the northeast side of this large cluster and a very bright (magnitude 6.5) star about one cluster diameter to the northwest. Look for a distinct "chain" of tiny 10th magnitude stars that bridge from the center of the cluster to this brighter star. Note: this brighter star is NOT part of the 2000 light year-distant (same as Messier 25!) cluster.

Object 17 - Messier 6 - Open Cluster in Scorpius

(Can't Bypass This One...."in the neighborhood!") Be sure to re-visit this beautiful open galactic cluster in Scorpius while you are so close....it's only a bit to the west and sure to be a hit under dark skies! Almost the same size (25' arc) as Messier 23, above, but containing only about 50 stars, this 5.3 magnitude object is discussed in full in our "GO TO" TOUR guide to Scorpius under the GUIDES tab here on this webpage. I have put this galactic cluster in our current Sagittarius Guide as a comparison of the MANY Milky Way arm clusters that are all located at near the same distance from our Earth.

Object 18 - Messier 7 - Another Scorpius Comparison Galactic Cluster
Also in Scorpius, this one is so much different in appearance that it bears a second look in comparison to the very dense and rich clusters of Scutum and Sagittarius. M-7 contains only about 50 stars, all visible in our scopes at very low power, but these stars are spread over a LARGE on-degree area, and thus the true "cluster" nature of this object is all but lost in telescopes other than low power, wide field small ones . For more information see the Scorpius GUIDE here on ASO under the GUIDES tab.

Objects 19 - *THE TRIFID NEBULA* (Messier 20) and associated galactic cluster Messier 21
This object is a treat in nearly all telescopes used in very dark sky conditions. Located in the same wide field (about 1 degree) field of view, look for the galactic cluster M-21 as a grouping of stars about 12' arc diameter; the stars range in brightness from 8th to 12th, with about 10 of them brighter and the

rest dim. Look for a very yellow star with a close companion that should appear somewhat grey in color...you can't miss this pair - the yellow star is magnitude 9 and the grey one only a magnitude dimmer; use medium-high (about 125x) to view this pair.

The famous TRIFID NEBULA (M-20) is only about 3/4 degree from this nice cluster to the southwest (see accompanying photos). This remarkable object is familiar to all of us from photographs, as three equally-shaped and equally-bright wedges of "pie" of blue and red. Indeed, on a very dark night, the larger telescopes "may" be able to hint at this true color of the nebulosity. Often overlooked, right in the middle of this wonderful object, are three stars - magnitudes 6.9, 8 and 10.5. Our smaller telescopes can only resolve the TWO brightest ones, however.

With high magnification and steady seeing, the 6-inch or 8-inch might reveal the third dimmest star at about 5.5" arc from the brighter pair. The human eye cannot, unfortunately do this object - nor its rival, Messier 8, the "Lagoon Nebula" justice.

The photograph below, taken with the magnificent 120-inch telescope at Lick Observatory, has been slightly digitally altered to match the overall appearance of The Trifid nebula in telescopes ranging from 8-inches to 12-inches and above. Note the conspicuous dark "lanes" which trisect this bright nebula into its three components; the 4-inch and larger telescopes are capable of showing these three rifts easily on moonless, dark nights.

Smaller telescopes WILL show the three distinct nebula concentrations, but will fail to show some of the intricate detail within this wonderful object.

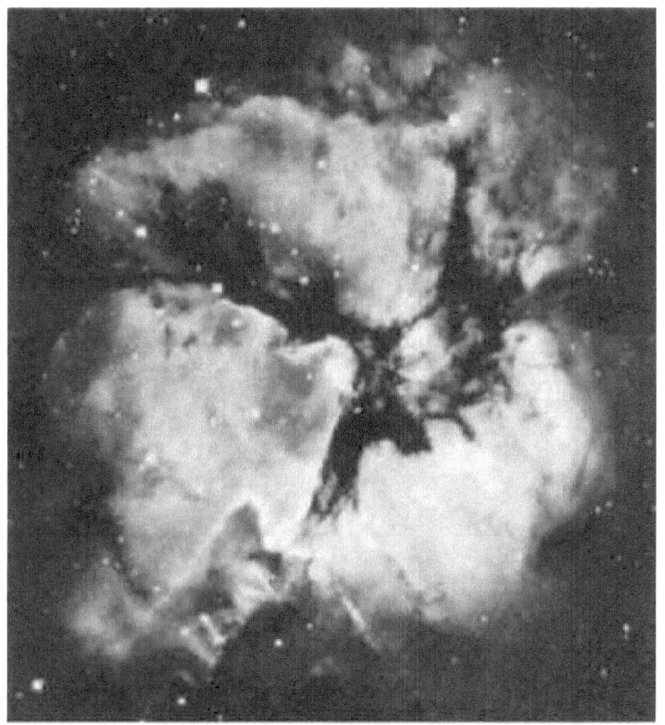

MESSIER 20
The "Trifid Nebula"

Recent studies suggest that this nebula and associated stars are located at almost the same distance from us as the fabulous Messier 8, Lagoon Nebula, and hence are likely to be part of a greater association of objects within the rich Sagittarius arm of our own galaxy.

Object 20 - THE LAGOON NEBULA! - Messier 8

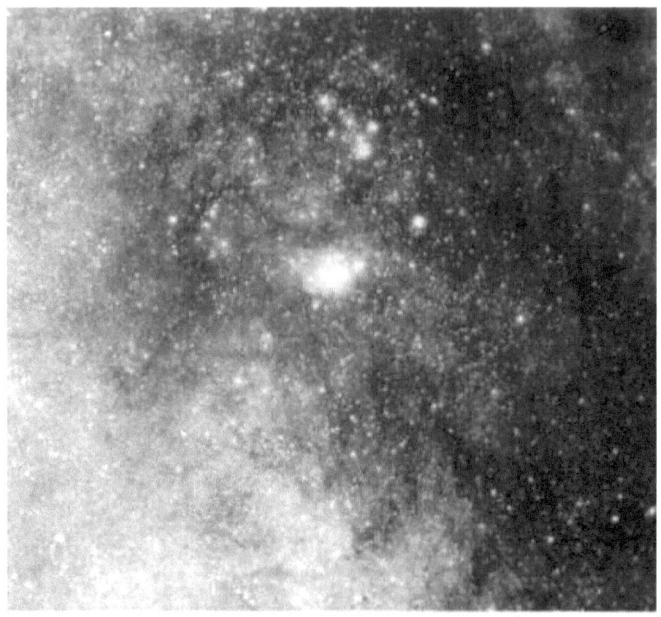

THE GREAT SAGITTARIUS STAR CLOUD
Showing the Lagoon Nebula (center) and Trifid (above)
Photo: 200mm f.3.8 telephoto, 20 minutes, SO 410 film
Arkansas Sky Observatory

The magnificent view afforded by the Sagittarius star cloud (seen to the left in the photograph above) provides us with a small glimpse of one inner arm of our congested and star-studded galaxy. Within that arm is the wealth of nebulae, stars, clusters and objects that are not unlike our own solar neighborhood....only more of them packed more densely together. In this photograph you can clearly see the Trifid Nebula (messier 20) and below it the huge and very bright "Lagoon Nebula," or "Omega Nebula" (as coined by William Herschel as he could see the Greek symbol for "omega" in the nebulous cloud). This very bright cloud of stellar gas envelopes a tiny star cluster - NGC 6530, to the nebula's east, which contains about 25 very faint stars. In small telescopes, the "omega" shape might be missed and the object will likely appear as an oval with some mottling, or dark detail. Visible to the naked eye and clearly distinguishable in your finderscopes, Messier 8 shows a tremendous amount of dark and bright detail....in addition to a wonderful "sprinkling" of stars....in the 6-inch and larger scopes. Together, this swarm of stars and associated nebulosity make this one of the summer's highlights from your dark sky sight! Look for the two brightest stars (both 9th magnitude) to be at the western edge of this nebula, and very near one-another, like a double. Another interesting feature is the very dark nebula (dark dust cloud) B88 which can clearly be seen in a wide field with a 6-inch or larger on a VERY dark night; this dark cloud is nearly in the center of M-8, just a bit to the west.

Perhaps it is not so much this marvelous Lagoon that is so intriguing in this part of Sagittarius as is the phenomenal array of stars, clusters and associations that abound here. Take the time.....turn off the motors....put in the widest field eyepiece you can find....and just scan until you lose yourself in the remarkable Milky Way centered on Messier 8. This is undoubtedly one of the star-richest areas of sky of all and worthy of repeated visits, and prolonged stays while you explore to your heart's content.

WANDERING ABOUT....YOUR NEW "USER OBJECT" IN SAGITTARIUS

And you really thought I would keep my word about "....no variable stars." Well, we can't let this one go by. The remarkable variable star RY Sagittarii is so easy, so quick and so readily accessible, that you should add this star to your list of "USER OBJECTS" for your skyk program or programmable keypad. If nothing else, you can have a finder chart and a link to the AAVSO observing charts and information. And, certainly RY Sag is visible in all telescopes.

This is one of the most unusual of all variable stars, in that it is normally about magnitude 6.5 to 7.0, but suddenly and totally unpredictably.....it will dim dramatically to less than 12th magnitude where it will stay for as much as 200 or more days! As a "R Coronae Borealis" type star, RY Sagittarii is the best of southern such stars for observing. When brightest, it is easily within reach of all our telescopes....when suddenly dimming, even the 3-inch scopes can reach it and monitor its wild fluctuations until it returns to its "normal" near-naked eye brightness! The following sample light curve gives you an idea of the normal brightness (highest point) of the star and its sudden demise to very faint magnitudes that may last for nearly a year.

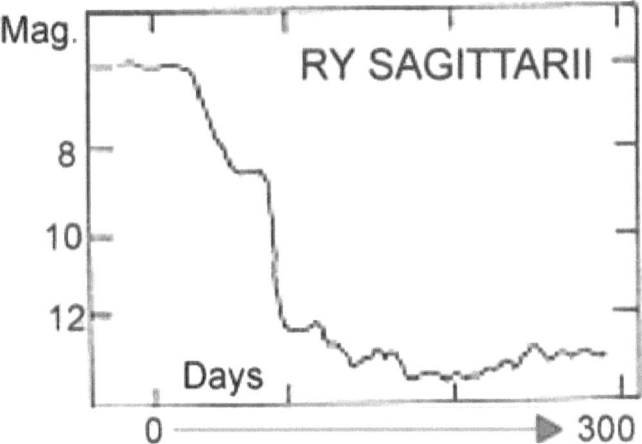

The following link will patch you into the AAVSO star chart "a", which is the low power, wide-field finder chart for the star; when at brightest, this star chart will suffice for comparison star magnitudes; however, when dimming either the "a" or "g" (medium and narrow field, respectively which show fainter and fainter stars) should be used to follow the irregular fluctuations of this interesting cataclysmic star.
https://www.aavso.org/apps/vsp/ . Note for these charts, simply type in the NAME of the variable at top to generate your choice of chart.

The coordinates for RY Sagittarii to enter as a User Object are: R.A. - 19H 13m / DEC - (-)33 37. On AutoStar, go to: "Select/ Object [enter]...." scroll down to "User Object" [enter]. Now enter the coordinates given above for "RY Sagittarii" - , using the number keys on AutoStar. After entering the coordinates and pressing "Enter" yet again, scroll down one and you can list the magnitude of the object as "6" [Enter].

Remember that your keypad, App or sky program will vary from the techniques for Autostar....but all are logically the same. Please read your User Guide or HELP to learn the process to produce your own User Object library for your computerized scope!

Johannes Hevelius' Sagittarius
Originally in splendid color

*   *   *

"Our galaxy, the Milky Way, is one of 50 or 100 billion in the Universe. And with every step, ever window, that modern astrophysics has opened to our mind, the person who wants to feel like they are the center of everything ends up shrinking....."

Neil deGrasse Tyson

Chapter 32

SCORPIUS

Watching from a Distant Earth the Dreaded Arachnid that Felled the Great Hunter

Our Constellations guide for Scorpius takes us into the realm of mighty legend, where a tiny armored creature lurks in wait to do mortal harm to the sky's Great Hunter. We enter into the realm of the beautiful Milky Way galaxy as we peer very near the actual **galactic center** and marvel at the array of wonderful deep sky objects, dense clouds of gravitationally-bound stars, dark obscuring and light-absorbing space dust and wonders beyond imagination.

Although we have explored many "globular clusters" in previous constellation discussions, these discussed in detail in my GUIDE "Observing Globular Clusters" on the Arkansas Sky Observatories' website, www.arksky.org. The constellation of Scorpius provides us with two of the most interesting and CONTRASTING globulars of all such clusters, **Messier 4** and **Messier 80**, both outstanding objects but strikingly different from one-another in many ways.

In addition, you will be visiting, via your telescopes on board computer or PC sky program, the sky's most intense X-ray source from deep space! This is quite a gem to show (well…it really cannot be seen well but it is fun to point out 'where it is') and tell at the next star party!

You will begin your "GO TO" journey into

Scorpius via the bright star "***Antares***", a vividly reddish star that has been known as the "*Rival of Mars*," and indeed the two are frequently very close to one another, not just in brightness and their blood red colors, but also as we see them in space. The ECLIPTIC (the zone of twelve *Zodiacal* constellations through which all the planets, moon and sun appear to move from Earth) passes through Scorpius, and hence Mars frequently "visits" its rival - the heart of the scorpion.

These and other fascinating objects of Scorpius will be discussed following - first in a concise list with exact coordinates and magnitudes and then individually - with specific details of what your telescope.....of any size....can expect to show you....and what you can expect to NOT see. So many times casual observers are disappointed when they attempt to find an object beyond their telescope's capability....or perhaps let down from what the photographs of the great observatory telescopes have shown them in the textbooks, magazines and advertisements. Using our "GO TO" guide allows you to know right off what you can expect to see of each object.

Each GO TO object is discussed for your telescope regarding the type of conditions necessary for you to view it optimally for discern the very faintest details....double star challenges for each size telescope .....magnifications and aperture necessary for most objects, and much, much more. This is YOUR complete guide to pique you curiosity to further explore the stinging scorpion of space. Use these guides as an excellent reference for your next star party itinerary, or a beginning for further study into the thousands of objects visible in this part of the sky.

Truly, as reports continue to come in on these extensive Constellation Study Guides, they will most definitely put your computerized telescope to work for you in the most efficient and enjoyable way possible!

We hope you enjoy these comprehensive GUIDES to touring the constellations via your AutoStar or other sky program and its computer-driven telescope. Each new installment will appear frequently, complete with diagrams, charts and illustrations that you will find nowhere else. Please let us hear YOUR feedback and your observations of each and every constellation after YOU have toured its vast reaches of our skies!

Introduction

The constellation SCORPIUS (pronounced "SKOR-pee-us") is one of the few constellations that nearly all who see its outline (even novice skygazers and the general public) immediately accept the fact that its stars do, indeed, trace a very good likeness of a celestial scorpion. The constellation is HUGE and its bright stars and wonderful celestial objects are positioned toward our line of sight into the center of our spiral galaxy, the "*Milky Way*." Hence, there is a plethora of wonderful and easy deep sky targets.....so many that I was forced to work my way down from 52 selected objects to "only" 20.....because, like in Sagittarius we had to limit the list to only the most conspicuous or interesting. But once you begin exploring within the boundaries

413

of this magnificent constellation, you will quickly realize that "**20 objects**" simply does NOT do this chunk of sky justice!

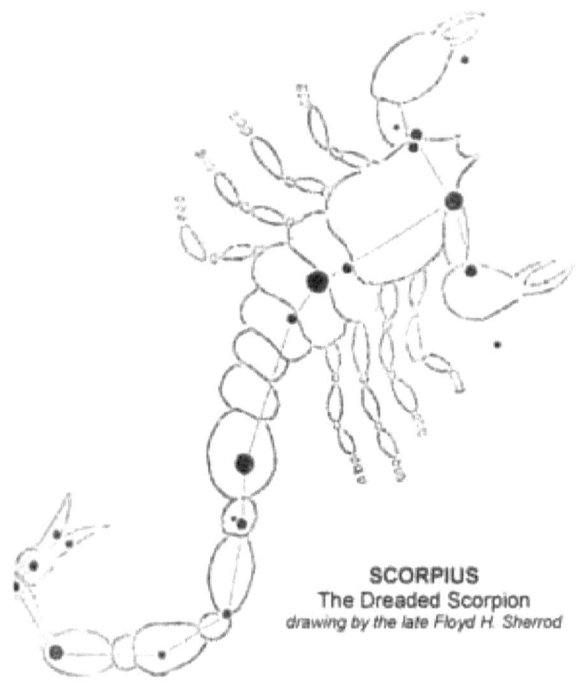

SCORPIUS
The Dreaded Scorpion
drawing by the late Floyd H. Sherrod

Dominant in the earliest Greek and Roman mythology, and sanctified through writings and art, is the legend of the mighty and monstrous scorpion and the seemingly immortal and invincible giant among "heroes", Orion, of winter sky fame. In fact, you might note at this point that Scorpius is almost EXACTLY in the opposite side of the sky from Orion! A mere curiosity? Not if you were an ancient Greek or Roman writer.

First, I must mention that even Greek writers and poets have their disagreements, and their legends and lore of the starry sky are no exception. One more colorful and perhaps heart-warming stories concerning the great scorpion is that of the young Greek *Phaeton*. In his braggingly boyish fashion, arrogant and ego-driven, this young man chose imminent death by leading his stead-driven chariot into the burning sun. Only, right before the sure-fateful encounter, Phaeton's horses which drew the mighty chariot were surprised and frightened by a huge scorpion that arose in the sky and diverted their paths away from the sun. For the valiant and saving effort, the mighty scorpion was placed prominently in the sky, signifying the place where the sun would have been on the day of Phaeton's suicidal journey. So in this heart-warming (pun intended) story, the Scorpion was our hero....saving a handsome young lad (not to mention the horses) from certain death.

But in the earliest Greek mythology and most embedded, is the tale that casts the dreaded scorpion in a much different....darker....light. For it was the great arachnid and his poisonous and feared sting that brought to his knees *Orion*, the Mighty Hunter and Hero of our nighttime winter skies. Orion was invincible, or seemingly so, until it was revealed by his adversaries that he had one weakness: a soft patch of exposed skin behind his sandal by the Achilles tendon. So, after surviving battle after battle and defeating celestial monsters and evil by the dozens, Orion was caught quite off guard one evening after defeating the "bad guys" as the scorpion sneaked up behind him...and, yep: stung him right there. After much pain and suffering, Orion finally succumbed to the sting and the gods placed him fittingly among the stars.....exactly

OPPOSITE the Scorpion that killed him. So the parade and chase continues today....Orion half-way across the sky in pursuit of villains to subdue while the mighty Scorpius engages in a chase that NEVER gets it any closer than it has always been to Orion. Once bitten is enough......

The ancient skywatches of Persia and Arabia, from whom we thankfully have been given most of the colorful names and legends of the stars and constellations, also saw a scorpion in Scorpius. It would seem that the dreaded scorpion would be quite familiar feature of the baked middle eastern lands. The oldest such connotation by the Persians is "*Girtab*", which translated literally means "*the stinging one.*" It is interesting to remember that the constellation of Libra, to the west of Scorpius, was once associated with this Scorpion of the Sky, with the now-Libra stars of *ZUBEN EL GENUBI* ("the southern claw") and *ZUBEN EL SCHEMALI* ("the northern claw") being the lower and upper, respectively, "pinchers" of the scorpion.

In Babylonian and earliest Mesopotamian lore, the constellation was also configured as "almost" a scorpion, but with a torso of a human being. Tombs and ritualistic material from very early Egypt also recognizes the scorpion, particularly through the legend of *Horus* who was slain by a scorpion, but nurtured back to life by his mother, *Osiris*, who used the magic of the gods for revival and protection. Hence, this SAME myth still persists today in Egypt and surrounding lands as homes, temples, buildings and even cross-country journeys are "protected" from the same magic through inscribed images that are displayed as emblems of immunity to bad fate.

Only in China did our earliest skywatchers NOT designate a scorpion to this clearly-defined outline; rather they saw it as "the dragon who devours the sun," surely a reference to perhaps a solar eclipse that must have occurred at one time in that constellation, during which the sun seemed to suddenly vanish (eaten, of course by the dragon) and then only to reappear! (regurgitated...apparently it did not agree with his delicate digestive system). Later, this famed dragon became known as the "*Azure Dragon*" and still persists today, as a kind and good spirit that is rising from the Celestial River (the Milky Way - see "GO TO" Draco) as he prepares to come to Earth with his blessings. It is strongly believed that this myth from the East somehow made its way thousands of miles across the oceans and influenced the pre-Colombian Toltec people of now-Mexico in the creation of THEIR myth of '*QUESTZACOATL*' a half-man/half-scorpion legend that is remarkably similar to the Chinese Azure Dragon.

YOUR SCORPIUS CONCISE DIRECTORY OF OBJECTS –

This mighty constellation is so rich in beautiful objects, stars, clusters, nebulae - from naked eye, to wide field, to telescopic - that is was very difficult to select (and limit) the number of objects for our "GO TO" tour.

I have chosen the finest 20 objects in Scorpius to showcase for your "GO TO" TOUR; in this case, all objects listed below will be visible in all telescopes (some naked eye) from the 3-inch to the 8-inch; of course larger apertures may "show" an object a bit closer and "better," but frequently a wide field and low power view is more desirable than aperture. This is the case for MANY of these objects in Scorpius since it is embedded in the very star-cloud-rich areas of our Milky Way galaxy. Indeed, I strongly encourage your to step away from the telescope often and scan the beautiful open skies and star fields of the Scorpius-Sagittarius regions with a good pair of 7 x 50 or 10 x 50 binoculars. It is truly like taking a trip into space, with one brilliant star-filled cluster after another. Even the binoculars will show dozens of deep sky objects that are not visible to large telescopes in other areas of the sky.

The constellation's very low declination for observers in the northern hemisphere make it difficult for observations in latitudes higher than 40 degrees north; so when the skies are crisp and dark and you can "see all the way to the horizon" on those rare nights....don't let it go by: LOOK at Scorpius! Conversely, our observers south of the equator are blessed with favorable near-overhead views of this constellation and the brilliant clouds of the Milky Way nearby.

As with all of the "GO TO" tour constellation studies, I recommend a good star atlas and/or chart which will list all the finest objects, constellation-by-constellation. One very handy reference guide is the *PETERSON FIELD GUIDE TO THE STARS AND PLANETS*, which features complete lists with declinations, right ascensions, magnitudes, and all pertinent information for you to expand your observing horizons beyond this brief GUIDE.

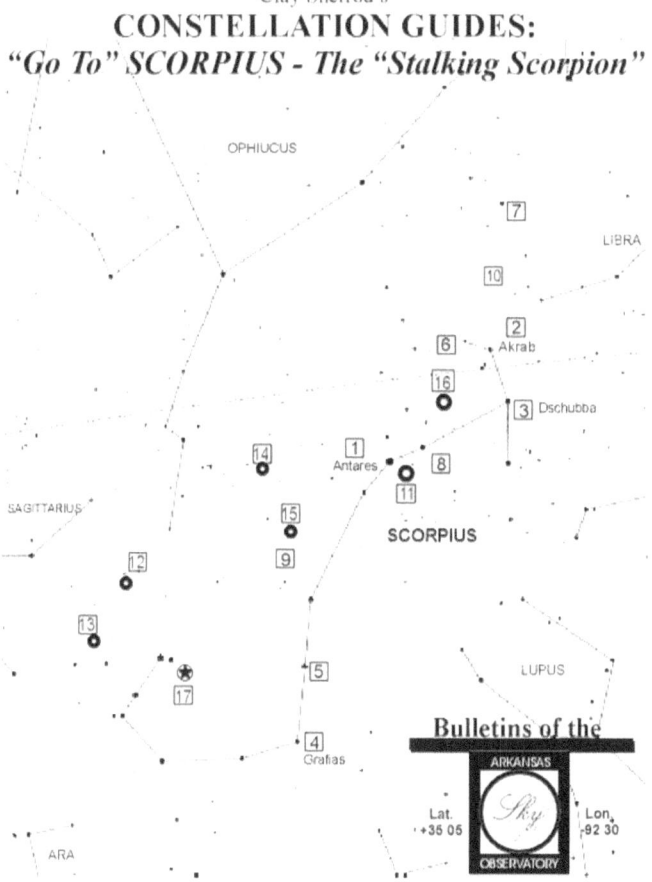

Clay Sherrod's
CONSTELLATION GUIDES:
"Go To" *SCORPIUS* - The "*Stalking Scorpion*"

The constellation tour Star Chart above (click on and save to a file on your PC; then open it and re-size to fit the page and print for a very handy at-the-scope star chart) will get you started on your journey for this constellation.

Following is the complete 20-object list for your "GO TO" TOUR of Scorpius; you may wish to find the majority of the objects from the App, program, or AutoStar-type Library (for example, you can easily pull up "Object/Deep Sky/Messier Object/..type in '80'...." and then press "Enter", followed by "GO TO" to access my favorite all time globular cluster. On the other hand, if you want to experiment and become a "better AutoStar user" try entering the exact R.A. and DEC coordinates of that object as described above after holding down the MODE key.

The exact process for bringing up a celestial object for your telescope to GO TO will vary with the sky program, telescope keypad, or App that you are using, but all fundamentally work the same logic to acquire your object as well as build your User Object Library. Always consult your HELP or User Guides of your specific program for the exact steps in using tabs and keystrokes to pull up and GO TO your objects!

OBJECT 1: very bright star - Antares (alpha Scorpii) - R.A. 16h 26' / DEC (-)26 19 - Magnitude: 0.92, brilliant red
OBJECT 2: fine bright double - Akrab or Graffias (beta Scorpii) - R.A. 16h 03' / DEC (-)19 40 - Mags: 2.6 & 4.9
OBJECT 3: what gives?? - Dschubba (delta Scorpii) - R.A. 15h 57' / DEC (-)22 29 - Magnitude: variable, mini-nova?
OBJECT 4: nice wide double - Zeta Scorpii (sometimes "*Grafias*") - R.A. 16h 51' / (-)42 17 - Magnitudes 3.6 & 4.8

OBJECT 5: wide nice double - Mu Scorpii - R.A. 16h 49' / DEC (-)37 58 - Magnitudes: 3.1 & 3.6 naked eye!
OBJECT 6: QUAD star - Nu Scorpii - R.A.16h 09' / DEC (-)19 21 - four stars, two close pairs.
OBJECT 7: 8-inch test double - XI Scorpii - R.A. 16h 02' / DEC (-)11 14 - Magnitudes: 4.8 & 5.1(tough!)
OBJECT 8: tough 4-inch test double - Al Niyat (sigma Scorpii) - R.A. 16h 18' / DEC (-)25 29 - Mags: 2.8 & 9
OBJECT 9: classic variable star - RR Scorpii - R.A. 16h 53' / DEC (-)30 30 - Magnitude: 5.4 to 11.7, 280 days
OBJECT 10: *strongest X-ray source*!! - Scorpius X-1 - R.A. 16h 17' / DEC (-)15 31 - Magnitude: 13
OBJECT 11: fine globular cluster - Messier 4 (ngc6121) - R.A. 16h 21' / DEC (-)26 24 - Magnitude: 7.4 (a must-see)
OBJECT 12: galactic star cluster - Messier 6 (ngc6405) - R.A. 17h 37' / DEC (-)32 11 - "Butterfly Cluster"
OBJECT 13: nice star cluster - Messier 7 (ngc6475) - R.A. 17 51' / DEC (-)34 48 - large and bright!
OBJECT 14: fine globular cluster (in Ophiuchus) - Messier 9 (ngc6333) - R.A. 17h 16' / DEC (-)18 28 - Magnitude: 8.5
OBJECT 15: tiny globular cluster - Messier 64 (ngc6266) - R.A. 16 58' / DEC (-)30 03 - very compact, bright center!
OBJECT 16: best globular cluster - Messier 80 (ngc6093) - R.A. 16h 14' / DEC (-)22 52 - very tight - compare to M-4!

OBJECT 17: diffuse nebula - ngc6302 - R.A. 17h 11' / DEC (-)37 03 - Magnitude: 10 (a must-see: the "bug nebula")
OBJECT 18: galactic cluster - ngc6231 - R.A. 16h 51' / DEC (-)41 43 - magnificent brilliant white cluster! (a must-see)
OBJECT 19: very nice cluster - ngc6242 - R.A. 16h 52' / DEC (-)39 25 - Magnitude: 8.8, same field as ngc6231!
OBJECT 20: distant globular cluster - ngc6388 - R.A. 17h 33' / DEC (-)44 43 - nice, bright, very small!

A VISUAL GUIDE TO OUR DEEP SKY OBJECTS IN SCORPIUS –

Object 1 - Very Bright Star ***ANTARES*** (Alpha Scorpii)
Only the "red drop of winter" - *Betelgeuse* - is more scarlet than the brilliant reddish Antares. From the Greek to mean "*the one who rivals Mars*", this reference eludes to the deep red color and brightness that is nearly an exact match for the Red Planet when at its closest (perihelic) opposition....which the best always take place in summer months and frequently VERY near Antares. The opposition of Mars in 2001 is no exception when both stood near one-another as a shining red pair.

Both Betegeuse (Orion) and Antares are perhaps the largest stars of our Milky Way (that we know of); although Betelgeuse is a bit large, Antares boasts a diameter.....get this: *600 MILLION MILES ACROSS!* That compares to our sun at slightly less than one million miles diameter. There is a faint (magnitude 9) companion star to Antares that is a full 3" arc nearly due west of the bright star; even though this may SEEM easy for a 3-inch and larger scopes, it clearly is not because of the brilliance of the main star. With larger telescopes, you MAY be able to glimpse this fainter, overshadowed star under high magnification, about 250x to 300x.

In wide field telescopes, such as an APO refractor, compare the color of Antares to nearby stars of spring and summer, notably VEGA (clearly blue-white), DENEB (definitely yellowish), as well as ALTAIR (orange-yellow).

Object 2 - A Fine Double Star *"Akrab"* or "Graffias" (Beta Scorpii)....a very good double star for smaller telescopes!
The latter name "Graffias" is from the Greek and sometimes associated with ZETA Scorpii, so some confusion exists in the actual name. It is interesting, however that "Graffias" signifies the "star of the crab," a creature that was not clearly distinguished from the Scorpion in ancient Greece.
At magnitude 2.5, Beta has a nice bright companion star magnitude 4.9 about NNE from the star a full 14" arc seconds distant. It makes a wonderful pair at very low power, even with the wide field refractor. Using MEDIUM magnifications in all telescopes, look for the wonderful color contrast of these two stars. The brighter star has always looked distinctly yellow to me on a steady night, with the fainter star seeming blue-green. Most observers see pretty much the same contrast, although some eyes see more blue or more green in the secondary star.

Object 3 - Delta Scorpii - "Dschubba" - What a ride this star is taking us on!

Normally this is nothing more of interest than a somewhat variable star that is also fairly bright and comprises the "head" of the scorpion. However, beginning fittingly at the start of the 21st century, Dschubba erupted almost "nova style" from its normal 2.3 magnitude to attain a brightness that continues to persist at almost the same magnitude as Alpha....Antares! Dschubba is an "early "B"-type star, so this type of minor nova activity is unusual for such a high-energy star. Photographs show some nebulosity coming from this star and possibly "connecting" to other stars nearby, indicating the possibility of similar outbursts in the past. This is a star that CERTAINLY bears watching. You can compare magnitudes easily with nearby bright stars with the naked eye. Look for any color changes using very low magnifications in your telescopes.

Object 4 - Zeta Scorpii - also known as "*Grafias*" (note only one "f" in this duplicated star name) I have included this star because it is a very wide and fine optical double star for low power, wide field telescopes and thus very fitting for the small telescope. It can be enjoyed even in a pair of binoculars, but the wide field telescopes provide one of the most striking star fields against these two stars imaginable. The two stars, magnitudes 3.8 and 4.2 are a full 7" arc apart in an almost exact east-to-west orientation. These stars are NOT a true double, only appear to be optically from our vantage point here on Earth.

OBJECT 5 - Nice very wide Double Star; great

wide field views of the sky - Historical Interest!
This is a beautiful field in the middle of the rich summer Milky Way! The double is made of two very wide stars, about 6 minutes (') arc (not seconds - ") , the brighter of the pair being about magnitude 3.1. Mu Scorpii is referred to in Polynesian cultures as two children who are bound together in faith in spite of abusive and tormenting parents. Be sure and scan the wide-field sky at your lowest magnification and try to spot ngc 6281 and ngc 6242, both very fine galactic star clusters located in the "Sagittarius Arm" of our galaxy.

<u>Object 6</u> - A Neat Quadruple Star - Nu Scorpii
Although this is not quite as outstanding nor famous, this is much like the well-known "Double Double Star" of Lyra (see the ASU GUIDE for Lyra under the GUIDES tab here on this website) for the complete Constellation Guide for Lyra). This is an interesting low power field in itself, with this complex star system on 1.5 degree east of Beta Scorpii (see chart). The first pair (denoted "A" and "C" stars, magnitudes 4.0 and 6.2) is separated from one-another almost the girth of the planet Jupiter seen in the same eyepiece, so it is easily resolvable with all telescopes; look for the orientation nearly NNE to SSW. Begin to raise the power on your telescope on the FAINTER of the two stars ("C") until you split it into two more stars (its companion is "D", magnitude about 7.7). This might be difficult for the small telescope, but should be relatively easy in the 4-inch and larger scopes. They are well separated in my ETX 125 at 227x. NOW, using the 6-inch and larger scopes and medium high (about 227x) power, look carefully at star "A" which should resolve into a VERY close but clearly

definable two stars, magnitude 4.6 and 5.6. The companion to star "A" is designated as "B". I would be very interested, particularly from 4-inch scope users, to find out the difficulty or the ability to split star "A" into its two components.

Object 7 - A true test for an 8-inch telescope –
Perhaps this star is asking "too much" from the 8" aperture. However....it can be done, even if only partially. This very, very close 4.2 magnitude star CAN be resolved at high (i.e., 300x and above) power. Look very carefully and very close to the brighter star at this magnification to the NORTHWEST and touching the main star. If nothing else, the resolution of the large 8" should allow for the star to be at least "elongated" in a NW-SE direction, indicating that the Airy Disks are attempting to be resolved. Use a very steady night; anything less and you are wasting your time. If you can split this star, and you are SURE that you have zeroed in on the correct star, I would like to hear of your observations. At separation only 0.6" arc or less at the present time, there is no reason to even try with smaller apertures.

Object 8 - Very Nice Double - Al Niyat (sigma Scorpii)
This star signifies the hard crusty exoskeleton of the rugged scorpion, the panel directly covering his heart, as the name implies "heart shield." This star is magnitude 2.9, very bright in the telescope, and exhibits a very faint companion a fair distance (20" arc, about half the diameter of Jupiter in the same eyepiece), almost DUE WEST of the brighter star. This white (brighter star) and blue pair is a challenge for the ETX 90 because of the brightness

of the main star; if it were not for that it could be seen small quality telescopes but that would be a very tough challenge. It is relatively easy in the 6- and 8-inch scopes at moderate (about 150x) power. NOTE: one of my favorite all-time globular clusters, Messier 4 (see below) is located within only one degree SE of this star; user your finder to locate the cluster.

Object 9 - A Classic Variable Star: RR Scorpii –
An ideal variable star for all telescopes!
RR Scorpii is a classic "Mira-type" variable star which actually grows in diameter to cause its seen brightness increases; as it subsequently shrinks back down in size we see a diminishing brightness here on Earth. That total cycle takes the star from naked eye brightness (!!) of 5.2 all the way down to telescopic-only brightness of about 11.5. Even at that low brightness, the small telescope can easily keep up with the entire cycle on a very dark nights. Only 1 degree WSW from Messier 64, a bright globular cluster, this is an easy star to find your first time out. To aid in locating - and more importantly ESTIMATING using selected comparison stars - click on the link below to the American Association of Variable Star Observers and download to a file the accompanying chart; save it to file, click on it to open the file, resize to fit your page and then print out....other charts are also available for comparing when the star is very bright ("a" charts, wide field for acquiring brighter stars) and very dim ("c" chart for when the star is dimmest, showing stars to magnitude 11 and below).

Link to chart, RR Sco:

https://www.aavso.org/apps/vsp/ . Note for these charts, simply type in the NAME of the variable at top to generate your choice of chart.

With a total cycle of 279.7 days, this is an easy star to keep up with. Note from the light curve shown below that the star reaches a VERY SHORT peak at maximum and then fades rapidly but predictably to a very short, but somewhat longer minimum brightness, only to increase again.

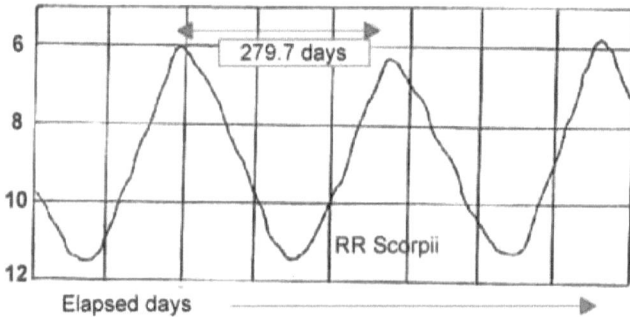

Since the maximum can fluctuate between less than 6th magnitude to up to 5.2, it is important for observers with moderate telescopes to monitor this star about every week; merely compare the star's brightness to one of the many comparison stars (or extrapolate between two or more) and write down your estimate. Send that estimate about once a month to: http://www.aavso.org so that they can include YOUR observations in the scientific database about this variable star.

Object 10 - A MOST UNUSUAL OBJECT
Powerful X-Ray Emitter Scorpius X-1!

To say that this is a "must see" is most definitely an overstatement since you are likely to NOT see anything when you GO TO this object!

Nonetheless, are YOU going to go through life without saying that you have SEEN the "....*MOST POWERFUL EMITTER OF X-RAYS IN THE UNIVERSE?*" Even better, wouldn't you like to actually show this object to lots of people? Well, now you can.

What you won't see is a mysterious burst of X-rays coming from the northern realms of Scorpius; this object is the most powerful source of X-rays known and first discovered in 1962 in a space flight. For identification purposes, use the accompanying star chart that I have put together; although it does not reach the magnitude 13.1 which this mysterious source appears to be....it will at least get you close enough to say "...my eyes have seen it; I may not know it, but my eyes have actually seen this object." In the chart below, the wide field chart (large one) shows the northern part of Scorpius, with the stars Antares and Dschubba clearly marked; the small square is the area of X-1; note that the small white-on-black chart at upper right is REVERSED to correspond with the image as seen in a Maksutov or Schmidt Cassegrain. The faintest star shown in the small chart is about magnitude 13.0.

At a distance of about 1600 light years, Scorpii X-1's nature is still not clearly understood; it most certainly is a "double-something," as deduced from careful observations since 1962. It is possibly a high energy NEUTRON STAR being encircled by a cooler white dwarf star that has one side of it (pointing toward the neutron star) strongly heated and energized creating a hot disk or shell about the size of planet Earth.

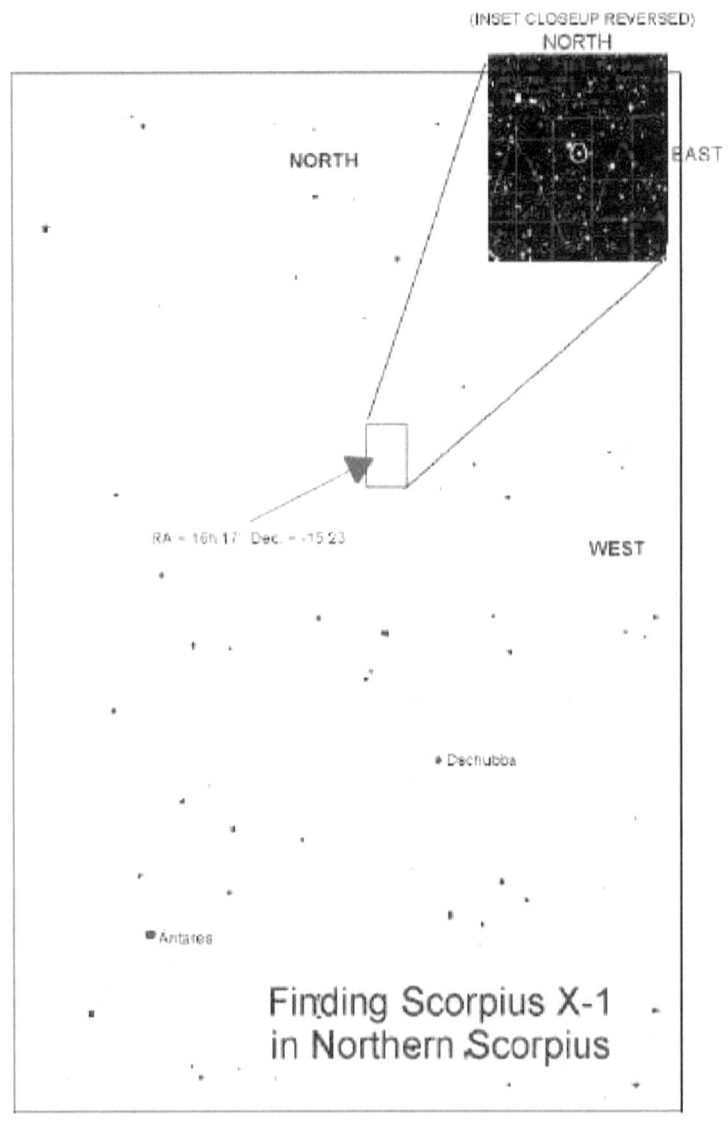

Finding Scorpius X-1 in Northern Scorpius

careful observations since 1962. It is possibly a high energy NEUTRON STAR being encircled by a cooler white dwarf star that has one side of it (pointing toward the neutron star) strongly heated and energized creating a hot disk or shell about the size of planet Earth.

Remember, this object is DIFFERENT: it is not the reward of "seeing something spectacular" with this one...it is the fact that you are looking at and showing to others what might be the most active source of X-radiation in the Universe! Sometimes the imagination has to supplement the realization. This is one of those times.

Object 11 - A Fantastic Globular Cluster –
Messier 4

You owe it to yourself to do a very detailed comparison between Messier 4 - a VERY bright and spread-out globular just west of Antares - and Messier 80, another globular but extremely small and tightly compacted toward its center. The differences in these two are fantastic and worth looking at time and time again. You might guess by its size (14' arc, half the size of the moon!) that M-4 is relatively close to us compared to tiny (3.3' arc) Messier 80 which is five times more distant. Interestingly the two of them are about the same brightness, with M-4 being magnitude 6.7 and M-80 at 7.4....however, M-4 will appear to be the DIMMER of the two objects. This is because the total light ("integrated light") is spread out over the large total area; thus the more compacted object will appear much more concentrated in luminosity. Messier 4 is only 15. degrees due west of bright Antares, and sure enough the small APO scopes can get them BOTH in the same low power field of view, offering absolutely stunning views of this portion of the summer Milky Way. Messiers 4 and 80 are my personal two favorites among the globular clusters. Messier 4 exhibits a wide spread of nearly equal magnitude 13 or slightly brighter; there are maybe 30 or 40 such stars which you might think, because of that brightness, would be tough to resolve in our scopes....they are not. Even with the 4-inch, many peripheral stars can clearly be seen at about 120x; the 6-inch shows good star images all the way to the center and the views in the 8-inch are outstanding, with stars appearing as a "splat" of star dust filling a medium power eyepiece. Unlike many globular clusters, M-4 does NOT exhibit a very condensed core or central nucleus, but rather a nice even spacing of stars throughout its large expanse.

Objects 12 and 13 - A Fine and Close Pair of Galactic Clusters, Messiers 6 and 7

Ideal for the smallest of telescopes and presenting perhaps the finest of all shots of the Milky Way star clouds, Messiers 6 and 7 team up near the tail (five degrees north) of the Scorpion to provide an outstanding low power wide field vista. Both can be seen as "faint fuzzies" to the naked eye on a very clear and moonless night. Messier 6 is large, about the size of the moon's disk in the sky. In the small refractors (which are the ideal instruments for this cluster, by the way!), up to 60 stars to magnitude 10.5 can be clearly seen in this packed large cluster over a field of nearly one degree! Although the entire cluster can be seen in wide field eyepieces of the 3-inch and larger scopes, there is no advantage to aperture here. With a wide field, observers can begin to imagine the shape of a winged "butterfly" for M-6....hence the nickname: "Butterfly Cluster" that appears on your named ojects in the sky library directory. Messier 6, at a distance of about 1300 light years is the more distant of the two galactic clusters.

Messier 7 is only 3.5 degrees southeast of Messier 6 and appears larger, has more stars and some associated nebulosity which CAN be seen on a dark night at the very low magnifications. Unlike a pretty "butterfly" of M-6, Messier 7 outlines pretty much a "square" shape in very low power instruments, showing about 25 stars brighter than magnitude 9.2! For observers with an 8-inch or larger telescope, use that larger aperture to look for a very faint GLOBULAR CLUSTER in the same field of view as M-7! NGC 6453, magnitude 11, is only 20' arc northwest of the brightest star of this group; it is very small and unresolvable, but will appear like a "fuzz ball" in the 8-inch.

Object 14 - Messier 9 - A Fine Globular Cluster in Ophiuchus –
Here for comparison purposes
Messier 9, in the constellation of Ophiuchus, is discussed in that "GO TO" GUIDE (see GUIDES/ Constellations/ Ophiuchus). Please use your GO TO function to slew over to this nice globular - it is a classic shape, size, and morphology - to use as a comparison for Messiers 4 (discussed above) and 80 (discussed below). You will see clearly just how different both M-4 and M-8 are from the "norm" of globulars. This is a great exercise and a really good "talking point" comparison for star parties and educational sessions with the telescope.

Object 15 - A Wonderful Globular Cluster – Messier 62

While comparing the "typical nature" of Messier 9 (above), be sure to include this globular in the judging as well. Messier 62 is also unique. Located about 7 degrees southeast of bright Antares, a small telescope will see the cluster as a very nice bright glow in which a very pronounced central "core" of brightness is contained. This is very evident in a 4-inch, although typically NO stars can be resolved clearly with this telescope. The 6-inch begins to hint at some stars around the periphery, but the 8-inch will only break down a very few more. It is a tough object. Its total magnitude is 8.3, but that brightness is packed so tightly in the center that little can be resolved. Remember that the AVERAGE brightness of its 30 brightest stars is only magnitude 16! It can be seen clearly in small scopes and is well worth searching out. Expect it to look like a very faintly illuminated tennis ball far out into the sky in all scopes.

Object 16 - The Fine Globular Cluster Messier 80

Discussed above, this is a really nice and compact little globular, well worth spending some time on comparing to Messier 4 (above). M-80 appears very much like M62, only about 1/3 smaller; curiously, the stars around the fringes of M-80 are more readily seen than those of the larger cluster. The 4-inch is about the smallest scope that will see peripheral stars (on very dark nights and medium-high power) in this cluster. You MUST observe the area from this globular WESTWARD with very low power and wide field in slow scans....it is here that we observe **William Herschel's** *"hole in the heavens"*, a pronounced dark area of the Milky Way where the background stars are obscured by the dark dust of the Milky Way. On a very dark night in remote areas, you will plainly see these areas "where stars ain't...." as it was put to me once.

Object 17 - The "Bug Nebula" - NGC 6302
As in tiny bug. This little diffuse nebula measures only 2' arc long by 1' arc wide, about three times the size of Jupiter in the same medium power eyepiece. It is too small and faint (magnitude 10.0) for small scopes, but begins to show somewhat in the 4-inch as a "bug-like" shape. Look for a faint glowing outline amidst a very bright star field; it will stand out against the stars clearly; some magnification (about 30x per inch) is required to really get a good look at it. It is on the named objects of your Deep Sky commputer library list. Let us know your impressions of this odd object....collect it, or squash it?

Objects 18 and 19 - Gorgeous Galactic Clusters – NGC 6231 and NGC 6242
These two glittering gems are like pendants hanging from fine necklaces of space. They are absolutely breathtaking on a very dark night, but unfortunately located in very far southern skies for high northern latitudes. Both clusters are excellent objects for all size telescopes, but wide fields and low powers are preferred. NGC 6231 contains about 25 stars to magnitude 9, scattered over an area about half the size of he moon; thus about 15x per inch aperture of your telescope is ideal. Larger telescopes can detect more; the 6-inch should reveal about 90 stars total, with some 120 or so visible in the 8" aperture. Not

so many are visible in the much fainter and more distant NGC 6242, but it is still a nice object. More compact and with fainter stars, only about 20 total stars can be seen with the 6-inch or larger scopes. About half that many can be discerned in the 4-inch. Expect NGC 6242 to be about half the size of NGC 6231. In very wide field eyepieces, take the time to scan between these two clusters; you are looking at an absolutely incredibly star-rich area of our Milky Way's Sagittarius Arm. You could spend an entire night looking through this region with new objects coming into the field with every turn.

Object 20 - NGC 6388 - Small but nice globular cluster
This is an overlooked pretty, yet small, globular cluster, VERY far south in Scorpius. At magnitude 7.1, if the object was higher in the northern sky it would be certainly one of the Messier objects. It is actually a bit larger than M-80, at almost the same distance in space and only a slight bit dimmer. As with M-80, the 6-inch will show some outer sprinkles of stars; the LX 90 offers little gain even with the larger aperture. With the smaller scopes, expect a nice compact and bright fuzz ball due south of the Scorpion's "stingers"

WANDERING ABOUT....YOUR NEW "USER OBJECT" IN SCORPIUS

How can we resist.....you already KNOW what the recommended new "User Object" will be, don't you? Even if you can't SEE it...it still needs to be in your database for your next neighborhood star party or big astro-bash in the mountains. Show people something *they cannot see*, only feel (and not really sure of that either!).

That's right, for our Scorpius User Object, let's key in the coordinates of Scorpius X-1, the most powerful X-Ray emitter in the Universe (so far as we know). Get the coordinates for OBJECT 10 (above) And let's load this conversation-starter into our User Object database. Take your device and key in the particulars for "x-1 Scorpii." This will your FIRST radiation source to your computer library of interesting and unusual "User Objects."

On AutoStar, go to: "Select/Object [enter]...." scroll down to "User Object" [enter]. Now enter the coordinates given above for "X-1 Scorpii", using the number keys on AutoStar. After entering and pressing "Enter" yet again, scroll down one and you can list the magnitudes of both stars [Enter]. One scroll more after that and you may type in your brief description of the object....(I like: **"Warning! Radiation!"**) [Enter again].

While in this area, visit CYGNUS, to the north, among my favorites of constellations. Abeautiful constellation, yet it is not full of "show material" like Scorpius. In this wonderful *Northern Cross* we will explore the large planetary nebula, the Dumbbell Nebula, the wonderfully exciting colored double star Alberio, the North American Nebula, and much, much more.

* * *

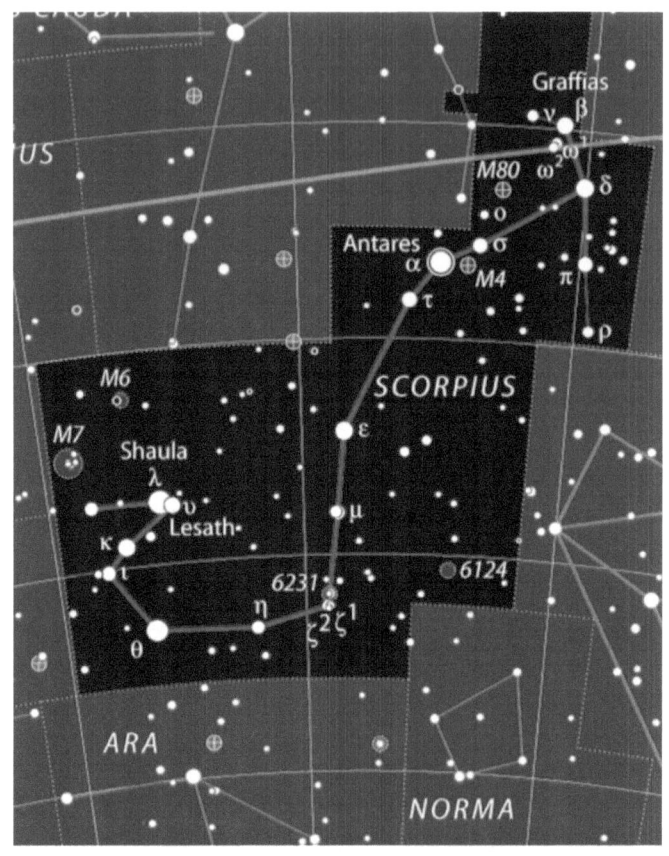

Sky chart of the Scorpius constellation.
Adapter from and courtesy of
IAU and Sky and Telescope

Chapter 33

TAURUS
....the brilliant field of stars that's bullish by nature....

The striking constellation of TAURUS, the "Bull," is a wonderful and brilliant constellation benchmarked by the bright fiery red star *ALDEBARAN* and the two closest star clusters and asterisms, the *HYADES* (Mel 25) and the *PLEIADES* (Messier 45). Because this star-rich constellation is located immediately west of the star clouds of the winter Milky Way, it is embedded in wonderful star groups and associations, and wonderful deep sky objects.

Clay Sherrod's
CONSTELLATION GUIDES:
"Go To" TAURUS: the "Bull"
With crimson eye and mystic horns toward Orion

Because Taurus precedes the brilliant winter constellation of Orion in the sky, the advent of the bright red star ALDEBARAN is a benchmark for the thrilling views of winter skies. This "eye of the bull" clearly marks the asterism - and physical star cluster - the HYADES, a very easy-to-differentiate "V" shaped asterism of brilliant stars. As a matter of fact, this cluster is a wonderful method through which you can determine the LIMITING MAGNITUDE of the naked eye, binoculars and even the finderscope that is installed on your telescope using this cluster. (see my discussion of this topic at the ASO Guides - Frequent. The diagram below shows my revised star catalog, with appropriate magnitudes of brighter stars of this cluster which may be identified in optics of 50mm and smaller, all the way to the naked eye. You can double click on this image and save to file and then print and laminate for a very handy reference to use in determining your limiting magnitude on any given night!

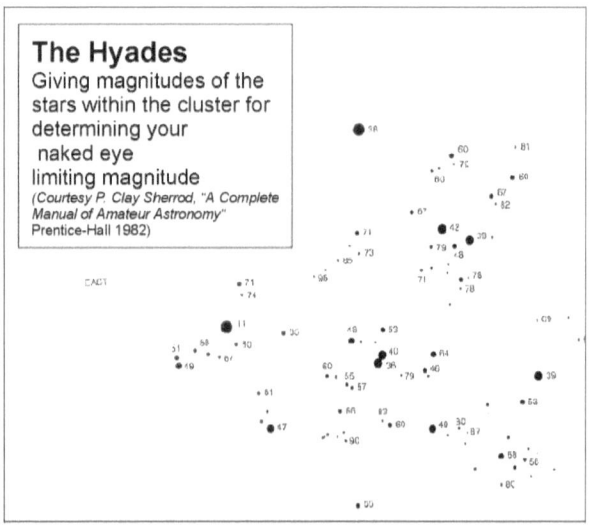

Likewise, to TEST critically the limiting magnitude

of your telescope, you can use the next chart of the PLEIADES, also conveniently located in Taurus. There are two charts below: the first one is the appearance of the "*Seven Sisters*" as it would look in good binoculars or a very wide field, low power telescope; the second chart is an enlarged view of the very narrow field of view around the bright star *ELECTRA*, which can be accessed at OR.A. 03h 45m . DEC +24d 07m, or merely center this bright star in your finder crosshairs to locate. Use this second chart to note the very faint stars embedded in and around the small "triangle that you see of brighter stars....the stars go all the way to about magnitude 14.7, which CAN be attained with the 8" and larger telescopes.

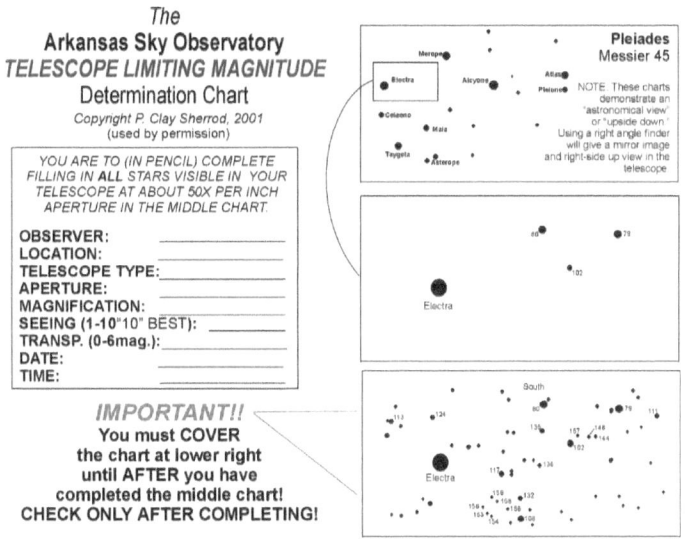

Use the Above Chart to Determine the Limiting Magnitude of Your Telescope

But, in addition to these wonderful star clusters which will be discussed individually in our "GO

TO" tour objects, the field of Taurus offers splendid viewing, whether one uses the naked eye, a good pair of 10 x 50 binoculars or a huge telescope. Unfortunately I am able to only select a few of the finest objects to discuss in our Taurus "GO TO" Guide; I very much encourage all telescope users to obtain the three-set copy of *Burnham's Celestial Handbook for* reference on each and every constellation; there is no finer reference work for deep sky viewing to be found.

Also note that there are program tools now available for you to install on your computer to CONVERT the epoch 1950-1960 coordinates listed in the Burnham reference tables DIRECTLY to any epoch to account for annual precession. These can be can installed within minutes....all you must do is merely type in the coordinates exactly as they appear in Burnham's and enter, with the resulting 2001 conversion done instantly for use with your sky program GO TO function.

Once scanning the pages of Volume 3 of Burnham's, you will quickly see the quandary one is put into when selecting the "best" deep sky objects and particularly the "best multiple" and unusual stars for discussion. There are no less than 100 fine multiple stars within reach of most amateur telescopes, many of which are fascinating objects. A few have been selected here. In addition we of course have our two closest star clusters as noted above and discussed later....AND, lest we forget, the very FIRST Messier Object....the "*Crab Nebula*", M-1.

Taurus is one of the most-referenced of all

mythological constellations and is one of the original 48 constellations as cataloged by the Greek philosopher/astronomer Ptolemy in the 2nd century A.D. and has the distinction of being the second of the "*Zodiacal Constellations*," the twelve star patterns through which the ECLIPTIC - or apparent paths of the sun, moon and planets as seen from Earth. In most legend, it has always been assumed or inferred that the horns of the great Bull are aimed in pursuit of Orion, the Hunter to the animal's east; indeed, as can be seen in the two old charts from star atlases of **Bayer** (1603) and **Bevis** (1750) the curved and ominous horns are most definitely aligned to a defending Orion (not seen in chart).

Look at this wonderful image from 1687 by Johannes Hevelius. Taurus, the Bull is obvious; to his lower right is Orion, appearing to be fighting back the charging bull, while Perseus flees in the opposite direction. To the upper right is the ever-calm Auriga, tending to his goats.

But, back to the Bull and mythology. It is interesting to note that we MAY be witnessing through historic legend more importance on the bull's HORNS than on the animal himself. This is true throughout Arabic, Oriental, Egyptian, Mesopotamian and even Greek, Roman and even Native American cultures, even without apparently literary exchanges between these early advanced cultures.

An animal's horn - whether it be from the bull, a ram, goat, or Walrus - was a dominant symbol of fertility and bountiful riches in these cultures for thousands of years. In addition, the tip of the bull's horns would have served as the precise calendrical indicator for the *Vernal Equinox* as it occurred at around 4000 BC, just at the height of the Egyptian empire and the emergence of Native American art and culture in the western hemisphere.

The constellation Taurus is also key to the Greek mythology of ***Europa***, daughter of the king ***Agenor***, and a great Bull. As the story goes, Princess Europa went to the seashore to gather flowers with her entourage of escorts, handmaidens of the kingdom. Being in love with this beautiful young woman (as he apparently was with EVERY beautiful young Greek woman) the great god Zeus had been watching this field trip from above and decided to "make his move" on the young woman. To disguise himself and venture near Europa, ***Zeus*** transformed himself into a magnificent white bull, and as such he mingled in grazing herd near the meadow where the flowers were being picked.

Being so beautiful, Europa did not scare any of the great animals of King Agenor's herd and they were like sheep to her glance, and while looking over the great and gentle beasts, she spotted the unusually wonderful and white one.....Zeus in disguise!

The story is very familiar to most of us that Europa turned her flowers into wreathes for the white bull, wrapped them around his horns, and begin to lead him around the meadow. He followed without incident, like a puppy following his mom. In play, Europe decided to mount the large bull and they galloped flockingly up to the seashore, where Zeus (aka, "the bull") suddenly plunged into the sea and carried the princess to Crete. Once there, Zeus transformed (much to the shock of Europa, I am sure!) from the bull into an eagle and became a bit too familiar with Europa, the end results being three sons, the first of which was the famous **MINOS**. It was Minos' wife, Pasiphae, who had an illicit affair with yet another bull and begat the famous "*Minotaur*", a half man-half bull creature to whom the Cretans continued sacrifices until the animal was completely destroyed by Theseus (no kin to *Roget's Thesaurus*).

Minos is said to have introduced the bull cult to the Cretans. He had Daedalus build a labyrinth in the depths of his palace at Knossus, which became the home of the Minotaur (offspring of Mino's wife Pasiphae, and a bull). Seven young men and seven maidens were ritually sacrificed to the Minotaur until Theseus killed it. Minos, in fact, was the title of the ancient rulers of Crete, and the story probably tells of their mythic origin. From this, and throughout even earlier cultures as attested by the symbolic imagery of horned bulls in the 40,000-year-old cave drawings in *Lacaux, France*, the association of the mighty BULL as royalty or even god-like has persisted. Indeed, many cultures throughout both time and location have worshipped through sacrifice the horned bull as a religious rite, a practice that does continue today.

The head of the bull and the horns themselves - and NOT the bull - are pretty much all that is represented by the stars of the constellation we recognize today. The left (southern) horn starts begins near The Hyades, of which Aldebaran is included (but NOT a part of the actual Hyades cluster!). It extends from *Aldebaran* to zeta Tauri, near the eastern edge of the constellation, a star known to the ancient Chinese astronomers as "Tien Kwan", or the "*gateway to heaven.*" The right (northern) horn angles up just west of the Hyades, from delta Tauri through tau Tauri and ends at *EL NATH*, or beta Tauri (we discussed this star as not being part of the large pentagram of Auriga!). As imagined in the Hevelius chart shown above, all else that is shown of the bull is part of his upper body and two very disproportionately thin front legs. It may be that the bull that we are supposed to image the bull as half-emerged in water since on his back he carries Europa across the sea to Crete.

TAURUS (pronounced "TAW-rus") is dominated by the Hyades so far as easy recognition of this star pattern; from mid-northern latitudes, the Hyades rises at about 9 p.m. local time on October 5, crosses the meridian (as high in the sky east-to-west as possible) about 3:45 a.m the following morning and does not set until that afternoon. Midnight culmination (highest point at midnight) occurs each year almost exactly on December 1.

THE STARS OF TAURUS –

For a very comprehensive and complete listing and cross-reference of Bayer, Flamsteed, SAO,double and other star information for the constellation of Cassiopeia, I highly recommend that you visit: http://ad.usno.navy.mil/star/star_cats_rec.shtml

OBSERVING THIS CONSTELLATION WITH BINOCULARS –

For those who wish to explore the regions of Taurus in binoculars I highly recommend a standard good quality 7 x 50 or 10 x 50 glass used in very dark, moonless skies away from artificial lighting. Remember to let your eyes become "dark adapted" for at least 15 minutes prior to searching out fainter objects. For a wonderful selection of binocular tour objects, visit http://www.dibonsmith.com/tau.htm , a tremendous guide by Richard Smith from his web page entitled "The Constellations."

GETTING STARTED –

As with every "GO TO" TOUR guide, each GO TO object in TAURUS is discussed for your telescope regarding the type of conditions necessary for you to view it optimally for discern the very faintest details.........magnifications and aperture necessary for most objects, and much, much more. This is

YOUR complete GUIDE to get you on your way to exploring the best (and few!) objects in this small constellation. The chart provided above from the Arkansas Sky Observatory and the subsequent detailed listing of "BEST" objects contains the finest or most interesting from my own observing experience and preference.

Use the attached star chart and the following Guide as an excellent reference for your next star party itinerary, or a beginning for further study into the thousands of objects visible in this part of the sky. To access and print the chart, double click on it and save the image to a file on your computer. Once saved, open the file and RESIZE this image to fit the normal paper format for your program and save again....then merely print out the chart on high quality paper for a field reference in this GO TO tour!

Every deep sky object and every double/multiple star will have a "PERFECT MAGNIFICATION". This is the magnification that you should use that will show the object as bright and with as much as detail with possible and still increase its size appreciably so that you can view it comfortably and unmistakably. The rule for determining "optimum magnification" is that: 1) too low power results in sky background glow detracting or diminishing the contrast against the deep sky object; 2) too high magnification darkens BOTH the sky background AND the object; 3) medium magnification can be achieved at which you have MAXIMUM contrast between the object and its darkened background sky. I have found through three decades of direct observing that about 15x per inch aperture (36x for the ETX 60/70; 55x for the ETX 90; 75x for the ETX 125; and, 125x for the LX 90).for deep sky observing is PERFECT for most objects. That being said, always remember that DOUBLE or multiple stars require whatever power you can crank out....the seeing conditions are the limiting factor here.

For my complete and comprehensive discussion regarding seeing conditions and sky transparency, see my guide on the ASO website.

With all deep sky objects, avoid attempting to observe when the moon is in the sky, even a very thin crescent, as its brightness in the sky will overshadow the very dim contrast afforded by even the brightest deep sky object; if you see the object at all against moonlight, you will NOT see the subtle outlying areas or the full detail of what is presented.

For detail descriptive lists of the great double stars within Taurus, and as with all of the "GO TO" TOUR constellation lists, I recommend a good star atlas and/or chart which will list all the finest objects, constellation-by-constellation. One very handy reference guide is the *PETERSON FIELD GUIDE TO THE STARS AND PLANETS*, which features complete lists with declinations, right ascensions, magnitudes, and all pertinent information for you to expand your observing horizons beyond this brief guide. For the many double and multiple stars, I again urge you to refer to the indispensable "*Burnham's Celestial Handbook*", Volume 3 for a complete abbreviated listing.

Truly these Constellation study guides will most definitely put your star program or telescope handbook to work for you in the most efficient and enjoyable way possible! As a matter of fact, MANY computer users are now programming their own "Tours" based on these guides, using each constellation as a separate GO TO Tour for the library that can be added in or deleted through the main edit screen on your PC or MAC computer.

We hope you enjoy these comprehensive guides to touring the constellations via your AutoStar or your computer-driven telescope. Each new installment is complete with diagrams, charts and illustrations that you will find nowhere else. Please let us hear YOUR feedback and your observations of each and every constellation after YOU have toured its vast reaches of our skies!

YOUR TAURUS CONCISE DIRECTORY OF INTERESTING OBJECTS –

As mentioned, Taurus awaits the curious and discovery-driven observer with a wealth of fine double and multiple stars (for a full discussion on double star observing and their "Position Angles" refer to my brief overview in the "GO TO" TOUR guide for Lacerta in the Constellation GUIDES of this website) and some outstanding galactic clusters and the most famous (albeit NOT the brightest nor most interesting!) planetary nebula of the skies.

As an **added bonus** to this installment of the Taurus "GO TO" tour a complete discussion of the star clusters Hyades and Pleiades is presented in depth to enhance the efforts of dedicated observers.

The most interesting **11 targets** in the constellation have been chosen for this TAURUS "GO TO" tour; as with all constellations, all objects listed below will be visible in most telescopes (some naked eye) from small APO refractors to larger observatory telescopes; of course larger apertures may "show" an object a bit closer and "better," but frequently a wide field and low power view is more desirable than aperture for FINDING the objects initially. Indeed, I strongly encourage you first FIND the target object, or its approximate location through your GO TO function with your lowest power and then - once IDENTIFIED positively - move up slowly in steps with magnification if necessary. Remember, not all objects "like" magnification. Sometimes better "field of view" (such as provided in smaller telescopes) is desired over light gathering and magnification of larger telescopes.

Note that your sky library may NOT have every object listed on every constellation GO TO tour....this is intentional. You can access some of the most interesting objects of the sky directly from their coordinates. It is quite simple as you merely follow the instruction for keystrokes or tab use for your particular planetarium program.

The constellation tour Star Chart shown previously will get you started on your journey for this constellation.

Following is the concise object list for your "GO TO" TOUR of TAURUS; you may wish to find many of the objects from the sky program Library (for example, you can easily go the Pleiades, Messier 45, if you pull up "Object/Deep

Sky/Messier Object/..then type in '45'...." and then press "Enter", followed by "GO TO" to access this rich cluster. If your telescope is accurately positioned and level, this should take you to a point nearly midway between *ALCYONE* AND *MEROPE*. On the other hand, if you want to experiment and become a "better computer user" try entering the exact R.A. and DEC coordinates of the Pleiades as given below after holding down the MODE key. You will find the accuracy of entered GO TO's to be somewhat less better those stored in your program, but the capability of acquiring unlisted objects is fantastic!

You can also access (this cluster for example) in other ways. It does NOT have an NGC designation, but it is listed under "DEEP SKY / STAR CLUSTERS" and also under "NAMED OBJECTS / PLEIADES".

You will access your FIRST GOTO target - (usually the brightest star in each constellation) - via the command "SETUP / OBJECT / STAR / NAMED....and scroll to "*ALDEBARAN*", then press "Enter" and subsequently "GO TO" to move your this bright star.

You may also access the constellation by: SETUP / OBJECT/ CONSTELLATION / Cassiopeia / Enter....GO TO, which will subsequently take you to the brighter star ALDEBARAN, which is the unmistakable red "eye" of the fierce and seemingly angry bull (with Europa on his back on his way to Crete, why should HE be angry??)

452

OBJECT 1: bright star - *ALDEBARAN* (alpha Tau) - R.A. 04h 36' / DEC + 16 31 - Mag: 0.85 - "The Eye of the Bull"

OBJECT 2: star cluster - *HYADES* (Mel 25) - R.A. 04h 17' / + 15 31 - Collective magnitude = 0.8, 40 stars (Aldebaran is NOT one of them!)

OBJECT 3: star cluster - *PLEIADES* (Messier 45) - R.A. 03h 47' / DEC + 24 07 - Collective magnitude = 1.4, 130 stars, "the Seven Sisters"

OBJECT 4: double star - 7 Tauri - R.A. 03h 35' / DEC + 24 28 - Mags: 6.2, 6.7, & 9.7 - primary pair tough even for 8", third star easy with 3" scope!

OBJECT 5: 5" test double star - 47 Tau - R.A. 04h 14' / DEC + 09 16 - Mags: 5.1 & 7.3 - A great double star test for a 5" and larger scope!

OBJECT 6: 3" test double - 80 Tau - R.A. 04h 30' / DEC + 15 38 - Mags: 5.9 & 7.9 - 1.6" separation, good test for 3" scope...nice stars!

OBJECTS 7: multiple stars in Pleiades! - Messier 45 - (use "DEEP SKY/ NAMED OBJECTS) - 5 nice multiple stars here! Use the following chart for "Object 7"

OBJECT 8: wild variable star - RR Tauri - R.A. 05h 40' / DEC + 26 23 - Mag: 9.1 to 13 - wildly fluctuating star...maybe a 95 day period!

OBJECT 9: nice galactic cluster - ngc1647 - R.A. 04h 46' / DEC + 19 04 - Mag: 6.5, very large with 25, 8th mag. stars, nearly due E. of Hyades. Low power!

OBJECT 10: galactic cluster - ngc1746 - R.A. 05h 04' / DEC + 23 48 - Mag: 6, actually TWO clusters, 50 stars all about 8th magnitude.

OBJECT 11: the *CRAB NEBULA* - Messier 1(ngc1952) - R.A. 05h 35' / DEC + 22 01 - Mag. 8.4, very large and very difficult object; not much detail.

OBJECT 12: planetary nebula - ngc1514 - R.A. 04h 09' / DEC + 34 46 - Mag. 10.5 - Very difficult nebula but 9th mag. central star should be easy!!

YOUR VISUAL GUIDE TO DEEP SKY OBJECTS IN TAURUS

<u>Object 1</u> - Our "Starting" Bright Star – "ALDEBARAN" (alpha Tau - pronounced: "al DEB-aron")

At a magnitude of 0.86 and fiery red, Aldebaran is the 13th brightest star of the sky and has been associated with many legends and designations throughout history, but perhaps none so famous and memorable as "the eye of the bull" in reference to the great bull raging horns-directed toward Orion. Highest in the sky in early December midnights, this star is positioned in the southern regions of the large naked eye cluster Hyades (see below), but is NOT one of the stars gravitationally bound by the stars within that group. Whereas the Hyades is the closest star cluster to Earth, Aldebaran is still closer at a distance of only 68 light years, or about half the distance to the bright cluster. *Claudius Ptolemy* - the "namer of the original" 48 constellations - perhaps borrowed from the ancient Babylonians in referring to this star as "the torch bearer" or the "...star leading all other stars" in their courses across the sky. Because Aldebaran is conveniently located amidst the ecliptic - the band of sky in which the sun, moon and planets pass from our line of sight here on Earth - it is occasionally OCCULTED, or "covered up" by the moon in its eastward path. It is one of only three first magnitude stars that can be so occulted.

If you are using the 8" or larger telescope, look for a true physical companion to Aldebaran, at magnitude 13.1 about 32" arc (about 2/3 the size of Jupiter in the same eyepiece....use medium power) to Aldebaran's east and just a bit south. Block the bright glare of Aldebaran to see this clearly red dwarf star at its best. About four times that distance from Aldebaran to the NORTHEAST is an 11th magnitude "optical double" star that is not a true companion to the bright star....this star should be seen in even a 3" scope under dark skies.

Object 2 - HYADES - Mel 25 - (pronounced: "HIGH-a-deez") The Closest Star Cluster to Our Solar System!
This system of only 40 odd stars seems more dense than the actual numbers indicate. Remember that this wonderfully large star cluster is VERY close to us (some 130 light years distant....the closest of any deep sky object other than stars) and that MANY of the stars that are actually distant background objects to the naked eye and binoculars that appear to be part of the cluster are what we refer to as "line-of-sight associations" and not part of the 40+ star cluster, of which all stars are bound gravitationally to one-another. In addition there are a few stars - Aldebaran of course is the best example - that are also FOREGROUND objects in between us and the more distant cluster. In any case, the chart provided below shows the NAKED EYE stars (in RED) that ARE visible without optical aid for those with very dark high altitude skies, no moonlight and good eyesight. Those stars indicated with a BLUE number are fainter stars that will be seen in standard binoculars or a good finderscope. Extensive testing over the years has revealed that a trained observer

under such circumstance CAN detect stars in our modern artificially light and polluted skies as faint as magnitude 6.2! The following chart I have prepared provides star magnitudes WITHOUT a decimal point (to avoid confusion of the decimal as a faint star!); hence a star showing as "59" is actually magnitude 5.9.

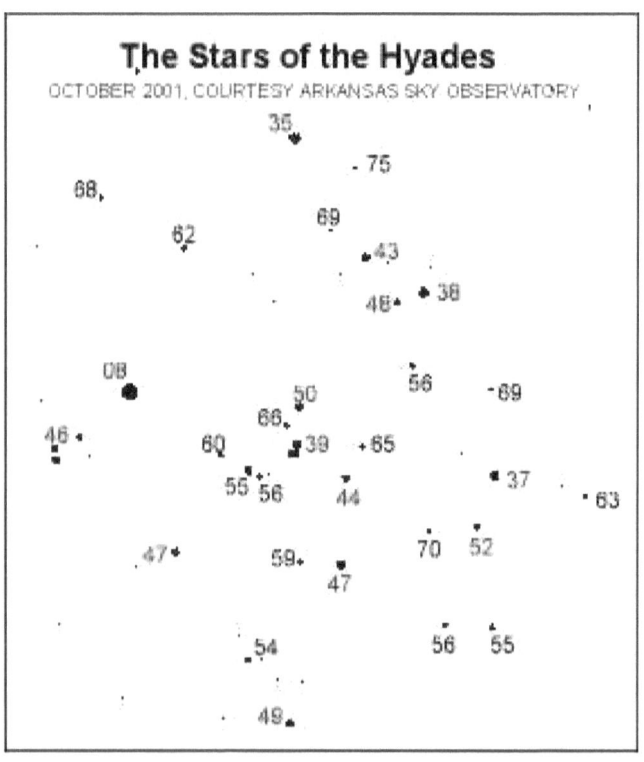

As we mentioned concerning Aldebaran's placement in the paths of planets, sun and moon, the Hyades stars are similarly - and much more often - occulted in spectacular fashion - by the moon; this is particularly rewarding for low power wide field observing (and timings of star disappearances!) when the moon is in its waxing stages and it is the dark, un-illuminated limb of the moon that will intercept the star's light and subsequently result in that light "blinking" immediately out of sight!

The Hyades, from the constellation's character standpoint, is predominantly the stars which comprise the HEAD and HORNS of the bull, extending almost 5 degrees east-to-west. It is the unmistakable "V" shape seen naked eye when peering upward in late night skies on fall evenings, the Pleiades looming just north of this larger cluster. To the trained eye there should be about 26 or maybe 27 stars which can be seen without the aid of a telescope or binocular, these being from magnitude 3.34 to 6.65.

In early Roman times, the grouping was known to the layman as "little pigs," apparently confined to a small pen, whereas in oriental texts and very old Greek references, the grouping was significant in association with rains, storms and violet weather. Greek mythology did not ignore the beautiful Hyades, designating them as the daughters of Earth-holder *Atlas* and his wife *Aethra*....the same maidens who cared for the all-important newborn Bacchus who would later become the god of drink and merriment, and thus VERY important in the social scheme of things.

The Hyades is best observed in binoculars; the naked eye cannot reveal the rich nature of the star field in and around the cluster; on the other hand telescopes cannot span the needed 4-degrees of sky necessary to bring the entire group into one wide field of view. A standard pair of good binoculars on a good night will reveal ALL 40+ stars to this cluster.

Object 3 - The *PLEIADES* - Messier 45 / the "Seven Sisters" - (pronounced: "PLEA-a-deez") Our Second-Closest Star Cluster to Earth!

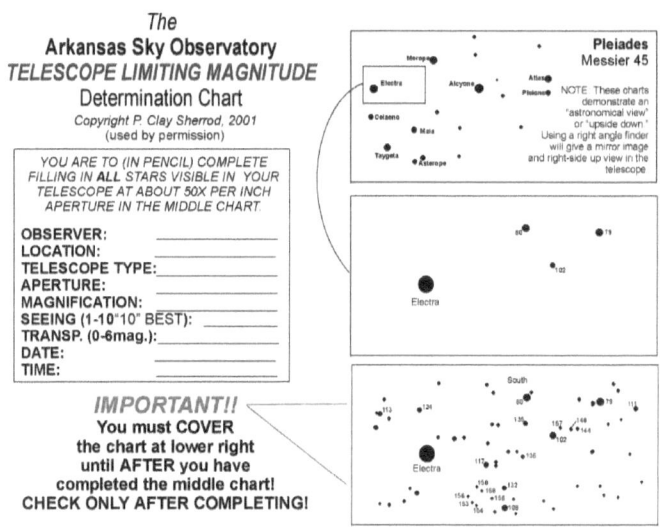

This finder/locator chart is identical to that provided above for the Hyades, and will serve to assist you in determining which stars can be seen 1) naked eye; and 2) in binoculars. The RED stars are those visible (under ideal conditions) with the naked eye, while the BLUE stars are those visible in binoculars; for telescopic tests of limiting magnitude, refer to the extensive test charts shown very early in this "GO TO" tour guide.

The photograph below was taken with five-inch APO wide field telescope, only 120 seconds in exposure. It shows pretty accurately the view that you might expect of the Pleiades on a very dark mid-winter's night through a good pair of binoculars or very low power, wide field telescopic field. This is one object where the smaller telescope really shines over larger more narrow field instruments!

The *Pleiades* is second only to the Hyades (above) in proximity of deep sky objects; whereas the neighboring Hyades is only 130 light years distant and more sparsely populated in stars (only 40), Messier 45 contains a whopping 625 or so stars from magnitude 2.8 (ALCYONE) to less than 14th magnitude (see the "Limiting Magnitude" chart discussed previously in this "GO TO" tour). The actual number of these stars which have been confirmed as physical members of the cluster is about 250, but more of those seen are surely associated as well. Its distance is well known at 410 light years and the cluster is actually drifting as a group through space in a southeast direction as viewed from Earth. Within the Pleiades are MANY multiple stars, five of which are featured as

"OBJECTS 7" (below) in our GO TO guide for Taurus. In addition, the darkest of nights and the finest of wide field optics WILL reveal the famous "*Merope* Nebula" surrounding that bright star (see chart above to identify Merope). It can be seen with about 20x in a 3" telescope; increasing magnification will merely make this object appear vague and ultimately disappear from lack of contrast against the background sky. A moonless and very dark night are required and only attempt to observe this nebulosity when the Pleiades is nearly overhead.

How many Pleiades stars can YOU see with the naked eye? Many observers (before light pollution and smog) were routinely able to plot as many as 13 to 15 of the brighter members; known also as the "Seven Sisters," the "Hen and six chicks" from the Bible, and by a series of seven distinct dots enclosed in a circle by the Native Americans of North America as well as the *Aztec* and *Toltec* Central American cultures, today ONLY SIX of these stars appear prominent...so what happened to the seventh? Being a very YOUNG star cluster (about 20 million years old....just a pup in the celestial scheme of things), it is possible that one of these active and energetic stars has stabilized a bit and decreased in intensity in recent centuries. For more information on observing this wonderful star cluster, visit my observing guide in previous chapters of Volume Two, *The Constellations*.

The distinguishable shape and brightness of this unique cluster have predictably given rise to much lore and legend by association. The earliest Chinese celebrate the "*Feast of Lamps*" in honor of the "seven lamps" of the sky, while native Americans worshipped the stars as part of the famed *Devil's Tower* in the plains of Wyoming. In Biblical references, the Pleiades have been linked to the plagues that infested Egypt, while the earliest Greek stories tied this innocent group of stars to such events as the sinking and eventual disappearance of the fabled land of Atlantis.

As mentioned with the Hyades and the bright star Aldebaran, the Pleiades is ALSO frequently occulted by the moon, as very spectacular and rewarding sight for even seasoned stargazers; such events are usually announced far in advance on various Internet calendars of celestial events and in popular publications such as Sky and Telescope and other periodicals.

Object 4 - A Nice but Difficult Triple Star - "7 Tauri" - Two easy.....the third, well: TOUGH Seven Tauri is listed as a triple star by Vehrenberg and others, but is noted as only a double by **Burnham**. This is an unusual situation for a triple star in that the two brightest components are the most difficult to differentiate, while the fainter third component (magnitude 9.7) is an easier target about 23" northeast of the brighter 5.9 magnitude combined brightness of the two primary stars. The star 7 Tau has a VERY close companion at 0.6" arc nearly in the same direction (Position Angle 23 degrees, or about NNE) as the fainter star (Position angle 58 degrees) refer to the chart below to locate these stars relative to one-another. The two primary and closest stars CAN be resolved on the best night with an 8" or larger telescope; the third star is easy with the 3" and larger apertures,

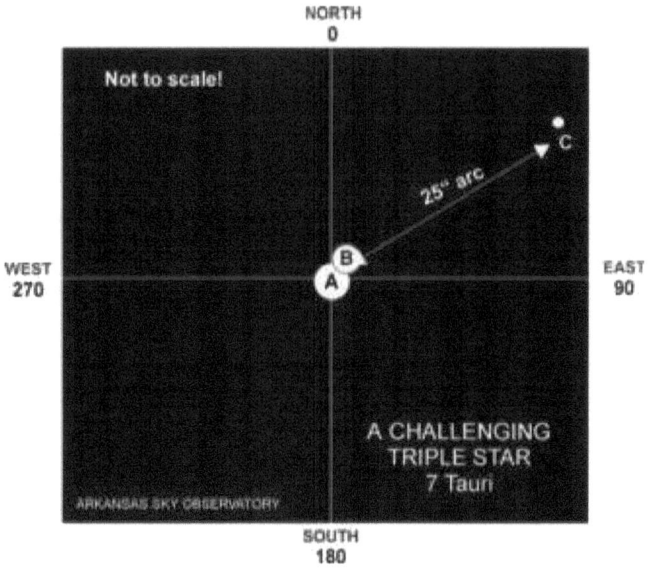

Object 5 - An Easier, but Still Not "EASY" Triple Star - 47 Tauri - Great test if you think your 5" or 6" scope has killer optics!

Can your 5" scope surpass Dawe's criteria? Divide your telescopes aperture in inches into the value "4.54" and that gives you the theoretical resolution in arc seconds (") that your telescope can resolve double stars of equal magnitude under perfect seeing conditions. For a 10" scope, for example, that would yield a value of 0.45" arc. In a 5" telescope two stars would have to be separated TWICE that much (or about 0.9" arc) to be cleanly split under the same conditions. The star 47 Tauri has a separation of just about that: 1.0" arc for its two primary (brighter) components and is a tough object. Look for two stars extremely close, almost "touching" nearly north-south-oriented; the fainter star (magnitude 7.3) will be almost due north of the brighter star (magnitude 5.8). This will be a test for scopes in the 4" through 6" range. A third component, magnitude 12 and barely visible in those telescopes but easy in the 8" and larger, is located southwest of this pair a full 30" arc.

Object 6 - Okay, Another Double: 80 Tauri in the Pleiades - This one is a challenge for the 3" aperture telescopes!
Here is a very nice double star for the 3" aperture telescopes and larger; it is an easy target at medium-high (150x) magnification in the 5" and larger instruments, but will be tough in the 3" with a separation of only 1.6" arc, somewhat wider than the theoretical limit of 1.3" for a 90mm scope. This will be an easy star to locate at magnitude 5.9, with the magnitude 8.9 companion very close and just east of due north from the brighter star.

Objects 7 - **Double Stars** in the Pleiades and Hyades!

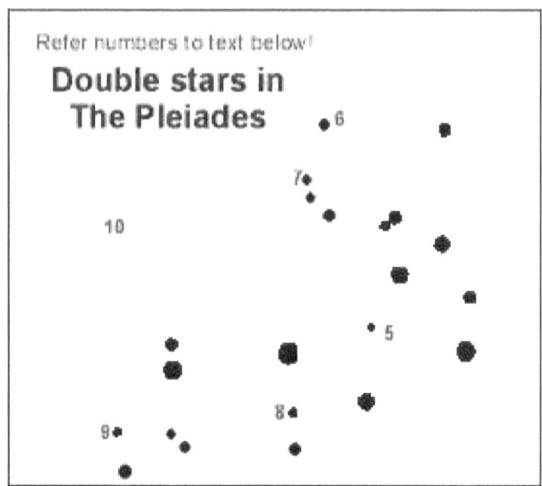

We have explored the Hyades and Pleiades already. These clusters we have seen are an excellent indicator of the ability of the naked eye to see faint stars, and to judge the limiting magnitude of optical instruments as well as the human eye. We have noted the proximity, ages and motions and discussed the faint nebulosity that is still "hanging around" from the Pleiades emergence through its primordial gas and dust cloud. With perhaps as many as 500 stars in the Pleiades and 50 in the Hyades which likely are bound through gravity in these clusters, many are sure to be "multiple stars" or double stars as we have just seen in the preceding objects....in addition to being part of the entire "multiple group" of the clusters themselves! I have selected some true physical double stars in each of these clusters (including again 80 Tau, discussed above). Use the chart above to locate the fainter stars in the Pleiades and attempt to find the brighter and easier stars that I have in the first list (Hyades) via the COORDINATES (R.A. & DEC - epoch 2001) that I provide at the end of each star description. Note that most of the first four (HYADES) are easy targets for the smallest telescopes; these are TRUE binary stars,and not just merely aligned to one-another in our line of sight as might be other close by stars. The second five stars (PLEIADES) are more challenging, fainter and closer for good challenges with your telescopes! Note that the number preceding each multiple star coincides with the number in RED on the chart above!

Following are all easy and bright stars to locate in the Hyades!

1) - **Theta2** (magnitude 3.4) and theta1 (magnitude 3.8) - wide separation of 337" arc, theta2 just below and to the east in PA 346°. Easy in all telescopes. (R.A. 04h 26' / DEC + 15 58)

2) - **kappa2** (mag. 4.2) and kappa1 (mag. 5.3) - an easily resolved binary with separation of 5.3" arc in PA 328° (nearly north-south orientation). (R.A. 04h 25m / DEC + 22 15)

3) - **Sigma2** (magnitude 4.8) and sigma1 (magnitude 5.2) - very wide separation of 431" arc - another wide binary in PA 193°, or almost north-south orientation. (R.A. 04h 39m / DEC + 15 47)

4) - **80 Tauri** is a difficult double star which is currently at nearly its maximum separation (see OBJECT 6, above). (R.A. 04h 30m / DEC + 15 38)

Now for some harder/fainter ones! Use the Pleiades chart above for these five:

5) - **Burnham 536** - (magnitudes 8. 9.5, 8 - triple) - very close (0.6")primary pair oriented N-S, third component SW wide 39" arc

6) - **Burnham 537** - (mag. 8.5 & 10.5) - very close (0.9") stars in N-S orientation; tough for 5" scope, not easy in 8"

7) - **Stuve 449** - (mag. 8.5 & 11) - wide pair (6.8") but secondary star (NW of primary) is pretty faint....good test for 3" and 5" scopes!

8) - Struve 450 - (mag. 7 & 9) - nice double and easy, with 6.1" separation; fainter star is due west of primary

9) - **O. Struve 64** - (mag. 6.5 & 6.4) - tough at 0.6" for 8" scope; look for equal pair in nearly N-S orientation!

Object 8 - A Very Wild and Erratic Variable Star: RR Tauri

As can be immediately seen by the light curve plotted below which covers a span of time covering 250 days, this is no ordinary star. **RR Tauri** will be a very difficult star to monitor throughout its entire light cycle with all but the 5" and larger telescopes since at minimum it dips into the low "12's" in magnitude and sometimes as faint as magnitude 13.1; however any 5-inch should be able to following this star's peculiar outbursts and declines very well.

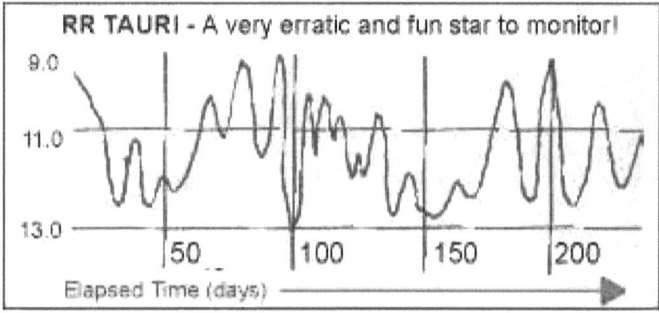

Note that the sudden increases to maximum as well as the downfalls to minimum light can take place in a matter of only a day or two! There is an indication of some degree of cyclic nature to RR Tauri's light changes of about 95 days, but that does stretch the imagination somewhat; at a distance of 3000 light years, the star might be similar to the "Orion-type" very young stars that are attempting to achieve thermal equilibrium.

To locate this star download the chart from: https://www.aavso.org/apps/vsp/ . Note for these charts, simply type in the NAME of the variable at top to generate your choice of chart.

which is the "a" finder chart for the star "U Orionis" and also includes the position for your finderscope for RR Tauri. This and other fine charts are available from the American Association of Variable Star Observers (AAVSO) with whom you can make contact via www.aavso.org to learn more about how to observe and contribute your estimates for the hundreds of variable stars for which amateur astronomers contribute nearly exclusively for each year. When the star reaches its faintest point, you will need the "g" or very narrow field and higher magnification chart which provides comparison stars down to magnitude 14. Note that this latter chart is also a "reversed" chart that the AAVSO has released to match the field of view of your telescope, with NORTH up and EAST to the right.

Merely right click on the chart and save it to file on your computer; call up the file and resize it to fit your standard 8.5 x 11 inch paper size and re-save in that new format. Then simply print out each of these charts, laminate for outdoor use and observe away! (see my complete discussion on variable stars in the Observational Guides Section here on ASO.)

Object 9 - Very Large Open Cluster - NGC 1647 - Just East of the Hyades
This is really an ideal object for binoculars or for a very low power wide field instrument. NGC 1647 can easily be found by "sweeping" a wide field of view about 3 degrees to the east of the center of the Hyades; the true "cluster nature"of this 40-star group is lost in high powers and in larger telescopes. The grouping is extremely large (40' arc

- larger than the apparent diameter of the moon), so the widest field of view is necessary. Most of the stars are 8th and 9th magnitude, so this is an excellent low power object for small telescopes.

Object 10 - Galactic Cluster NGC 1746 / 1758
Here is a very irregular cluster of stars that actually shares its stars between TWO designated "ngc" listings. NGC 1746 is technically the "western part" of this cluster and NGC 1758 takes up the eastern side. At magnitude 6.0, this brightness is deceiving since that total light is spread out over a huge 45 arc minute area, 1.5 times larger than the moon's disk as we see it in the sky; when you view this with VERY low power, you will realize why this "object" has two designation, with two conspicuous clumpings of 8th and 9th magnitude stars to either side of center. In total there are about 50 sparsely scattered stars in this cluster. This is another cluster that really requires a very low power instrument affording the widest possible field of view.

OBJECT 11 - **THE CRAB NEBULA**! Messier 1 –
The "bright Star of 1054"
Why did **Charles Messier** ever star his "list" of faint fuzzy objects now known as the "Messier Catalog?" The answer lies in an event that took place over six centuries before the French comet hunter was ever born.

Rising with a very thin crescent moon on the warm summer morning of July 5, 1054 a young Pueblo native American awoke to the haunting howls of many distant coyote, the pups screaming in discordant tones over the wails of their guardians. Throughout the canyon of the *Anasazi*, creeping through the remote crevices and sweeping through the distant hollows, more and more the cries of wildlife alerted the young man that something was different than it was the morning before when quiet met the impending dawn. To his east the bright fiery red Aldebaran, known to him as another name now forgotten in time, was gently hovering low in the now-brightening deep skies. It was a sight very familiar to him, the pointed shape of a cluster that would become known as "Hyades" in some future time describing exactly the pattern of the many projectile points routinely made by members of his group.

But there was a visitor to all this....heralded by the frantic coyotes, and emblazoned upon the sky with the accompaniment by the waning moon: there was a new star, once that outshone the familiar "red eye" and one that rivaled it in color and intensity, yet so bright that it nearly overshadowed the brilliance of even the moon. Shadows were cast upon the ground by this magnificent sight. So impressed, the Pueblo youth left an artist's image of what he had seen in red ochre smeared via his shaking finger against the flat writing surface of stone in a nearby overhang in the canyon.

The image still remains in Chaco Canyon today, a large circle that seems to encompass the crescent shape of the moon, perhaps indicative of the fact that the unknown object WAS brighter than the moon itself. And he was not alone. The spectacle was recorded around the world, notably by *Sung Dynasty* astronomers who observed and documented carefully the new visitor to the skies - a "guest star" in their records. With the suppressions of medieval European thought, there is unfortunately NO record of any observations of this "Supernova" from western cultures, although there are references that might indicate some Roman and mid-eastern early historic records of the stellar explosion.

What we see today, and essentially what **Charles Messier** first saw on September 12, 1758, is preserved in the photograph above, taken as a 120 second exposure with the ASO astrographic telescope.

Today the explosion presents itself as a planetary nebula. This huge cloud of dust and gas is the primary remains of a great stellar Supernova that result in the virtual annihilation of a great star almost 1,000 years ago. Although photographs today will reveal SOME changes from that seen above since the gases are now flowing outward at the incredible rate of 600 MILES PER SECOND (!), our view remains pretty much as it appears in that remarkable photo. Very large (10" and above) telescopes will reveal some of the faint filaments that you can see streaming from the center of this epic event. However, smaller instruments with modern CCD imaging can easily capture much of the fine structure that is surely representative of the catastrophic nature of this explosion some ten centuries ago at a distance of over 6300 light years away. One cannot look at this photo and not be overwhelmed by the cataclysmic force that created this remarkable view.

In my opinion, the "Crab Nebula" as we know this remaining cloud from the supernova event of 1054, is one of the most difficult of all the Messier objects; the unknowing observer can be peering directly at it and actually not even realize there is an object in the field of view, since its faint 9th magnitude "integrated brightness" is spread out over a large 5' x 3' arc area. It thus has a VERY low overall surface brightness. It can be seen in a 3" telescope under very dark skies, but expect no detail at all until using the 10" and larger telescope.

It is recommended to use higher powers on this object, contrary to most practice when observing deep sky objects; I have always found that a magnification of about 35x to 40x per inch is actually ideal for this nebula as well as other planetary nebulae. Always observe this difficult object only when the moon is totally absent from the sky; nebula filters most definitely help on this planetary, but not to the degree that they do on brighter ones such as the Dumbbell (Messier 27) and the Ring (Messier 57).

At the center of the nebula, and the resulting phenomenon from the stellar explosion is a very faint starlike object (magnitude 15.9) that ejects tremendous "pulses " of energy in both radio and X-radiation at the rate of precisely 30 times each second. So accurately and intense are these beats that such a source (and there are many now known) has come to be known as a "Pulsar."

The only reason that Messier even started his famous catalog was because he had confused this faint object for a comet that he was attempting to "re-find" from the night before....as he searched for the comet, he stumbled across this "faint fuzzy," and realized that it appeared very much as do very faint newly-discovered comets prior to brightening and forming the characteristic features such as a tail. He explained, "....*I endeavored to find others....in that astronomers would not confuse these [nebulae] with comets as we see them when they first begin to shine...*" And the rest is astronomical history.

OBJECT 12 - A Rival Planetary Nebula That Messier never Found - NGC 1514
This is a much fainter, more distant, certainly older and much more difficult object for small telescopes, the planetary nebula ngc1514. This interesting planetary has done a switch on amateur astronomers in that the STAR which created the shell of gas (similar to that of the Crab Nebula) is actually much easier to observe visually than is the gas shell that was ejected! The central star of ngc1514 is

magnitude 9.7 (visible easily in a 3" scope) while the faint strands of nebulosity are a very faint 10.8. Thus, it is possible to identify the star from this object, but not the gas planetary nebula unless using a telescope 8" or larger and in very dark skies.

WANDERING ABOUT....YOUR NEW "USER OBJECT" IN TAURUS

Back to the Pleiades: we have all seen the famous blue nebulosity that surround most of the brighter stars of this remarkable cluster, the most dense and "observable" being that encompassing the bright star Merope.

This nebulosity (only that enveloping Merope) is also designated with its own number, "IC 349" and is fairly easy to detect in small wide field instruments on VERY dark nights. I have actually detected it in my 10 x 50 *Nikon* astronomical binoculars, and it can routinely be see in richest field telescopes utilizing the lowest possible power and widest field of view. The fine photograph of this incredible nebulosity associated with nearly every star of the Pleiades and particularly Merope (arrow) below was recorded with a 3-minute unguided exposure from the wide field 13" f/4 camera at the Arkansas Sky Observatory in 1983, with the wispy clouds of nebulosity appearing as though an artist passed through the pool of gas with a broad paintbrush. The spurious reflection image at left is a flare produced in development of the original cooled emulsion negative.

All of this gas cloud that is still visible, and certainly photographable, is what remains

from the actual formation of the entire 500-member group of stars. This primordial gas is perhaps not at all unlike that of the equally famous Orion Nebula, except that more gas and less stars remain in the latter object. Eventually we might imagine that all of this remain gas will be "accreted" or gravitationally-tugged into Merope as well as the other nearby stars until none remains. In color photography, amateur astronomers, even utilizing piggyback photography via telephoto lenses can capture the true BLUE coloration of the young energetic and hot stars and the whitish glow of this surrounding cloud of gas.

The star Merope and IC 349 are our "GO TO" tour User Object for this installment of this constellation guide. The 2001 epoch coordinates of the star Merope are:
R.A. 03h 46m.2 / DEC. + 23d 24m

Of course you can use the chart above to find the star quite simply with your finderscope, but it will be quicker to access if you add this star and nebula as yet another of your growing library of User Objects!

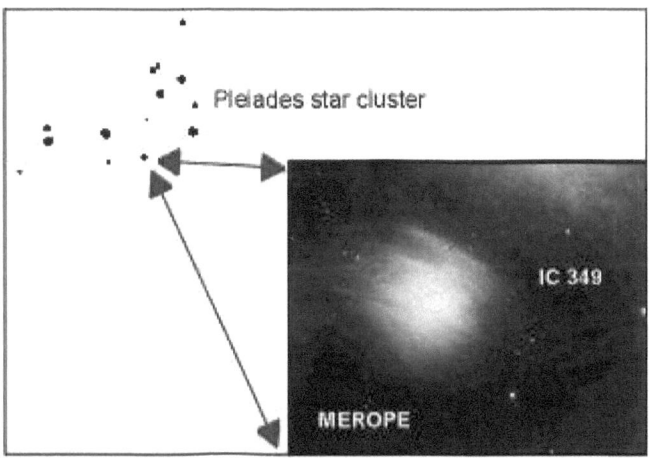

On AutoStar, go to: "Select/Object [enter]...." scroll down to "User Object" [enter]. Now enter the coordinates given above for "Merope", using the number keys on AutoStar. After entering the coordinates and pressing "Enter" yet again, scroll down one and you can list the magnitude of the object as "4" [Enter].

Remember that your keypad, App or sky program will vary from the techniques for Autostar....but all are logically the same. Please read your User Guide or HELP to learn the process to produce your own User Object library for your computerized scope!

* * *

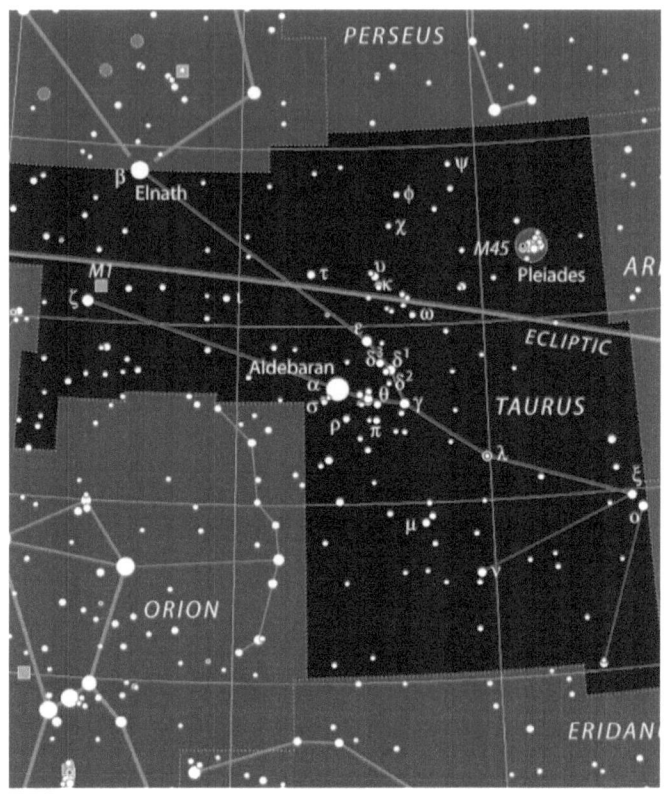

Taurus constellation chart
Adapted from the original by
IAU / Sky & Telescope

Chapter 34

TRIANGULUM
A Ancient Tribute to the Fertile Nile Valleys of the Pyramid Builders

This sounds like our first geometry lesson in our *Constellation* tours. This constellation represents one of only a few non-animal/human star patterns of the recognized 88 constellation groups.

Other than its fabulous "*Pinwheel Galaxy*," Messier 33, this small constellation is an often overlooked searching ground for deep sky splendors. Tucked in between the large constellations of *Andromeda* to its west and *Perseus* to the northeast, Triangulum offers several very nice and colorful double stars and at least two brighter "NGC-category" spiral galaxies. So all is not lost in this little corner of space. The constellation is made up of fairly faint stars to the naked eye, so use the wide field finder chart below to locate it relative to the familiar "strings" of stars in Andromeda (to the right in the chart) and the "*keep-on-truckin' man*" of Perseus (to the upper left in the chart).

Note that the stars which form the namesake of this constellation are naked eye....but just barely. The only named star in the group, Alpha Trianguli, is "*Metallah*", or "*Caput Trianguli*" (the Latin form of the original Arabic), is only magnitude 3.49 and forms the "point" of the sharp triangle pattern. Beta Trianguli is magnitude 2.9 (or given by some as 3.1 and in either case is another example of the "beta star" of a constellation outshining the "alpha" member!); Gamma Trianguli is a full magnitude fainter, at magnitude 4.0; both Gamma and Beta for the opposite angle points of the less-than-conspicuous triangle.

Surrounding Beta Tri are two additional stars that make a nice triplet in binoculars and very low telescopic magnifications: Delta Triangulum is to the star's north and slightly fainter (still naked eye) and "7 Tri" is opposite Beta from Delta, or a bit west of due south.

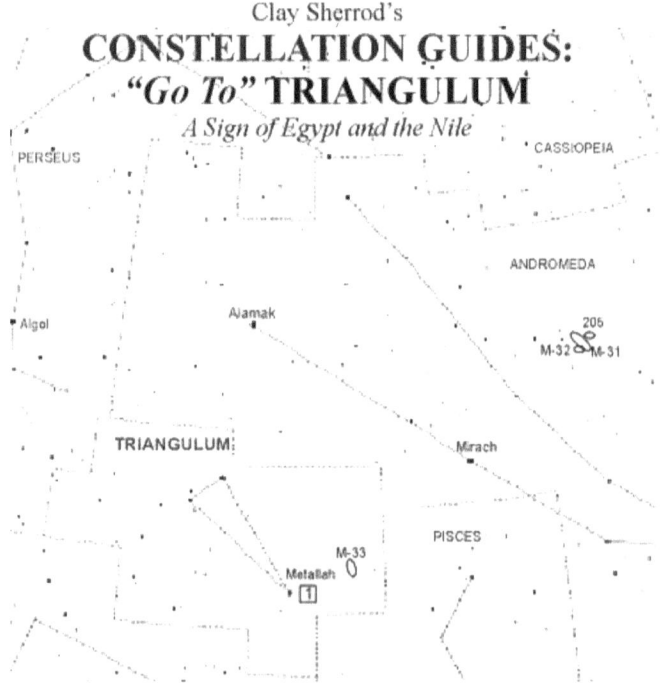

The three stars making the triangle are fairly close to our own solar system, with Alpha being only 65 light years distant, followed by Gamma at about 100 light years with Beta trailing farther away at nearly 140 light years distant. They are NOT close enough to one another to be physically associated as are the stars of Orion or the Hyades in Taurus.

History of the Little Triangle

Oddly enough, Triangulum is one of the original 48 constellations as set out by **Ptolemy** in the 2nd century A.D. Odd because it does not really fit the pattern of serving to honor myth, legend, and important Greek and Roman gods of antiquity. It is - and always has been - a "triangle." When you look upon this tiny triplet of barely-naked eye stars, it must come to mind that it should perhaps have been included in SOME OTHER constellation, like Andromeda, Cassiopeia or Perseus.

But upon closer examination, perhaps the symbol of the triangle was just as important to pre-Ptolemaic philosophers as were the gods and mythology of old. By the time that Ptolemy had designated this and other constellations by name, the Egyptian pyramids were long built and even new, smaller, ones were being constructed. The strength and grandeur of the pyramids has always captivated modern man and the reasons for their existence remains a mystery until today. So, perhaps Ptolemy was paying tribute to the engineering masterpieces of his time (and still today!).

For a brief summary of the naming and history of the constellations, see introductory chapters in this Volume Two.

Also - again an Egyptian influence - the TRIANGULAR SHAPE of the small asterism of stars is suggestive of the earliest Greek letter for "delta," and has since been long associated with the fertile **Nile River delta** so vital to the advance and survival of the earliest Egyptian people. Likewise, the three stars have become associated (without any explanation as to origin) to the island of Sicily - marked in a triangle by its three conspicuous promontories which shape the island into its characteristic shape.

So, unlike the inanimate (non-animal to be more specific) constellations such as *Argos* (a ship), *Sextans* (a marine sextant), *Microscopium* (the microscope) and so on.....the "Triangle" is a constellation of original antiquity....the others have come since the 17th century.

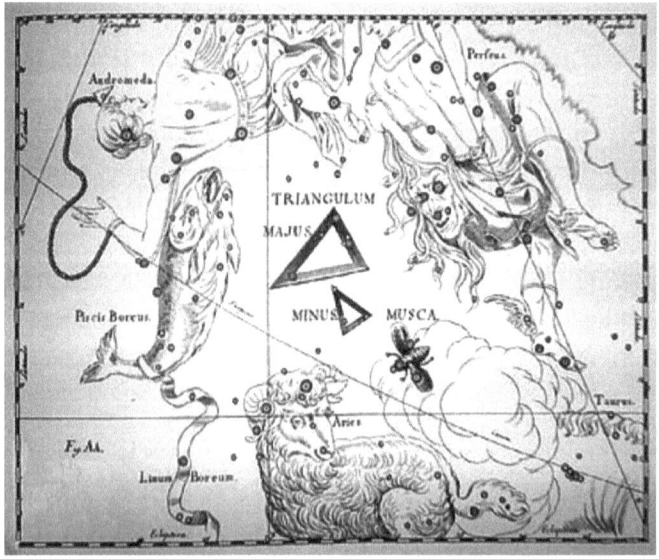

In Hevelius' drawing (1690) of Triangulum, note that there are two triangles. While the ram, Aries, rests below, the menacing gaze of evil Medusa and her snake head being held by Perseus is seen at upper right. The strong mythological influence of the triangle - plain and simply a "triangle" - can be realized in the portion of the **Hevelius** star chart shown, with Triangulum sandwiched clearly between Andromeda to its west and Perseus to its east and north. What a triangle is doing between the two lovers......who knows.

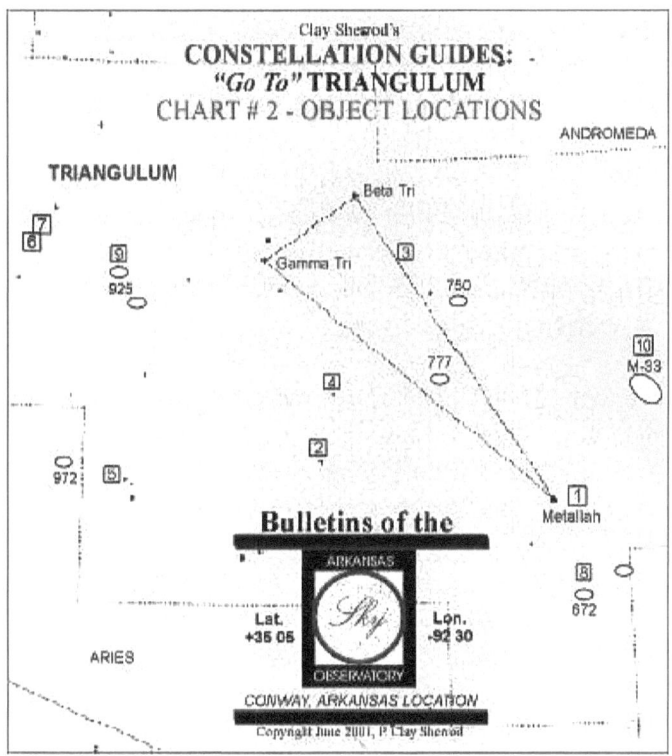

Above is a closer view of Triangulum and the nice variety of objects within it for our "GO TO" tour. Other that about eight or ten variable stars, two dozen multiple stars and ngc/IC objects that are much too faint for amateur telescopes, what is shown in the chart is pretty much the entire menu of objects available for interesting viewing in this constellation. Nonetheless, those objects shown are quite nice and many have fantastic facts associated with them!

There are MANY galaxies in the realm of Triangulum, but most of these NGC objects are too faint for our telescopes and, indeed, many are beyond the reach visually of even very the largest telescopes except photographically. Hence, these galaxies (and some other faint objects of other types) are not discussed in this guide. Throughout our "GO TO" guides, only objects that are within reach of telescopes ranging from 60mm to 200mm aperture are typically discussed; occasionally you will find noted an important object or "position" that is of historical or other interest that CANNOT be seen in our telescopes.

As with all "GO TO" tour guide star charts, use as a nice quick desk reference or good field chart as you proceed through the Triangulum tour.

Also with every "GO TO" tour guide, each GO TO object in Triangulum is discussed for your telescope regarding the type of conditions necessary for you to view it optimally for discern the very faintest details.........magnifications and aperture necessary for most objects, and much, much more. This is YOUR complete guiude to get you on your way to exploring the best (and few!) objects in this constellation. The following listing of "BEST" objects contains the finest or most interesting from my own observing experience and preference.

Use the attached star chart and the following Guide as an excellent reference for your next star party itinerary, or a beginning for further study into the thousands of objects visible in this part of the sky.

YOUR TRIANGULUM CONCISE DIRECTORY OF INTERESTING OBJECTS –

The charts shown above will assist you in becoming familiar with the neighborhood within the little triangle. I have chosen the finest (or most interesting) **10 objects** in this TRIANGULUM "GO TO" tour; as with all guides, all objects listed for this constellation will be visible in most telescopes (some naked eye) from the small beginner telescope to large observatory instruments. Of course larger apertures may "show" an object a bit closer and "better," but frequently a wide field and low power view is more desirable than aperture for FINDING the objects initially.

Indeed, I strongly encourage you first FIND the target object, or its approximate location through your GO TO function with your lowest power and then - once IDENTIFIED positively - move up slowly in steps with magnification if necessary. Remember, not all objects "like" magnification. Sometimes better "field of view" (such as the wonderful wide fields provided by the modern wide field APO refractors) is desired over light gathering of larger telescopes and higher magnification.

The rule for determining "optimum magnification" is that: 1) too low power results in sky background glow detracting or diminishing the contrast against the deep sky object; 2) too high magnification darkens BOTH the sky background AND the object; 3) medium magnification can be achieved at which you have MAXIMUM contrast between the object and its darkened background sky. I have found through three decades of direct observing that about 15x per inch aperture (36x for a 3-inch) 55x for the 4-inch; 75x for the 6-inch; and, 125x for the an 8-inch) for deep sky observing is PERFECT for most objects. That being said, always remember that DOUBLE or multiple stars require whatever power you can crank out....the *seeing conditions* are the limiting factor here.

With all deep sky objects, avoid attempting to observe when the moon is in the sky, even a very thin crescent, as its brightness in the sky will overshadow the very dim contrast afforded by even the brightest deep sky object; if you see the object at all against moonlight, you will NOT see the subtle outlying areas or the full detail of what is presented.

The high northern declination of Triangulum and its very small size results in EVERY object within its borders being tightly restricted between declination +25 degrees (and only a VERY small area that far south!) and about +37 degrees, with most area and objects confined between 30 degrees and 35 degrees north declination.

Triangulum is a "fall" constellation for the northern hemisphere, rising at the end of dusk very far north of due east from mid-northern latitudes about 9:30 p.m. local time on August 7; on that same night the constellation "culminates" (or passes through the celestial meridian) at about 4:30 a.m. the next morning. Midnight culmination occurs each year on about October 20.

As with all of the "GO TO" TOUR constellation lists, I recommend a good star atlas and/or chart which will list all the finest objects, constellation-by-constellation. One very handy reference guide is the *PETERSON FIELD GUIDE TO THE STARS AND PLANETS*, which features complete lists with declinations, right ascensions, magnitudes, and all pertinent information for you to expand your observing horizons beyond this brief GUIDE.

Note that your handbox or App will NOT have every object listed on every constellation GO TO tour....this is intentional. You can access some of the most interesting objects of the sky directly from their coordinates. It is quite simple as you merely enter these coordinates as described in the HELP or User Guide for your particular sky program.

The constellation tour Star Chart shown here will get you started on your journey for this constellation.

Following is the concise object list for your "GO TO" tour of TRIANGULUM; you may wish to find the majority of the objects from the sky Library (for example, for Autostar, you can easily go to the huge galaxy Messier 33 if you pull up "Object/Deep Sky/Messier/..then type in '33'...." and then press "Enter", followed by "GO TO" to access this remote globular cluster. On the other hand, if you want to experiment and become a "better computer user" try entering the exact R.A. and DEC coordinates of that object as described above after holding down the MODE key. You will find the accuracy of entered

GO TO's to be somewhat better than those stored in sky program but the capability of acquiring unlisted objects is fantastic!

Also, remember that you can access objects in several other ways in this constellation as well: (keystrokes will vary with program)

1) **by constellation**: key in SETUP / OBJECT / CONSTELLATION / [scroll to Triangulum] / "Enter" / "Go To"
2) **by key star**: key in SETUP / OBJECT / STAR / NAMED / [scroll to "Metallah"] / Enter / Go To;
3) by Messier Object: key in SETUP / OBJECT/ DEEP SKY / MESSIER / [type in "33"] / "Enter" / "Go To"
4) **by NGC object** (same as #3 above, except key in "NGC" rather than "Messier" and type in "598"
5) **by object name**: key in SETUP / OBJECT / DEEP SKY / NAMED [scroll to "Pinwheel Galaxy...."] Enter

So here is your Triangulum concise list of objects by name and celestial coordinates:

OBJECT 1: brighter star - METALLAH (alpha Trianguli) - R.A. 01h 50' / DEC + 29 20 - Magnitude: 3.6
OBJECT 2: great double star - Iota Trianguli - R.A. 02h 10' / DEC + 30 04 - Mags. 5.4 & 6.9 - great colors!! Easy.
OBJECT 3: really good double - Epsilon Tri. - R.A. 02h 00' / DEC + 33 03 - Mags: 4.4 & 11.3 - great challenge!
OBJECT 4: fabulous double - Struve 232 - R.A. 02h 12' / DEC + 30 10 - Mags: 7.5 & 7.5 - wide,

great for all scopes!
OBJECT 5:   3-inch test - Struve 269 - R.A. 02h 25' / DEC + 29 38 - Mags: 7.5 & 10 - great double star, close!
OBJECT 6:   another test star - Struve 285 - R.A. 02h 36' / DEC + 33 12 - Mags: 7 & 7.5 - nice equal pair, close!
OBJECT 7:   variable star - R Trianguli - R.A. 02h 34' / DEC + 34 03 - long period (266 days) Mag. 5.7 to 12.4!
OBJECT 8:   spiral galaxy - ngc672 - R.A. 01h 45' / DEC + 27 11 - Mag: 11.8 - interesting in larger scopes only
OBJECT 9:   spiral galaxy - ngc925 - R.A. 02h 24' / DEC + 33 22 - Mag: 11.4 - very large, faint! 6-inch+
OBJECTs 10:   "*Pinwheel Galaxy*" - Messier 33 (ngc598) - R.A. 01h 31' / DEC + 30 24 - also ngc604 inside!

A VISUAL GUIDE TO OUR DEEP SKY OBJECTS IN TRIANGULUM

Object 1 - Our "Starting" Brighter Star – "*METALLAH*" (alpha Trianguli)
A very close spectroscopic binary star, Metallah is a star very much like our own sun, except about 12 times more luminous. At a distance of only 65 light years, it is the closest of the few brighter stars within Triangulum, although at magnitude 3.6, it is not as bright as Beta Trianguli which shines at magnitude 3.1. Alpha Tri makes the "pointed" end of the long triangle of this constellation.

Object 2 - Iota Trianguli - A Wonderful Double Star for all telescopes!!

This is a wonderful double star for low powers in all of our telescopes, although about 50x to 75x should be used with small telescopes. This is a well known star for its remarkable color contrasts and one that you do NOT want to pass up seeing. Most observers see two distinct colors: the brighter star (magnitude 5.4) is distinctly yellow, while the fainter star (magnitude 7.0) is clearly deep blue or green-blue. This star can be resolved with a 3-inch, but medium powers are desired to best observe the distinct colors. In all scopes be sure to use about 20x per inch aperture for best views. The companion star is just north of due east of the brighter star, separated by 3.8" arc, easily resolvable if enough magnification is used. This is a splendid sight in the 6-inch and larger scopes.

Object 3 - Another Good Double Star – Epsilon Trianguli - Good separation, but FAINT companion!
Here is a wonderful test object for the keen-sighted 3-inch telescope user and a pretty darned nice object for all telescopes! Epsilon Trianguli (also "3 Tri") is a nice double with a wide magnitude contrast! it is easily found about one-third the distance from Beta Tri to Alpha Tri, along that side of the triangle. The primary star is a brighter magnitude 5.4 while the companion - a nice 3.9" separation (far enough to resolve, but close enough to make it "fun!") - is a faint magnitude 11.1, quite a challenge for the smaller scopes on a dark night. In all telescopes, best views and best chances to see the companion are with medium high (about 25x per inch to 35x per inch aperture). Look for the faint star almost exactly SOUTHEAST of the brighter star. (Position Angle 118 degrees - see discussion of Position

Angle in my *Constellation* guide *Lacerta* for details regarding double star observing and determining "Position Angle")

Object 4 - A Double You Do Not Want to Miss –
Twin Stars: Struve 232
Here is a "double star" that will make you think you are seeing double.....Struve 232 is a pair of EQUAL magnitude 7.5 stars, both brilliant blue-white! The stars are oriented in an almost EXACT EAST-WEST orientation, making the view even more pleasing....many have likened the view of this star at high power as "...like the headlights of a car far in the distance." The separation of this wonderful star - and its relative brightness - allows for it to be viewed easily in all telescopes. However, the colors of the stars are best appreciated at magnifications reaching 25x per inch aperture. Perhaps this star is BEST seen in a 4-6 inch telescope, affording a nice field of view at about 150x plus nice contrast of their brilliant white against the dark sky.

Object 5 - The Stars are Getting Closer! Tough Double for small scopes - Struve 269
Below, find a "Position Angle Chart" for both objects 5 and 6, Struve 269 and Struve 285. BOTH stars are good tests for a 3-inch and might require a 5-6 inch on some nights; neither is a simple shot with either telescope! Both stars have companions at exactly 1.6" arc, seemingly easy resolution for the larger telescopes but just wider than the threshold resolution of 1.3" arc for a 3-inch. Struve 269 has an easy-to-find 7.5 magnitude primary star, very orange in color. Look for the companion (use the chart below to identify where to look in your telescope, this being oriented exactly as your

telescope with the right angle view - north at top and east to the right) in Position Angle 345 degrees....or just WEST of due NORTH. The companion is 10th magnitude, so you "might try" this star with a 3-inch at very high power on a steady night!!

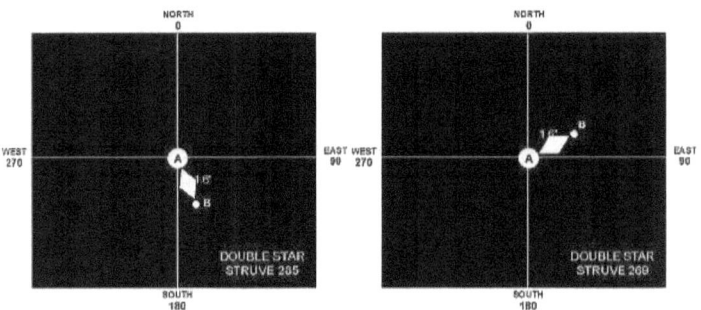

STRUVE 269 and STRUVE 285
Two difficult double stars for the ETX 90 Telescope

Object 6 - The Other Double Star (see above chart)
- Struve 285
Like Struve 269 above, Struve 285 is a tough double, but MUCH more interesting as far as "pretty views" are concerned. Here is a pair of equal magnitude stars - 7.0 and 7.5 - VERY red in color! Look for the slightly fainter star (use the chart above to help determine where to look!) only 1.6" arc away in Position Angle 166 degrees, or just about SOUTHEAST of the brighter star. Powers of about 25x per inch minimum are recommended, and even more (up to about 40x per inch) will reveal the beautiful colors of this pair.

Object 7 - Nice Long Period Variable Star –
R Trianguli

Like so many long period variables, R Trianguli is an "M-class" Red Giant star. Its variation comes from actual pulsating of its size due to expansion from the nuclear processes, and then subsequent contraction as gravity attempts to pull the mass of the star back toward its center. So dramatic is this celestial tug-of-war with this star that it will vary in magnitude from a bright 5.7 to as faint as 12.5 in only 266 days! This remarkable change can be monitored and even plotted by those using very small telescopes and certainly with larger telescopes. Reports of your observations are always valued by the American Association of Variable Star Observers (AAVSO) and you may download and print FINDER/COMPARISON charts for free from their site.

To locate the star to begin your observing, use the "a" finder chart for R Trianguli at https://www.aavso.org/apps/vsp/ . Note for these charts, simply type in the NAME of the variable at top to generate your choice of chart. This will allow you to see the star relative to the entire constellation and many of the brighter stars for comparison when the star is at its brightest. However, as the star fades and you need fainter comparison stars, use the "g" narrow field chart with star magnitudes down to 13.0 and fainter.

Both charts should be downloaded and saved to a file (just name the file "R Tri"); open this file and the chart will be HUGE! Resize of course to fit your standard 8.5 x 11 inch page and RE-SAVE the file in your new format. Then you may print each of these charts and laminate or put in a sheet protector for use in the night air.

Object 8 - A Faint Spiral Galaxy - NGC 672
For lack of a better description, this is a "rugged" spiral galaxy; the 6 or 8 inch telescope "may" show some clumping of this object, this being the conglomerated star groups that are lumped tightly in a couple of the spiral arms. Although this faint galaxy is a medium sized object (5.5' by 1.5' arc), it is very faint and seen best only on VERY dark nights; wait until the object is nearly overhead for viewing. It is possible to see this galaxy in a 4-inch telescope as a very small dim glow in about 150x magnification if your skies are very dark. Look for this very faint galaxy just south of Alpha Trianguli.

Object 9 - Another Faint Spiral Galaxy - NGC 925
This object - like ngc672 above - is a very "rugged" spiral galaxy, exhibiting bright clumpings in the spiral arms from huge star clusters. The clusters cannot be seen, but the clumping CAN be seen in an 8-inch or larger. This, like ngc672 is a very faint galaxy, but it is much larger - 9.5' x 4.3' arc - and hence perhaps easier to spot in a 4-inch, though a challenge. This galaxy forms an isosceles triangle with Gamma and Beta Tri., being just EAST of Gamma.

Object 10 - The Famous "Triangulum Galaxy" or "Pinwheel Galaxy" - Messier 33

You are looking at perhaps the MOST DISTANT object in all of space that can be seen with the naked eye! Messier 33 is just a bit farther from us than its more famous cousin, the *Andromeda Galaxy*, Messier 31, found only a few degrees WEST of M-33. The Hipparcos satellite surveys have changed the cosmic distance scale, and has resulted in "moving out" the distance to M33 to about 3.0 million light years. Most sources give a distance of 2.3 to 2.4 million light years, but the Sky Catalogue 2000 has more than 2.9 million light years.

Compare that distance to slightly neared Messier 31 at a distance of some 2.3 million light years; both are part of the same gravitationally-bound "Local Group" of galaxies (see *The Constellations*, Vol. Two)

The Two Most Distant Objects as seen with the naked eye are Messier 33 (Pinwheel Galaxy) at lower right and the Great Andromeda Galaxy (Chapter One, The Constellations) above. While it takes an average night to quickly see the smudge of the Andromeda Galaxy, the very dark skies out of urban areas are a must to glimpse the huge glow of the Triangulum Galaxy, the Pinwheel.

Credit for the first records of M-33 is given to astronomer **Hodierna** in about 1654; like so many celestial "discoveries" of the time, this beautiful galaxy was independently rediscovered by Messier in 1764. **Sir William Herschel**, who otherwise carefully avoided Messier's objects in his survey, assigned it the number HERSCHEL V.17, on September 11, 1784. Also because of the cataloging of Herschel, the brightest NGC object embedded within this galaxy - NGC 604 - has received an

NGC number of its own. This interesting "clump" of celestial material is PART OF Messier 33 and is situated in the northeastern part of the galaxy; it is the bright knot near the top of the image shown below in the NASA photographs; the photo at right shows the region very close through the eyes of Hubble, resolving over 200 young hot massive stars (of 15 to 60 solar masses) which have recently formed here. This region in the Pinwheel galaxy is MUCH like our own Milky Way's Orion region, a stellar birthing place and nursery for young hot stars.

Messier 33 gives our small telescope users a BIG advantage over the larger telescopes. Because M-33 is so large - 65' a 35' arc (twice as large as the moon's apparent diameter long and just the same size as it wide!) - and so faint, this object is best viewed always with very wide field and low power instruments; in standard 10 x 50 binoculars on dark mountaintops, this thing is huge and bright! It appears as a "thumbprint of a glow". I always thought that binoculars were "the way to go" when viewing this close galaxy UNTIL I had a chance to see it through a small quality refractor....just outstanding! Just as with the large wide field needed with observing Messier 31, the wide field refractor provide the sharpest contrast and widest fields imaginable; combined with their very fast focal ratios, the images are truly outstanding. With any 3-inch, use as wide a field as possible, preferably the 32mm Plossl if possible; the 6-inch and larger scopes warrant an even wider field eyepiece; my 35mm Televue Panoptic gives and incredible 1.4 degree field of view with the 8" scope and takes in all of the Pinwheel, though not M-31. On most telescopes, I have found that the 32mm Televue Plossl giving it some competition with somewhat better star images across the field.

This is an excellent object for piggyback astrophotography (see my discussion about piggyback photography at the *ASO GUIDES - Frequent*) requiring about a 200mm telephoto lens, dark sky and some stacked CCD frames of about 60 sec each.

Just to the north of direct center of this large galaxy is NGC 604, which is NOT a separate object at all (see the photograph and locator arrows above); this is actually a huge cloud of dust, gas and stars embedded in one of the larger spiral arms of M-33, very similar to the Orion nebula here in our Milky Way. A 6-inch on the darkest of nights can SEE this, if you know exactly where to look; it will look as an incredibly faint "ragged" fuzzy star at about 200x, due north about 1/2 field of view in the 26mm Plossl. It shows up better (brighter) with no additional detail visible in an 8-inch at about the same magnification, although it can be seen easier in this scope at somewhat lower powers.

WANDERING ABOUT....YOUR NEW "USER OBJECT" IN TRIANGULUM

What did I tell you....this tiny constellation is just PACKED with good objects; other than the fabulous Messier 33, there are NOT that many interesting deep sky objects, but there are plenty more doubles to be observed for the curious. Consult the *Burnham's Celestial Handbook*, Volume 3 for additional double and multiple stars - as well as several other good variable stars for our telescopes.

One such multiple star will be our TRIANGULUM USER OBJECT for this constellation "GO TO" TOUR. This is the quadruple star "**Struve 158**" this being the 158th multiple star cataloged by the pioneering visual double star observer **Otto Struve** throughout the early 20th century. Struve 158 is a difficult star to locate....once found though, it is unmistakable - there will be two faint stars, one ("A") at magnitude 8.5 and the other ("B") slightly dimmer at magnitude 9.1 (just for starters!) The separation of these two stars is ONLY 2.1 second of arc, but still easily resolvable in a 4-inch telescope; the trick here is to make sure of the identify at low power and then gradually move up to higher magnifications to achieve both images (about 30x per inch minimum).

Locate Struve 158 at: R.A. 01h 44m ; +32 degrees 55m.

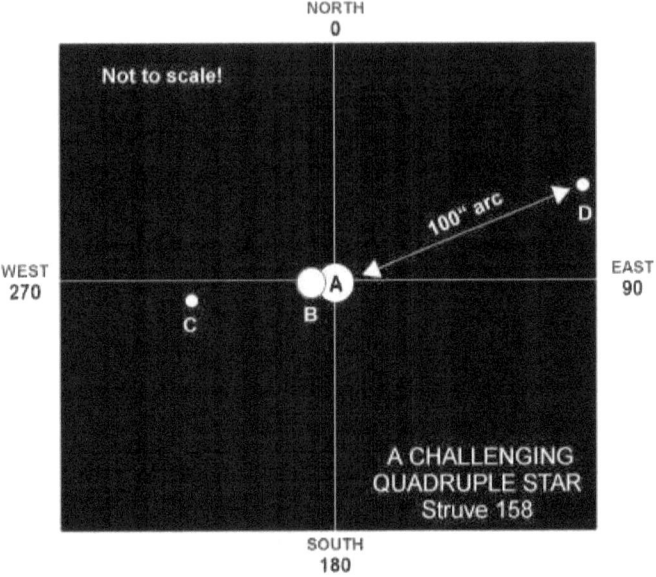

As shown in my diagram "B" is in Position Angle 267 degrees, or almost exactly DUE WEST of the brighter star; that should aid in its identification. The brighter "A" star has another companion, "C" with a VERY faint magnitude 12.5 (likely NOT visible in a 3-inch but attainable in a 5 inch and larger) in Position Angle 256 degrees, ALSO nearly DUE WEST! However, look for this very faint star (use about 40x per inch in the 5" and 8" scopes) much farther away, some 50" arc, or about the diameter Jupiter would look in the same eyepiece. There is yet a FOURTH member of the multiple Struve 158, this being not quite so faint as "C", at magnitude 11.4, hence easily visible in the ETX 90. This "D" star is TWICE the distance (100" arc) from "A" that "C" is, but is found nearly opposite the brightest star, in Position Angle 70 degrees, just north of EAST (see the finder chart, previous page).

For AutoStar, go to: "Select/Object [enter]...." scroll down to "User Object" [enter]. Now enter the coordinates given above for "Struve 158", using the number keys on AutoStar. After entering the coordinates and pressing "Enter" yet again, scroll down one and you can list the magnitude of the object as "8" [Enter].

The exact process for bringing up a celestial object for your telescope to GO TO will vary with the sky program, telescope keypad, or App that you are using, but all fundamentally work the same logic to acquire your object as well as build your User Object Library. Always consult your HELP or User Guides of your specific program for the exact steps in using tabs and keystrokes to pull up and GO TO your objects!

You will likely be glad you have added this object to your library of User Objects or PC sky program: first, you will likely NOT see all four stars at your first attempt and be tempted to move onto something easier ("....I know! let's look at the Pinwheel Galaxy!"); second if you DO see all four stars then you will surely want to show off your trophy catch!, and third, if you cannot resolve all four stars, you will want to remember this challenge when you move up to a larger telescope

*   *   *

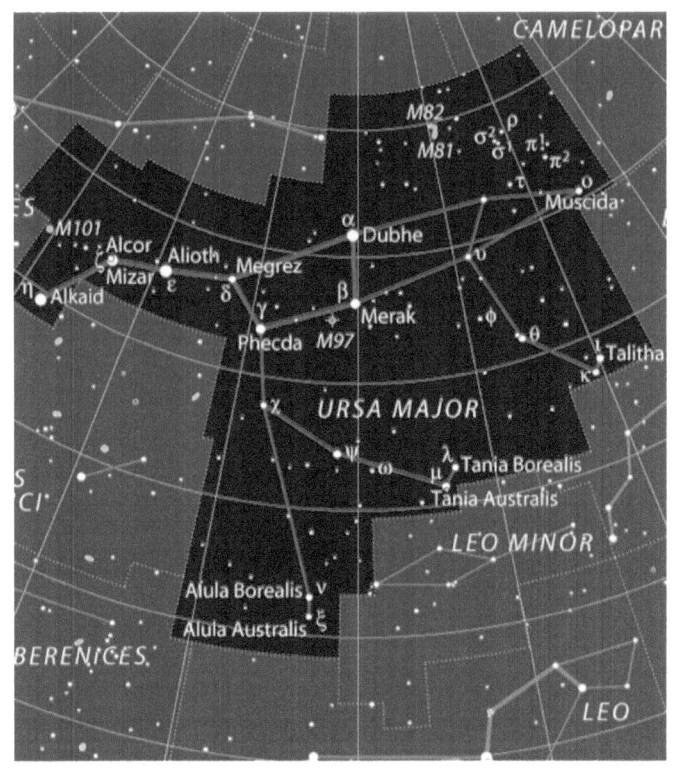

Sky chart for Ursa Major
And surround areas

Courtesy IAU and Sky & Telescope

Chapter 35

URSA MAJOR
Touring the "Greater Bear"

As a result of a hundreds of requests urging me to write a "guided tour" through various constellations via the GO TO function of our modern computerized telescopes this chapter/installment was the first in a series of OBSERVATION GUIDES which was to be entitled:

"GO TO [-*constellation name*-]"
Touring the Constellations With Your GO TO Telescope"

The series, all of which was from the popular Arkansas Sky Observatories' website: www.arksky.org , has become incredibly popular and is used by beginners, schools and educators throughout the world.

This first-written installment, "GO TO URSA MAJOR", features "*the Big Dipper*" since it so prominent throughout all seasons of each; it features a "start" with an easy GO TO to the bright star **ALIOTH**, and then get rights into the meat of it, with a slew to the beautifully-colored and complex multiple star system **ALCOR** and **MIZAR**, with discussions along the way and what to expect from each telescope size and type. We will view the multiple star systems, galaxies, and all deep sky phenomenon with exact descriptions of what the viewers with all telescopes should expect to see...and what to NOT expect to see!

The finest stellar and deep sky objects in the constellation will be featured....and - yes - there will be something for everyone and every telescope...Even naked eye and binoculars when appropriate!

Discussed are useful magnifications for EACH GO TO object, what type of night and conditions are needed to see certain details, double stars that can be resolved in each telescope model, and much, much more. It is your complete guide for your deep sky observing pleasure and a very handy tool for use at your next star party!

Needless to say, it WILL put your computerized telescope to work for you in a most efficient and enjoyable way!

I hope you will enjoy these comprehensive guides to "Touring the Constellations" which will feature all of the more prominent constellations, complete with diagrams, charts and illustrations. Please let us hear from you with summations of YOUR observations through these constellation tours!

Introduction

The wonderfully recognizable constellation **URSA MAJOR** is known by many names: throughout history the seven bright stars have been referred to as "*seven wise men*", *King Arthur's chariot*, the English *Plough of oxen*, and even the Egyptian *boat of the God Osiris*. In modern cultures these stars are commonly referred to as "*the Big Dipper*," appearing as a large ladle in the sky. It is one of the earliest recognizable star patterns by children throughout the ages.

Indeed, the seven bright stars and many others nearby are part of an actual star *ASSOCIATION*, stars held together and moving relatively together in space; the Ursa Major association is only *75 light years distant* from us here on Earth, and the group measures about 30 by 15 light years across and contains perhaps a total of 17 stars.

The images of the bright "dipper stars" of Ursa Major are a favorite with artists, among them **Van Gogh** who portrayed them often, and most notably on his work "*Starry Night on the Rhone.*"

The "*Great Bear*" designation for this constellation and star association is deeply rooted in both history and culture. Worldwide it has bee associated with the furry beast, from Aristotle, to the American Iroquois and Algonquin, while oddly the ancient Chinese saw virtually every imaginable connection to this star pattern - from scales, mountain peaks, to floating sky-lofted temples - *EXCEPT* a bear.

YOUR URSA MAJOR "GO TO" TOUR – THE CONCISE DIRECTORY

We will concentrate on 14 objects in Ursa Major for this quick tour; all are in reach of telescopes from 3-inches to 8-inches, yet each telescope will demonstrate uniquely different and challenging aspects of the objects. While we cover the 14 finest objects, there are MANY more exciting things to see in Ursa Major than those given here - hundreds of double and multiple stars, hundreds of galaxies and deep sky objects, stars of curious colors and motion - that can be found with your telescope.

Clay Sherrod's
CONSTELLATION GUIDES:
"Go To" Ursa Major - The "Great Bear"

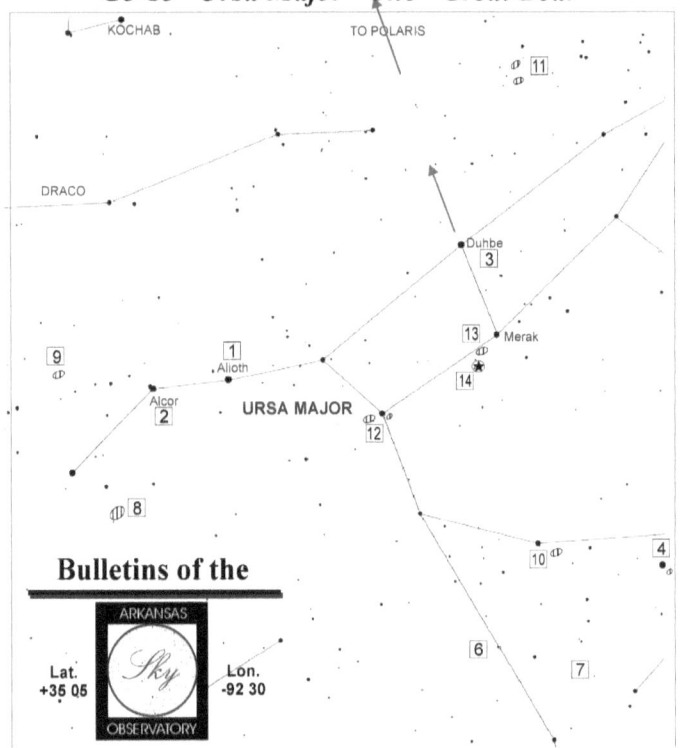

A finder chart for locating many of the GO TO objects in the constellation of Andromeda; if using a computer planetarium program, you are encouraged to plot the objects on your screen for higher resolution than this chart provides.

*   *   *

I recommend good a good star atlas and/or chart/and PC sky program which lists the finest objects constellation-by-constellation; you merely need to hold down your MODE key on the AutoStar or other telescope control for three (3) seconds and the RA and DEC coordinates appear for the telescope. Merely press "GO TO" and the cursor appears prompting you to enter the Right Ascension of the object if it is NOT listed among the objects in the AutoStar library; once the RA is entered, press "Enter" and the cursor once again prompts for the Declination coordinates (these coordinates for epoch 2000) are found in all good observing guides). Once those are entered, merely press "GO TO" once again and your telescope will slew to the position of the object!

Now, just use this constellation guide to point out the details of each object you wish to view.

The special star chart here will get you started, as it demonstrates the relative positions of all objects in this "tour" to the seven conspicuous stars of "the big dipper." NOTE that **Messier 51** (object #8 on this list) is actually in *Canes Venatici,* not Ursa Major, but is included here because it is a favorite for this portion of the sky (most people think it is in Ursa Major anyway....so that gives me the right to include it here....and there is NOT a constellation chapter for Canes Venatici anyway).

Following is the complete **14-object** list for your "GO TO TOUR" of Ursa Major; you may wish to find the majority of the objects from the sky program library (for example, for Autostar, you can merely pull up Messier 81 by going to "Object/Deep Sky/Messier Object/M-81....enter....GO TO" or...if you want to experiment and be a "better GO TO user", try entering the following coordinates as described for your particular PC program, telescope handbox or smart device App.

OBJECTS IN URSA MAJOR:

OBJECT 1: star - *ALIOTH* - R.A. 12h 52' / DEC + 56 14 - Magnitude: 1.8
OBJECT 2: multiple star - *ALCOR & MIZAR* - R.A. 13h 22' / DEC + 55 11 (Mizar) - Magnitudes: 4.0 & 2.4, respect.
OBJECT 3: multiple star - *DUHBE* - R.A. 11h 01' / DEC + 62 01 - Magnitude: 1.8
OBJECT 4: multiple star + galaxy - *TANIA AUSTRALIS* (MU UMa) + ngc3184 - R.A. 10h 19' / DEC + 41 45
Magnitude: 3.1 (galaxy is 10.3m)
OBJECT 5: multiple star - *ALULA AUSTRALIS* (XI UMa) - R.A. 11h 16' / + 31 49 - Magnitudes: 4.3 & 4.8
OBJECT 6: multiple star - 57 Ursa Majoris - R.A. 11h 26' / DEC + 39 37 - Magnitudes: 5.3 & 8.5
OBJECT 7: one of the closest stars - LALANDE 21185 - R.A. 11h 01' / DEC + 36 18 - Magnitude: 7.6
OBJECT 8: face-on galaxy – Messier 51 ("*Whirlpool Galaxy*" - ngc 5194) - R.A. 13h 28' / DEC + 47 27 - Magnitude: 8.1
OBJECT 9: difficult galaxy - Messier 101 (ngc 5457) - R.A. 14h 01' / DEC + 54 35 - Magnitude: 9.5
OBJECT 10: bright symmetrical galaxy - ngc 2841 - R.A. 09h 19' / DEC + 51 12 - Magnitude: 9.3
OBJECTS 11: beautiful galaxies: Messiers 81 & 82 (ngc3031 & 3034) R.A. 09h 52' / DEC + 69 18

Magnitudes: 7.9 & 8.8 (81 is spiral, 82 is irregular)
OBJECT 12: very difficult galaxy - Messier 109 (ngc 3992) - R.A. 11h 55' / DEC + 53 39 - Magnitude: 10.7
OBJECT 13: very difficult galaxy - Messier 108 (ngc 3556) - R.A. 11h 09' / DEC + 55 57 - Magnitude: 10.7
OBJECT 14: nice planetary nebula - Messier 97 (the "Owl nebula") - R.A. 11h 12' / DEC + 55 18 - Magnitude: 11.2

A VISUAL GUIDE TO THESE OBJECTS –

Object 1 - Bright Star *Alioth*
We begin our journey through Ursa Major with a slewed GO TO to the bright star ALIOTH. Obviously, "epsilon Ursa Majoris" as it is also known is bright enough to locate with the naked eye, marking the "inside-most bright star of the "dipper's handle." Actually, at magnitude 1.79, Alioth is the brightest star of Ursa Majoris.... NOT Duhbe, which has the distinction of its ALPHA designation. Alioth derives its name from the Arabic and can be translated curiously as either "*bright eye*," or strangely...."*fat tail*." It is an "A" spectral type star, very similar to our own sun although some 80 times more luminous.

Object 2 - Multiple Stars *Alcor* and *Mizar* (Zeta Ursa Majoris)
This has the distinction of being the **first double star** ever recorded.....in more ways than one. Mizar is a double in itself, which was discovered in 1650....prior to that, legend has it that the pair with Alcor had them as the "horse and rider" to native North American people. The star purportedly was

used as an *eyesight test* for young warriors...those who could make out the "*rider*" (Alcor) atop his "*horse*" (Mizar) was deemed suitable for the trials of manhood. Alcor and Mizar present my absolute FAVORITE color contrasts of the sky, surpassing even the famed "*Albireo*" in Cygnus. Mizar is a distinctive **yellow-white color**, contrasting beautifully with Alcor, which shows itself as a **green** or **blue** star, depending on who is looking. Alcor, at magnitude 4.0, is the fainter of the two and is immediately east of Mizar; the pair is easy at the lowest power in the smallest telescopes. Medium power in the 4-inch and up will reveal "MIZAR B", a bit fainter than Mizar and appearing much closer than Alcor to the primary star; all three can be seen in the same field of view with the 6- and 8-inch scopes, with clear separation of Mizar, but with still enough field to accommodate Alcor with a wide angle eyepiece. The faint galaxy Messier 101 (# 9 on our list) is immediately EAST of Mizar.

Object 3 - Bright Star & Multiple Star *Duhbe* (Alpha Ursa Majoris)
A relatively easy star to see, a companion star encircles the bright star Duhbe at about 6 minutes arc. This 7th magnitude star reveals itself in all the telescopes, but ONLY with fairly high magnification to overcome the glare of the 2nd magnitude Duhbe. The name "Duhbe" is also Arabic, signifying the "*back of the great bear.*" Although Duhbe is itself a double, only very large telescopes can reveal this difficult double star. Although Duhbe system is only 3 times more massive than our sun, the system puts out an incredible luminosity almost 150 times brighter than the sun.

Object 4 - Double Star Tania Australis (Mu Ursa Majoris) and Nearby Galaxy ngc 3184
This is an incredible object for all telescopes and a good target for small apertures. Always observe this nice double with low magnification for an outstanding and curious difference in color! The main component of Mu is a very, very **red** *RED GIANT* star of the M-class, contrasting nicely to the more mainstream and **yellow** component. Mu itself is magnitude 3.1. Due WEST of this interesting star, don't miss the opportunity (in the same field of view of your low power eyepiece!) to see the small spiral galaxy, **ngc3184**. In very large scopes and photographically, this is one of the most beautiful face-on spirals there is, with delicate rotating spiral arms from a highly concentrated central nucleus. In the 6- and 8-inch it appears as a uniform round (very small) glow with a concentrated brightness in the center; the smaller scopes will show this galaxy a medium-high power, but typically will NOT reveal the nucleus.

Object 5 - Nice, but difficult Double Star Alula Australis (XI Ursa Majoris)
The beautiful name denoting "*First Spring*" is rivaled only by the beauty of this very close and equal magnitude double star. Both stars are about 4.5 magnitude and extremely difficult, except in an 8-inch and larger telescope. Under steady skies and when overhead (high power) this will show some elongation to the pair, and a good 6-inch might see a bit of a thread of darkness between these two stars. Both stars are very similar to our own sun. In about 15 years, these stars will appear farther apart from Earth (over 3" arc) and thus be easily resolved in all telescopes under high power.

Object 6 - Double Star 57 Ursa Majoris
This is an outstanding double star, and typically resolvable in even the smaller telescopes under very high power on a good night. They are solar-type stars at magnitude 5 and magnitude 8 and are separated by about 5.5 seconds arc. Because of the relative faintness of the companion star, this is truly a TEST for the smaller telescope, but it can be seen on very good nights.

Object 7 - Lalande 21185 - One of the Closest Stars Outside Our Solar System
This is a star that - in your lifetime - you can actually chart as it moves! It is so close to Earth, at only 8.3 light years (!) that it shows great "*proper motion*" as it moves against more distant stars behind it. It moves almost 5" arc annually due southward (due north of this interesting object is the nice double star "51 Ursa Majoris, so be sure and check that out while in the neighborhood). **Lalande 21185** is the fourth-closest of all stars outside of our sun. At magnitude 7.6, this might be a hard star to spot....but use the following chart to help you (magnitudes of some nearby stars are given). ALSO, you can find this star easily by remembering that it is **CRIMSON RED**, being a nice example of a true *RED DWARF* star. The grid marks on the following chart are 1 degree; the 6- and 8-inch scopes provide just under and over (respectively) a one-degree field with a wide field 32mm Plossl for reference; whereas the small APO refractors give so much field that you might be confused by all you see!

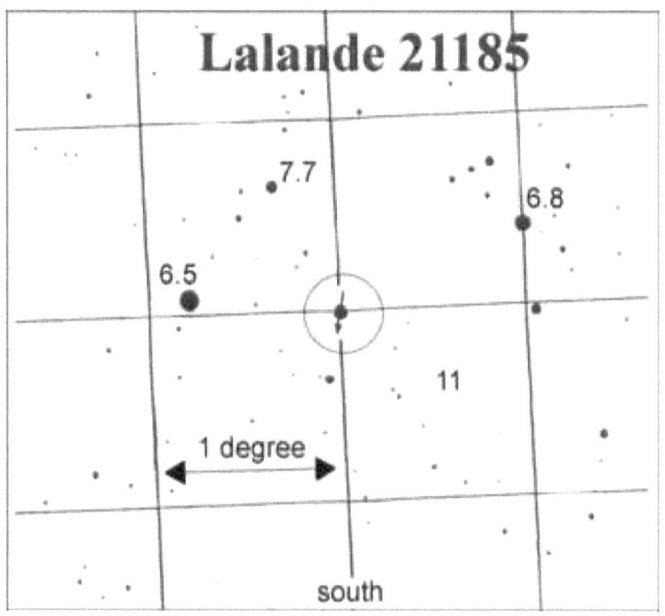

Object 8 - *The Whirlpool Galaxy* (actually in Canes Venatici)

Although this is a favorite, both visually and photographically, this is a pretty tough object UNLESS you are observing on a very dark night away totally from city lights. In suburban conditions, even the 6- or 8-inch will show little detail, other than the bright central core of this face-on galaxy and the other smaller "companion" galaxy that appears to be attached to one arm of this famous sight. In very dark conditions, even the smaller telescopes will clearly show both of these components at medium power and as well will show the one large spiral arm leading down to the "companion." Some detail other that this one arm and a clear dark "lane" between it and the galaxy center will be revealed to the 8-inch at about 120x under very, very dark conditions. HOWEVER...do not expect to see what **Lord Rosse** "saw" (uh-huh.....) in his massive 72 inch reflecting telescope over 130 years ago, as shown in the rare drawing shown below.

Messier 41
as drawn by Lord Rosse
with his 72-inch telescope
1882
from the Arkansas Sky Observatory Library

Clearly Mr. Rosse had much greater "mental resolving power" than perhaps he did "visual acuity" resolving power. Nonetheless, it is interesting to look at his clearly-recorded spiral nature of this object decades before the true nature of spiral galaxies and their structures were even eluded to!

The Whirlpool Galaxy – Messier 51 – from ASO 1-inch Celestron Schmidt-Cassegrain, 60 seconds

* * *

Object 9 - Galaxy: Messier 101
Due east of Mizar, this is a very difficult galaxy, even for an 8-inch under normal suburban conditions; it is shown clearly under dark sky conditions in even a 6-inch, however. It is not an object that should worry users of the smaller telescopes however, although it can be glimpsed as a "smudge" in some on very dark nights. It is a relatively large (7' arc) face-on galaxy which exhibits no features at all in any of these telescopes; however, large scopes with long-exposure photographs reveal a beautiful spiral structure exhibiting many huge star clouds within this

galaxies spindle-like arms. On the very DARKEST nights, 8-inch and larger may reveal globules of stars that are, in fact, images of these huge star swarms. Lord Rosse (see M-51 above) was the first to note t he spiral structure of this object in his giant 72-inch reflector.

Object 10 - NGC 2841, a Nice Elongated Galaxy
This may be the finest example of a tightly-wound spiral type Sb galaxy; in the 4-inch and larger scopes you can clearly see its elongated elliptical shape, like a stretched tiny star. This spiral has many, many distinct arms that can only be made out with the largest telescopes. The "whirls" can be glimpsed on very steady and dark nights with an 8-inch, but not easily. It is a large (6' x 2' arc) elliptical shape, very bright across its entire length. Even at magnitude 10.2, this galaxy CAN be glimpsed as a tiny smudge at medium power with the smaller telescopes.

Objects 11 - Famous Messiers 81 and 82
Messier 81 is a bright, compact spiral galaxy; its two outer, very delicate, arms may be glimpsed in fleeting moments with the 8" scope under extremely dark skies; the 6-inch fails to show these. Medium power (about 20x per inch) shows the best views in all scopes. This is the larger and brighter of the M81-M82 pair which can be framed nicely in the SAME FIELD of the smaller telescopes even under medium power. On a dark night, this provides a spectacular sight. Look for a white, very uniform brightness for M-81 and elongated in a NE-SE direction only very slightly. A brighter center 1/2 of this object can be seen clearly in the 3-inch in dark skies. Immediately about 1/2 degree NORTH

Messier 81

of M-81 is the slightly less bright Messier 82, a very **chaotic**, eruptive and much more interesting sight than M-81.

Messier 82

Even with the 4-inch and a power of about 160x, much "mottling" of this cigar-shaped patch of light can be witnessed; in the 6-inch and larger telescopes on a very dark night, the sight of this cataclysmic system is sometimes overwhelming. I recently had my best view of this from the Mountain with a 12-inch and a magnification of about 230x. Seven distinct bright "globular" masses could be discerned and much dark matter in between; at that magnification, M-82 almost filled 1/2 of the field of view. Although dimmer and a bit smaller than its M-81 companion, I have always found M-82 to be a much more rewarding object telescopically. The smaller telescopes will clearly show the cigar-shape of this galaxy under high power, although no discernable detail should be expected.

Object 12 - Messier 109 and Galaxy ngc3953
Messier 109, just east off the bright star *Phecda* in the "bowl" of the dipper, is the toughest of all objects on this list....forget sighting it with small telescopes and expect much difficulty and a disappointing view (if you DO see it) in the 4-6 inch scopes; even an 8-inch really does not do much for it either. I have put it here only because it is one of the many Messier objects, although this is controversial to Messier's original list of 100. This, curiously, is one of the largest galaxies on this list at a size of 6.5' x 3.5' arc, but it is very faint at nearly magnitude 11. It is a tightly-wound type Sb spiral, but only reveals its nature in long-exposure photographs through very large telescopes.

Object 13 - Galaxy Messier 108
Not a cake-walk either, but also very large (over 7.5' across!) this galaxy's brightness is spread over so large an area to render it only about magnitude 10.7. In many ways it is more difficult than M-109,

above. This galaxy in an 8-inch or larger telescope under very, very dark skies can be clearly made out to be pencil-shaped, and edge-on galaxy with a clearly-defined character in medium-high power; the 6-inch shows it well, but lacks the aperture to clearly show the 7' arc length andn detail of this curious object.

Object 14 - The Famous "**Owl Nebula**", Messier 97 Concluding our GO TO TOUR or Ursa Major is the famous planetary nebula (gas cloud shell remnant from a supernova explosion) Messier 97, also known as the "*Owl Nebula*." It's name comes from two dark ovals equally spaced off-center that resemble the two "black eyes" of an old hoot owl. This is a VERY large object, measuring 4' x 3' arc and thus its magnitude of only 11.8 (I think that official magnitude is a bit too low) is spread out to make this a most difficult object. I have seen the "black eyes" only a few times with the 6-inch....I have never been able to discern any detail with the smaller scopes, although the nebula is clearly seen.

The drawing on the following page shows what can be expected in all scopes at low power; magnifications of about 15x per inch give the best view, and only then are the "eyes" reveal in the two larger telescopes. The central star (the one which exploded, leaving the gaseous shell) is still there, at 14th magnitude, just beyond the reach of even the 8" scope. However, its presence allows contrast for viewing of the two darker "eyes" offset to each side of this central star.

THE OWL NEBULA / M-97
(shown with ETX / LX orientation)

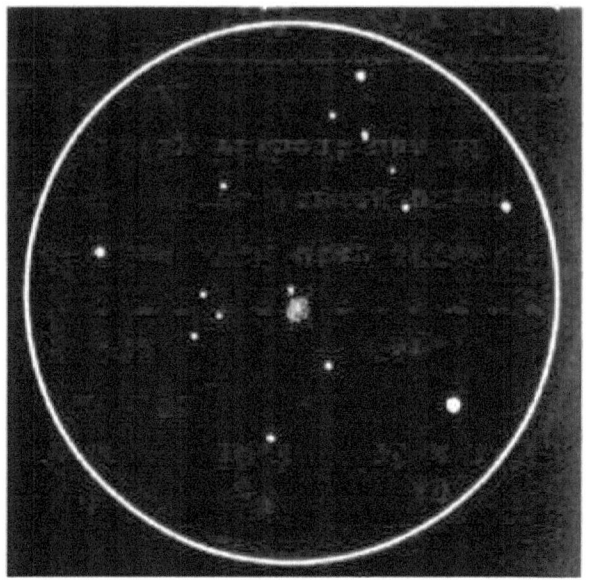

WANDERING ABOUT....Your Ursa Major User Guide

This brief GO TO tour of Ursa Major has revealed its most interesting secrets. But please do not stop here. Go ahead and locate many of the hundreds of other ngc galaxies that are present in this constellation.....there are several more interesting planetary nebulae that are within the reach of our telescopes. In addition, use this opportunity to actually LOAD A USER OBJECT onto your AutoStar or other telescope computer!

Your friends, family and fellow stargazers would be delighted if you could show them the "4th closest star" to our solar neighborhood, a fiery red one at that; merely go into "Select/Object/User Object....[enter]" an then add in the coordinates (R.A. and DEC given above) and a name/description of **Lalande 21185** and press "enter" to store your information into the AutoStar library!

The exact process for bringing up a celestial object for your telescope to GO TO will vary with the sky program, telescope keypad, or App that you are using, but all fundamentally work the same logic to acquire your object as well as build your User Object Library. Always consult your HELP or User Guides of your specific program for the exact steps in using tabs and keystrokes to pull up and GO TO your objects!

* * *

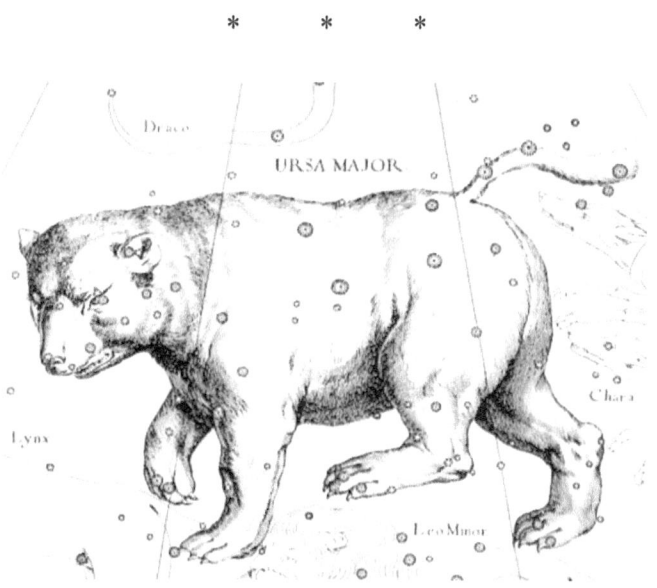

Ursa Major – The Large Bear – drawn by Hevelius

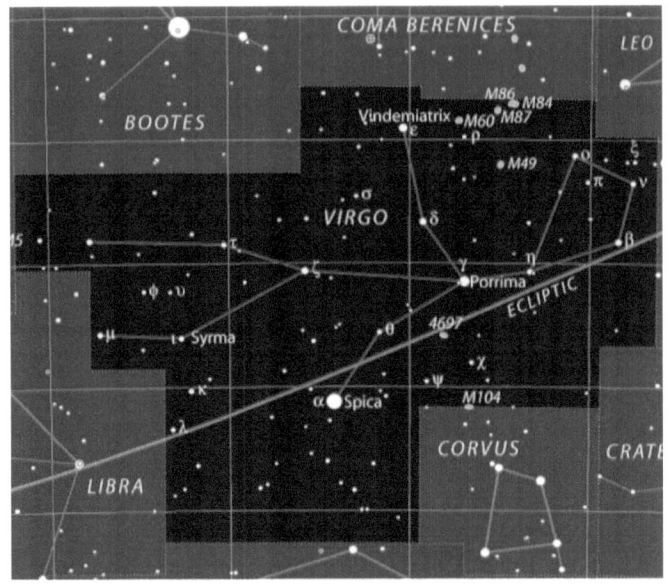

Virgo and surrounding constellations
For your final chapter of
The Constellations Tours

Chart adapted courtesy from the original
IAU / Sky & Telescope

Chapter 36

VIRGO

And the Maiden's Hope Chest Filled With Galaxies Galore....!

For our last alphabetical *Constellations* guide, we will continue into the far depths of the known Universe to examine the vast wealth of extraordinary galaxies and galaxy clusters that present themselves in this large region of sky.

In our last chapter on Coma Berenices, just to the north of Virgo, we examined scores of galaxies, both elliptical and spiral that grace the skies in that tiny constellations and star cluster known also as "Mel 111." Here, we extend that galactic search into our southern skies, low areas for most northern hemisphere observers, and unfortunately frequently ignore because of this low altitude!

But for deep sky enthusiasts, this is one area that should NOT be turned away from! Even in this brief tour, we will look at over *35 galaxies*! That is in addition to the nearly 30 galaxies from the Coma Berenices tour in Volume One. As with that previous article, I will be listing about **25 ngc** galaxies in "abbreviated form," that are in ascending Right Ascension order and by ngc listing. In addition, the brightest and most spectacular and interesting objects will be discussed in detail as usual.

Note from the sky chart included here that the CELESTIAL EQUATOR passes directly through the middle sections of Virgo. This is the reading "0" degrees on your properly adjusted declination setting circle. All angles NORTH of this equatorial line are positive ("+") and all angular measures (declinations) south of the celestial equator are negative ("-"); hence you will see references in this "GO TO" GUIDE to both "+" and "-" declinations for celestial objects.

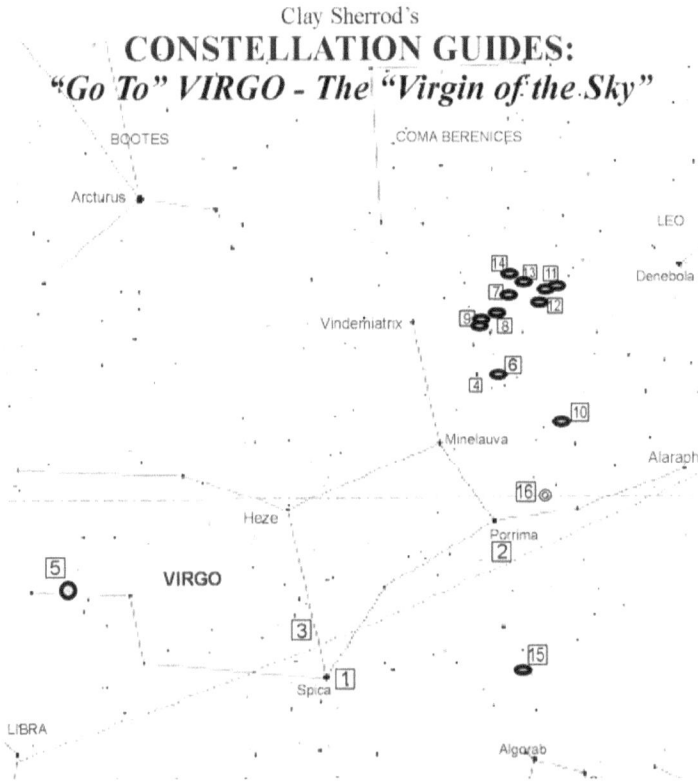

Clay Sherrod's
CONSTELLATION GUIDES:
"Go To" VIRGO - The "Virgin of the Sky"

Also note that the *ECLIPTIC* (the band in which all the planets, sun and moon appear to move relative to Earth) passes from southeast to northwest through this large constellation, passing only about 2 degrees north of bright SPICA.

The ancient Greek legends have given rise to our association of this star pattern (or perhaps originally with the purity of the WHITE star, Spica in all its brilliance in a star-void field. To them, and the Romans, the Greek goddess Astraea was immortalized among the stars in this constellation. The daughter of *Zeus*, *Astraea* was the last of the immortal Greek gods to leave the mortal Earth, after which no further direct contact between mankind and the gods would ever take place again.

Even in ancient Egypt, this star - and hence the entire constellation - was associated with a maiden and her purity and innocence, as it honored the Nile goddess Isis; indeed, the ancient ruined Egyptian city of Akhenaton and its temples appear to have been aligned and built in relation to the rising and setting of the bright star Spica in her honor. The Pharaohs knew the constellation as a "*devoted wife*," or the "*most dedicated wife*," while the Babylonian stargazers gave **Spica** and her starry regime the distinction of being the "*Queen of the Stars*." Even the Christian religion has honored the bright star as the Madonna.

It has always been my opinion that - taking nothing from the beauty of this star - that the isolation of Spica among the springtime stars and preceding the bright and rich Milky Way is the primary factor resulting in the undying attention and legend throughout history that is associated with Spica...... for the constellation itself is a very vague and uninteresting mildly-visible group of stars!

It is what you DON'T see in this area that is the most interesting. The "*COMA-VIRGO CLOUD OF*

GALAXIES" dominates the north-western realm of Virgo and the southern portions of Coma Berenices. Literally millions of galaxies seem to occupy this region, most at a distance of some 42 MILLION light years away. On a very dark, moonless night, merely scan slowly with any of our telescopes at very low magnification and wide field eyepieces to see hundreds of "faint fuzzies" that are islands in the sky harboring hundreds of millions of stars each. In Virgo alone there are at LEAST 50 galaxies visible in the smaller telescopes.....over 100 in the 6-inch telescope and above. The most "crowded area" of this huge "mega-galaxy of galaxies" is located at about 12h 24m in right ascension, just at the Coma-Virgo border.

Each GO TO object in Virgo is discussed for your telescope regarding the type of conditions necessary for you to view it optimally for discern the very faintest details.........magnifications and aperture necessary for most objects, and much, much more. This is YOUR complete guide to get you on your way to exploring the many hundreds of galaxies that are clustered in Virgo. The following listing of "BEST" objects contains the finest or most interesting.

As with our "GO TO" tour of Coma Berenices, there is also a bonus comprehensive listing of <u>ALL NGC GALAXIES</u> (they are on your AutoStar and most good PC sky programs) that can be seen in at least one or some of our telescope size ranges *(faint ngc objects that are NOT visible in at least an 8-inch telescope are NOT included in these "GO TO" tours.

Use this attached star chart and the following Guide as an excellent reference for your next star party itinerary, or a beginning for further study into the thousands of objects visible in this part of the sky. Truly these extensive Constellation Study Guides will most definitely put your handbox, sky program or smart device App to work for you in the most efficient and enjoyable way possible!

As a matter of fact, MANY computerized telescope users are now programming their own "Tours" based on these guides, using each constellation as a separate GO TO Tour for the library that can be added in or deleted through the main edit screen on your PC or MAC computer. In fact Smart Phones and iPads are now being used to run such on many telescopes via wireless such as Bluetooth and SkySafari.

We hope you enjoy these comprehensive GUIDES to touring the constellations via your GO TO program and its computer-driven telescope. Each new installment is complete with diagrams, charts and illustrations that you will find nowhere else. Please let us hear YOUR feedback and your observations of each and every constellation after YOU have toured its vast reaches of our skies!

YOUR VIRGO CONCISE DIRECTORY OF INTERESTING OBJECTS –

There are literally thousands of fantastic galaxies in Virgo and surrounding constellations, from wide field to telescopic. Because of the great number of galaxies, it was very difficult to select (and limit) the number of objects for our "GO TO" TOUR. Only the best and in some cases, most challenging, objects are chosen for this brief tour. The large number of galaxies made it prohibitive for me to include the myriad of wonderful double and multiple stars that I would normally have considered for this Guide. As can be seen with the photograph below (Hubble Space Telescope), there are galaxies virtually "crammed" into every field of view....the darker the skies, the better the view! The larger the telescope....the more the galaxies...!

I have chosen the finest (or most interesting) 17 objects in this VIRGO "GO TO" TOUR; as with all GUIDES, all objects listed below will be visible in all telescopes (some naked eye) from the 2.5 inch through an 8-inch; of course larger apertures may "show" an object a bit closer and "better," but frequently a wide field and low power view is more desirable than aperture for FINDING THESE GALAXIES!. Indeed, I strongly encourage you first FIND the galaxy, or its approximate location through your GO TO function with your lowest power and then - once IDENTIFIED positively - move up slowly in steps with magnification.

You will find that there is an "optimum magnification" with which to observe each and every galaxy. Too much "power" and you will darken the object as well as the field of view; not enough "power" and there might not be enough contrast between the dark background sky and the faint image of the "faint fuzzy" galaxy! Once your eyes are fully dark adapted, you will even be able to see fainter and fainter galaxies that at first might not even be glimpsed! I have found through practice with galaxies of this brightness and size that about 15x per inch aperture is PERFECT for good direct viewing once found.

With all deep sky objects, and particularly these small and faint galaxies, avoid attempting to observe when the moon is in the sky, even a very thin crescent, as its brightness in the sky will overshadow the very dim contrast afforded by even the brightest deep sky object; if you see the object at all against moonlight, you will NOT see the subtle outlying areas or the full detail of what is presented.

The convenient sky placement of Virgo lends itself well to very good and long-period observing for all telescope users both north and south of the equator. When rising about dark in the east (mid-spring) it will remain in the sky throughout the night, transiting the meridian at about midnight during that season. All deep sky objects and difficult double stars are ALWAYS best observed when they are located nearly overhead (or as high in the sky as possible), thus requiring the observer to look through the thinnest portion of the Earth's "lens" of atmosphere and haze. Hence, northern hemisphere observers will want to wait until the constellation transits the meridian to provide as high a view as possible; fortunately for those north of the equator, most of Virgo's faint galaxies are located "high" in the constellation, above the "0" degree celestial equator.

As with all of the "GO TO" tour constellation lists, I recommend a good star atlas and/or chart which will list all the finest objects, constellation-by-constellation. One very handy reference guide is the PETERSON FIELD GUIDE TO THE STARS AND PLANETS, which features complete lists with declinations, right ascensions, magnitudes, and all pertinent information for you to expand your observing horizons beyond this brief guide.

FOR AUTOSTAR type operated telescopes: Note that your AutoStar will NOT have every object listed on every constellation GO TO tour....this is intentional. You can access some of the most interesting objects of the sky directly from their coordinates.

The exact process for bringing up a celestial object for your telescope to GO TO will vary with the sky program, telescope keypad, or App that you are using, but all fundamentally work the same logic to acquire your object as well as build your User Object Library. Always consult your HELP or User Guides of your specific program for the exact steps in using tabs and keystrokes to pull up and GO TO your objects!

The constellation tour Star Chartwill get you started on your journey for this constellation.

Following is the concise object list for your "GO TO" tour of VIRGO; you may wish to find the majority of the objects from the sky Library (for example, you can easily go to the "Sombrero Galaxy" if you pull up "Object/Deep Sky/Messier Object/..type in '104'...." and then press "Enter", followed by "GO TO" to access this very nice and bright edge-on galaxy. On the other hand, if you want to experiment and become a "better computer user" try entering the exact R.A. and DEC coordinates of that object as described above after holding down the MODE key (or recommended key strokes for your pc program)..

Of course, for **named** objects such as the "Sombrero", you can also merely go to "Objects / Deep Sky / Named...." and then SCROLL downward until reaching the name "Sombrero Galaxy" and then press ENTER to GO TO that object!

OBJECT 1: very bright star - *SPICA* (alpha Virginis) - R.A. 13h 23' / DEC (-)10 54 - Magnitude: 1.0, beautiful white
OBJECT 2: double star - *PORRIMA* (gamma Virginis) - R.A. 12h 39' / DEC (-)01 11 - Mags: 3.6 **& 3.6, getting close!**
OBJECT 3: variable star - S Virginis - R.A. 13h 30' / DEC (-)06 56 - Mag: 6.3 to 13.2, 378 days
OBJECT 4: variable star - R Virginis - R.A. 12h 36' / DEC +07 16 - Mag: 6.2 to 12.1, 146 days - good star!

OBJECT 5: globular cluster - ngc5634 - R.A. 14h 27' / DEC (-)05 45 - Magnitude: 10.4 - tough, but visible in 3-inch!

OBJECT 6: elliptical galaxy - Messier 49 (ngc4472) - R.A. 12h 27' / DEC + 08 16 - Magnitude: 8.6

OBJECT 7: spiral galaxy - Messier 58 (ngc4579) - R.A. 12h 35' / DEC + 12 05 - Magnitude: 9.2, good object for all

OBJECT 8: elliptical galaxy Messier 59 (ngc4621) - R.A. 12h 40' / DEC + 11 55 - Magnitude: 9.6, round, starlike

OBJECT 9: elliptical galaxy - Messier 60 (ngc4649) - R.A. 12h 41' / DEC + 11 49 - Magnitude: 8.9, bright, easy

OBJECT 10: spiral galaxy - Messier 61 (ngc4303) - R.A. 12h 19' / DEC + 04 45 - Magnitude: 10.1

OBJECTS 11: two elliptical galaxies - Messiers 84 & 86 - R. - very nice pair!

OBJECT 12A. 12h 23' / DEC + 13 10 - Mags. 9.3 & 9.6: huge elliptical galaxy - Messier 87 (ngc4486) - R.A. 12h 28' / DEC + 12 40 - Mag. 9.2, very large oval!

OBJECT 13: bright elliptical galaxy - Messier 89 (ngc4552) - R.A. 12h 33' / DEC + 12 50 - Mag. 9.5 - round & bright

OBJECT 14: spiral galaxy - Messier 90 (ngc4569) - R.A. 12h 34' / DEC + 13 26 - Mag. 10.1 very elongated oval

OBJECT 15: "*SOMBRERO GALAXY*" - Messier 104 (ngc4594) - R.A. 12h 37' / DEC (-)11 21 - Mag. 8.7 - Interesting!

OBJECT 16: 3C 273 Quasar! - R.A. 12h 27' / DEC + 02 19 - Magnitude 12.8 - Tough, but WORTH the try!!

OBJECTS 17 THROUGH 41 –
NGC GALAXIES IN VIRGO (other than those listed above) listing of "ngc" galaxies in VIRGO in order of RIGHT ASCENSION (and NGC # order) visible in our scopes:
NOTE: These NGC galaxies are NOT detailed in the following "Visual Guide" as are those Objects 1-11 listed above...use the abbreviated descriptions as they follow the order:

NGC# / R.A. / DEC / MAGNITUDE / SIZE (in minutes arc -'-) / GALAXY TYPE (S=spiral, E=ellipt;, P=Irr.)

ngc4030 / 11 57 / -00 49 / 11.0 / 3.1 X 2.2 - spiral, face-on, 3-inch (barely) and above
ngc4216 / 12 13 / +13 25 / 10.4 / 7.4 x 0.9 - edge-on spiral, good for 4-inch and above, cigar-shaped!
ngc4365 / 12 22 / +07 36 / 11.1 / 1.3 x 1.0 - very tiny and dim elliptical, starlike. 6-8-inch+ only
ngc4429 / 12 25 / +11 23 / 11.2 / 3.3 x 1.0 - spiral, very faint, 8"+ only
ngc4438 / 12 25 / +13 17 / 10.8 / 8.0 x 6.3 - spiral, VERY large and face-on; visible in 3-inch, but tough
ngc4473 / 12 27 / +13 42 / 10.1 / 1.6 x 0.9 - tiny elliptical, but bright. Starlike in all scopes, fuzzy
ngc4517 / 12 29 / +00 21 / 12.0 / 8.9 x 0.9 - edge-on spiral, nice but too faint, tough even in 8 inch+
ngc4526 / 12 32 / +07 58 / 10.9 / 3.3 x 1.0 - small elliptical, fuzzy blob in larger scopes, starlike in small scope
ngc4527 / 12 32 / +02 56 / 11.3 / 5.3 x 1.0 - small spiral, only seen in 8" scope
ngc4535 / 12 32 / +08 28 / 10.7 / 6.0 x 4.0 - large face-on spiral, can be seen in ETX 90; nice in 6 inch

ngc4536 / 12 32 / +02 28 / 11.9 / 6.9 x 2.6 - large but VERY hard, even in 8-10 inch
ngc4546 / 12 33 / -03 31 / 10.0/ 1.8 x 0.8 - very tiny, but fairly bright elliptical, visible in all scopes
ngc4596 / 12 37 / +10 27 / 11.4 / 2.8 x 2.2 - barred spiral, very faint but barely seen in the 6 inch
ngc4636 / 12 40 / +02 57 / 10.4 / 1.4 x 1.3 - small starlike fuzzball, visible in 4-inch and larger scopes
ngc4654 / 12 41 / +13 23 / 11.2 / 4.2 x 2.2 - elongated spiral, very faint but can be seen in 5-inch
ngc4666 / 12 43 / -00 12 / 11.4 / 3.8 x 0.8 - near edge-on spiral, only visible in 8-inch and larger scopes
ngc4697 / 12 46 / -05 32 / 9.6 / 2.2 x 1.4 - small oval elliptical, but bright enough for small refractors and above!
ngc4699 / 12 47 / -08 24 / 9.3 / 3.0 x 2.0 - nice spiral, small and visible in all scopes, oval glow
ngc4753 / 12 50 / -00 55 / 10.8 / 3.3 x 1.1 - Spiral, tilted; visible as a smudge in 4-inch; not bad in 8-inch
ngc4762 / 12 50 / +11 31 / 11.0 / 3.7 x 0.4 - very thin edge-in spiral, very faint and tough in 8-10 inch
ngc4856 / 12 57 / -14 46 / 11.8 / 2.0 x 0.7 - unusual "squashed" oval elliptical; very hard even for 8-inch+
ngc4958 / 13 03 / -07 45 / 10.9 / 1.7 x 0.7 - elliptical, can be seen as small smudge in 4" scope and above
ngc5363 / 13 54 / +05 29 / 10.7 / 2.0 x 1.4 - tiny elliptical, but bright enough for 3-inch
ngc5566 / 14 18 / +04 11 / 10.4 / 5.6 x 1.1 - needle-like spiral edge on, bright enough for 3-inch, high power

ngc5846 / 15 04 / +01 48 / 10.5 / 0.9 x 0.9 - smallest elliptical you will see! Starlike....PERIOD.

....AND NOW ON WITH THE SHOW!! (refer to the Virgo Star Chart for all the objects described in detail on the "Guide")

A VISUAL GUIDE TO OUR DEEP SKY OBJECTS IN VIRGO

Object 1 - Very Bright Star - *"SPICA"* (alpha Virginis)
Our starting point for every "GO TO" TOUR is always (or usually!) the brightest star of the constellation or region and Virgo holds no exception. However, Spica holds special honors in that much of the ancient lore of the Virgo constellation is likely attributable to the STAR ITSELF, rather than perhaps the pattern of stars making up the asterism we know as "Virgo" (see above). It is the bright white-blue of this diamond-like star that is captivating. Add to that the fact that Spica appears to "stand alone" amidst an otherwise star-poor and uninteresting area of the sky to the naked eye.

Spica is the only *"true" 1st magnitude* star of the skies....it has the honor of having an exact visual magnitude of 1.00. Because of its proximity to the celestial equator, Spica is one of my favorite stars for testing the tracking accuracy of telescope drive systems during spring and early summer months. Center you polar-mounted telescope on Spica when it is on the meridian (the imaginary line stretching across the sky from DUE SOUTH to DUE NORTH) with a very high power eyepiece and let your telescope track it for about 10 minutes. Ignoring all drift in the east-west direction (right ascension), if Spica moves NORTH in your eyepiece, your polar mount is aimed too far to the WEST....if it moves SOUTH, then your polar axis is aimed too far EAST!

At a distance of about 240 light years, Spica has depleted a great amount of its primordial hydrogen, now depending on Helium as its primary food source. Although the seemingly unfailing associations of this star to an unblemished maiden or goddess has prevailed, the actual name "SPICA" has derived from the Latin of the same name: "...spicum" or "spiculus" meaning "the ear of grain." This association refers to the traditional Greek art in which the Virgin is holding a large handful of wheat in one hand, of which Spica is the principal "grain of wheat!"

Object 2 - "Porrima" - (gamma Virginis) - A Very Tough Double that is closing fast!
You "may" or may not be able to resolve this star. Look for TWO equal 3.6 magnitude stars closing in on one-another for gamma Virginis. *Porrima* is a double star with a period of 171 years. In 1910, the first pass of Halley's comet of the twentieth century saw with it Porrima's two stars at their widest separation possible, over 6" arc. It likely was a beautiful sight...but not for long as the secondary star continued its long loop around star "A" and is now closing fast. Look for the stars oriented in an almost E-W line to one-another and of equal magnitude. In the 6-8 inch scopes I can clearly see some separation of these two stars, but the 4-inch can clearly see an elongation of the two stars in that

orientation. But you must lwait....in 2007, the stars will appear so close together from Earth that they will appear only as ONE star! In reality, the stars on that year will still be 270 million miles apart from one-another in space. The star Porrima honors the goddess of Sage or Prophecy from the Arabian stargazers. However, the name is derived from the Arabic "Zawiat al Awwa", or the "angle" denoting the stars that comprise the heart of Virgo: Epsilon, Delta, Gamma, Eta and Beta. The stars will gradually increase their separation in every successive year for a while.

Object 3 - A Classic Variable Star - S Virginis
For those of you who have NOT tried their hands at observing and monitoring variable stars.....you really don't know what you are missing. Not only is it exciting compiling your own data and light curves from the star's magnitude changes, but YOU can contribute much to the science of stellar astronomy and astrophysics through your regular observations of stars like S Virginis. I have chosen this star (as well as the next one following) for this "GO TO" TOUR because of three factors: 1) the star's brightness changes go from 6.3 to 13.2 magnitude, easily observable in the 3-inch and larger telescopes all the way through its cycle; 2) its cycle requires 377.9 days to complete, thereby making it "easy" to observe, requiring an estimate only every TWO weeks!; and, 3) the star is easy to locate. To first locate the star, and to make estimates when the star is at its brightest, we use the "a" or finder chart. Click on the link to the American Association of Variable Star Observers (AAVSO) web site for this "a" chart and copy to file the chart found at: https://www.aavso.org/apps/vsp/ . Note for these

charts, simply type in the NAME of the variable at top to generate your choice of chart - merely open that file and resize the chart to print on a standard piece of paper. Now....once you have located this star using the "a" chart, you will want a MORE DETAILED, or "closer" chart providing fainter stars and more of them for estimating when the star is at dimmest, or midway between maximum and minimum. For that, the AAVSO "g" chart found at the same link will provide a more narrow field of view concentrating on a field of fainter stars for that purpose. For a complete discussion on the importance of variable star observing, the techniques for telescope users and a beginner's guide to select stars, you may find a thorough discussion from my "Observing Variable Stars" on the Arkansas Sky Observatories website: www.arksky.org .

Object 4 - Another GREAT Variable Star – Only requiring HALF the time! R Virginis

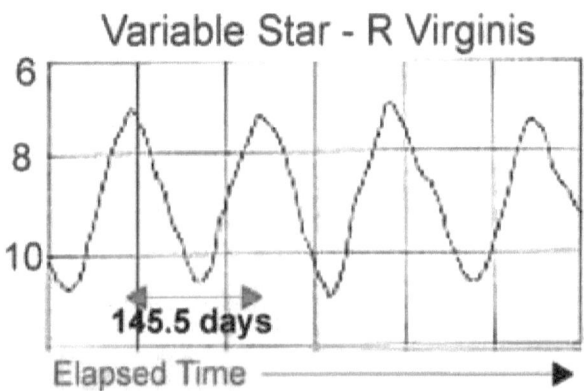

Also a long period variable like S Virginis above, R Virginis is also a classic variable star....mostly predictable, with impressive variations.....but this

one only requires HALF the time as S Viriginis! With a period of only 145.5 days, and a wide magnitude range from 6.2 to 12.1, this star can be monitored easily in the small refractors scopes as well!

The light curve shown above for R Virginis demonstrates its very rapid increase to brightness and equally sudden diminish from near-naked-eye visibility to a faint 12th magnitude! At its lowest, the star only "rests" at minimum for less than a couple of days! It is a "Mira-type" star, being a pulsating star that grows from pressure and subsequently diminishes in size from its own weight and gravity. Its short period is only HALF of a typical "Mira-type" star. The "a" finder chart from the AAVSO for this nice variable star is found at: https://www.aavso.org/apps/vsp/ . Note for these charts, simply type in the NAME of the variable at top to generate your choice of chart – while the more detailed "g" chart showing fainter stars to follow the light changes into minimum is located at the same link.

Object 5 - NGC 5634 - Our "Token" Globular Cluster in Virgo!
It may not be the brightest....or the biggest...and it certainly is NOT a show-stopper. But NGC 5634 is the ONLY globular star cluster in Virgo, and in this region of sky! With the overwhelming concentration of bright and large globular clusters in Ophiuchus not far away, this lone straggler has this lonely distinction. NGC 5634 (found on your AutoStar: OBJECT / DEEP SKY / NGC OBJECT....[scroll to] listings) is actually bright enough at magnitude 10.4 to be seen in even a small

refractor, but with some difficulty. Its entire size is ONLY 1.3' arc! Thus, it will appear like a very dim fuzzy star in that telescope; the 6-8 inch show this a bit better (and larger), but with absolutely NO star resolution. At 23,000 parsecs away, this is one of the most remote of our Milky Way's globular clusters.

Object 6 - Let the Galaxy Show Begin! Messier 49 - Elliptical Galaxy
At magnitude 8.6 this is one of the brightest, largest (about 5' arc) and easiest of all elliptical galaxies for amateur telescopes. I appears as a round glow in the smaller telescopes, but the ETX 90 will begin to show a VERY bright star-like center to this galaxy. In the 6-8 inch scopes the form is very interesting, as the circular glow of this shape dims very rapidly toward its edges with NO detail seen, except a clear star-like center. Like most galaxies of the Coma-Virgo group, this elliptical is an incredible 42 MILLION light years away. Because of its brightness, this is a great example of an elliptical galaxy for our smaller GO TO telescopes!

Object 7 - Messier 58 - a VERY Nice Spiral Galaxy
I re-visited this galaxy for the first time in years from Petit Jean Mountain on May 20. The sky was one of those 3-dimensionally dark events and the galaxies of Virgo were just incredible, showing MUCH detail in the LX 90 and presenting wonderful challenges in my ETX 125. Messier 58, a barred spiral, shows a distinct "bar" across its middle in the ETX 125 at about 125x, and clearly with 100x in the 8" scope. At medium power, about 75x to 100x, in the ETX 90 and ETX 125, you can barely "imagine" some very delicate outlying spiral

arms at the north and south peripheries....they ARE there! Use averted vision to occassionally detect this wispy arms! At low power in a good APO small refractor, expect to see a bright (magnitude 9.2) elongated oval with a very, very bright central one-half. There is a very distinct "lump" seen in a six-inch and larger scopes exactly SE from the center....this is actually a huge concentration of stars, like a cluster, in one of M 58's spiral arms!

Object 8 - Messier 59 - A Very Small, Starlike Oval Elliptical Galaxy
Look only about one wide field (1 degree) east from Messier 58 and find another elliptical galaxy, only this one if very faint and difficult; it is likely to be missed as a very faint star in a 3-inch and perhaps in a 4-inch as well to an inexperienced observer. This is a tiny elliptical of magnitude 9.6, but its size prevents it from being clearly distinguishable from a star! Curiously, in 1939, a "supernova" was seen in this galaxy at 12th magnitude....an exploding star that almost equaled the brightness of the galaxy itself!

Object 9 - A Bright and Easy Elliptical Galaxy – Messier 60
Messier 60 is a bit larger and thus easier to see with our smaller telescopes; even in the 8" it still appears like a very fuzzy star with no distinctive markings or features. A very faint spiral galaxy, NGC 4647, is found about 3' arc (in the same medium power field) to the NW of M-60, but is only visible in the 8 ilnch and larger scopes, and then very difficultly so. Messier 60 is magnitude 8.9, thereby making it an easy target for the small refractor scopes and quite a nice object in the 3-inch and larger

instruments. With a total number of sun more than ONE TRILLION, this is one of the largest of all galaxies known in the Universe....so LOOK at it!

Object 10 - A Large and Very Nice Face-on Spiral Galaxy! Messier 61

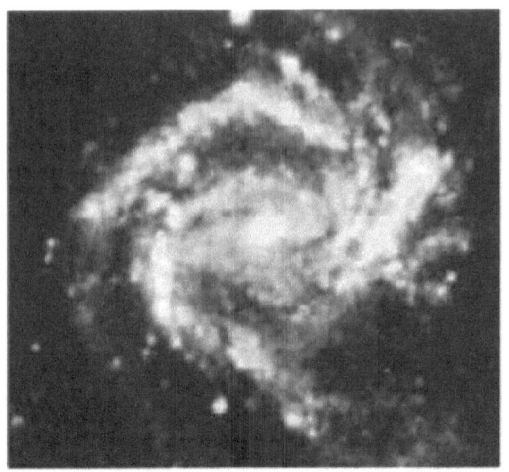

**Messier 61
Face-on spiral in Virgo**
Lick 120" Telescope Photo

From my dark skies of Petit Jean Mountain on May 20, I got stuck on this object....not finding it; both my scopes did just fine on their GO TO's....I got STUCK OBSERVING this magnificent object in near-perfect skies! This face-on galaxy has an incredibly bright nucleus, so bright that you might at first think you are seeing some sort of outburst in the center of this distant galaxy! Look carefully at the sketch adapted digitally to show what might e seen in a large telescope visually. The spiral arm at top is CLEARLY visible with a 6-inch at 125x, as is the huge "knot" just to the right of the center, this also being a spiral arm. These two features are absolutely incredible in a 10-inch at 100x; any more magnification and you begin to lose the spiral structure....any less and you cannot see it at all. The 3-inch showed the top arm and the knot, but not a clearly-defined spiral shape as the larger instruments. In the 10 inch, the lower (and left) spiral arm was also clearly visible. This is a very large (5.3' arc) and round object and is clearly visible in a quality small APO refractor, and actually shows some "graininess" from the alternating dark and bright arms. If you have a smaller telescope and wish to actually "see" a spiral galaxy....this is the one, far easier that Messier 52 (the "Whirlpool galaxy") although the latter is by far brighter. Messier 61 is oddly rated at only magnitude 10.1, although it certainly appears to be brighter.

Objects 11 - Two Galaxies - Messiers 84 and 86 –
Both Elliptical Galaxies
Only 17' arc apart, this is an outstanding low-power wide-field sight! M-84 is magnitude 9.3 and M-86 is magnitude 9.7, making them very easy in all scopes, appearing as two large out-of-focus stars at very low power. At medium (about 75x to 100x) magnification, they appear almost IDENTICAL in size and brightness, making for a beautiful sight. Keep in mind which is which! Messier 86 is the one to the east. For the 8-inch and larger telescopes, there are also two OTHER NGC galaxies (not listed in the concise directory above, as they are too faint for most scopes): ngc4388, a spiral only 16' arc south in a 20mm eyepiece; and ngc4402 only 10' north which is an edge-on spiral in which the dark

dust lane CAN be seen at about 150x in the 8" scope.....IF you use your imagination! Otherwise, look for M-84 and M-86 to be two large and bright oval "faint fuzzies" in our telescopes.

Object 12 - Messier 87 - An Incredible Elliptical Galaxy

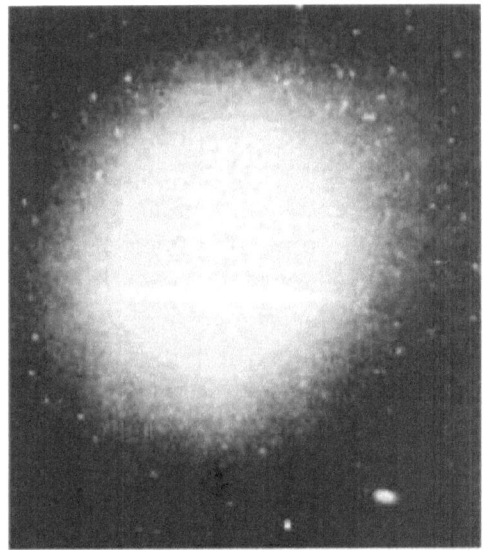

The Large Elliptical Messier 87
Palomar 200" Telescope Photo

Examine the 200" Palomar reflector's photograph of Messier 87 shown above.....see all those "larger" spots that appear to be outlying stars of this huge elliptical galaxy? They are NOT stars....those are **GLOBULAR CLUSTERS**! We obviously cannot see these in any of our telescopes, but it really puts the sizes of these galaxies into perspective. What you CAN see with M-87 is a fantastic galactic core, or "nucleus" that appears just like and 8th or 9th magnitude star right in the center of this uniform glow of **790 BILLION** suns. The galaxy itself appears only about 3.2' arc across, but is plainly visible in all of our telescopes at magnitude 9.2; like all the elliptical galaxies we have seen thus far in Virgo, it appears as a large oval glow....only this one has a very distinct nuclear center! This starlike core is clearly visible in a 3-inch, but may be lost even in high magnifications in smalll refractors.

OBJECT 13 - Messier 89 - A Very Tiny Starlike Elliptical Galaxy
Even at magnitude 9.5 visually, this is a tough galaxy, due principally to its extremely small size of only 1.3' arc diameter. It is unmistakable even in the small scope, but does not show any detail. The 8-inch and larger WILL reveal a very dim nucleus to this large elliptical, this appearing as a star of magnitude 10.5. Of all the galaxies in the Coma-Virgo Cloud, this one may be the closest to Earth.

OBJECT 14 - A Bright Spiral Galaxy - Messier 90
Only about 2 degrees northeast of Messier 87 (previous page) is found M-90, a very nice spiral tilted greatly toward our line of sight. It measures 7.1 x 2.2' arc in size, so it will appear much elongated. At its center is a stellar point of light, about magnitude 11, so this should be clearly distinguishable in the 4-inch and larger telescopes. This galaxy, although appearing very small through our telescopes, is actually about the same size and type as our own Milky Way galaxy of some 250 billion suns.

OBJECT 15 - "THE SOMBRERO GALAXY" – Messier 104

Okay, you've waited patiently for this one...but you may be disappointed. The photographs are tantalizing and teasing, and with such contrast you would THINK that some detail should be more visible. However, I also re-visited this galaxy in May 2001 with all the scopes and was a bit surprised. I have view this with wonder in a 40" reflector and with my 24" reflector many times.

The 6-inch shows a clear dispolarity - the top half being clearly fainter than the bottom half - of M-104. I could NOT see this effect in a 3-inch scope nor with smaller telescopes. However, the dust lane is barely visible in the 8" scopes, although NOT as clearly as I had remembered in scopes of this size. All in all, the most distinguishable feature is the difference in the two "halves" (see photo above) of this galaxy. Actually Messier 104 is an "added object" to Charles Messier's list....it is one of those I have eluded to in the past that he "should have seen" according to our way of thinking and thus we have included that in his original listing of "faint fuzzies." Messier 104 is seen nearly edge-on with a huge nuclear "bulge" that is clearly seen in the 200" photo above. That much is unmistakable even in the 3-inch scope. It appears as an elongated blur in the smaller instruments. One very distinguishing feature in an 8-inch that I noticed, however, is the "pointy" ends on each extremity of this galaxy; this was so obvious that it appeared as the points of two sharpened knives. With both the 6-inch and 8-inch I kept getting the impression that I could imagine further extensions from these "points" than actually exist. This is purely an illusion but I would be interested to hear from others who imagine similarly; I have never had this impression from larger telescopes. Looking at the above photograph, one would think that contrast alone would allow a much clearer and easier view of the huge dark dust lane.....it just isn't so, and it can only be seen with much difficulty, and - like said - is best imagined from the "huge" half section compared to the other side and not so much as a dark lane at all!

OBJECT 16 - **YOUR FIRST QUASAR!** –
Virgo3C-273
Get set to view perhaps the brightest yet most distant object in the universe....or at least one of them! The Quasi-Stellar Objects , or "Quasars" as they are coined are the most enigmatic of all celestial objects, with distances computed far beyond the most remote galaxies and emanating radiation at a rate more than 100 times the brightest and largest of galaxies.....yet we have no idea what they are. Only a dozen or so are actually known and verified.

To add to the curiosity, these objects are so distant

in the visible universe that they appear to be travelling as much as 37 percent of the speed of light!, or about 30,000 miles per second! The one in Virgo, like all the rest, appears nothing more than a very faint star visually. It is about magnitude 12.0, so it is visible in the 4-inch using a good star chart. Will appear as nothing but a very faint star....

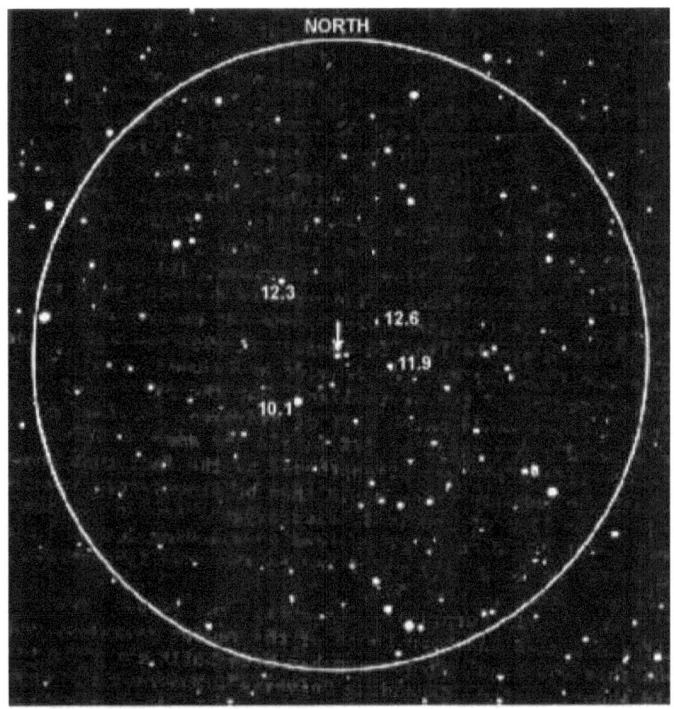

DISTANT QUASAR 3C 273
Finder Chart

As can be seen in the above chart, 3C-273 appears no more interesting than a very faint star, just on the limit of visibility of the 4-inch and even the 6-inch....but obviously there is more to this object than "meets the eye!" We do know now that this is actually a pair of stars, with at least 90 percent of all the energy we measure from these sources coming from the second, unseen "star." The actual "core" of the energy being emitted comes from something (something??) that is an incredible 7,500 LIGHT YEARS ACROSS! That means a pure energy source the size of a small galaxy! The jury is well out on this one, and likely not to come to a quick verdict. Options include and collapsing galaxy, with the gravitational collapse resulting in the outpouring of energy....perhaps ultra dense stars that have all their masses packed into something so small that you could put it on Long Island.....an "energy lens" shared by groups of hundreds of distant galaxy neighbors.....explosions on our dimensional wall created by black hole collapses in another dimension. Which explanation do YOU like?

At any rate, expect to see only a faint star at this location. Although I have included a starter finder chart here, download
https://www.aavso.org/apps/vsp/ . Note for these charts, simply type in the NAME of the variable at top to generate your choice of chart.
from the AAVSO and print this chart ("a" chart size) for a more detailed map with good comparison magnitudes. This object is also actually VARIABLE which is why the AAVSO has the chart on it.....varying from magnitude 12.2 to about 13.0. it is not a difficult object to find, and certainly one to provoke much conversation at your next star party....provided that YOU know how to "GO TO" it!

WANDERING ABOUT....YOUR NEW "USER OBJECT" IN VIRGO

Was there any doubt that I would select our **QUASAR 3C-273** in Virgo as our "User Object" for this Virgo Tour? You must have this object on your "quick list" as it amazes people at star parties, cook-outs and casual observing sessions with friends. A word of warning, however....make SURE that you are well-read on the subject before asking your AutoStar to "GO TO" it. Read up on this and the other Quasars in Burnham's Celestial Handbook, volume 3. It is amazing the theories and resources that are available on the subject; likewise, merely use your net search engines on "Quasar" (be prepared for televisions, microwaves, cars, flashlights and auction houses on that one) or better still, "Quasi-Stellar-Object" and see what information is now the most current.

The exact process for bringing up a celestial object for your telescope to GO TO will vary with the sky program, telescope keypad, or App that you are using, but all fundamentally work the same logic to acquire your object as well as build your User Object Library. Always consult your HELP or User Guides of your specific program for the exact steps in using tabs and keystrokes to pull up and GO TO your objects!

So with the addition of a Quasar, you are completing a set which includes a black hole, dark nebulae, curious multiple stars, variable stars and weird deep sky objects. All of these are wonderful for conversation-starters and crowd-stoppers at the big astronomical events!

The Author at one of his observatories in 1982.
There were no such devices as "GO TO" telescopes
And smart phones.

We found all of those wonderful sky objects the
hard way

www.ingramcontent.com/pod-product-compliance
Lightning Source LLC
Chambersburg PA
CBHW031810170526
45157CB00001B/26